CW01180345

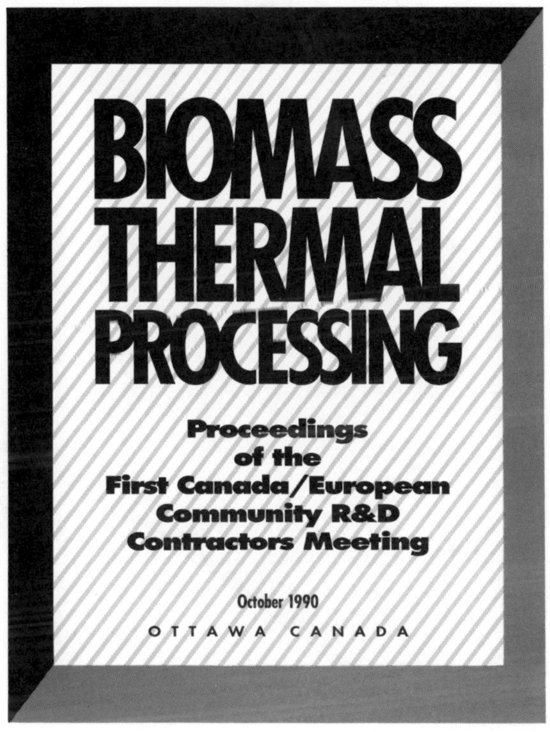

EDITED BY:

E. Hogan	CANMET – Energy, Mines and Resources Canada
J. Robert	CANMET – Energy, Mines and Resources Canada
G. Grassi	Commission of the European Communities
A.V. Bridgwater	Aston University

SPONSORED AND ORGANIZED BY:

Bioenergy Development Program – CANMET Energy, Mines and Resources Canada

Directorate General for Science Research and Development Commission of the European Communities

cpl PRESS

Published in the United Kingdom by:

CPL Press, a division of CPL Scientific Limited,
Science House, Winchcombe Road, Newbury, Berkshire RG14 5QX, UK
Telephone: +44 635 524064 Fax: +44 635 529322

for:

CANMET Energy, Mines and Resources Canada *and*
Commission of the European Communities

Printed by: The Chameleon Press Limited, London, UK

ISBN: 1 872691 50 1

PREFACE

Thermochemical conversion technologies offer the potential of converting biomass feedstocks into liquid and chars, either for fuels or as chemical feedstocks, helping to displace the current dependance on fossel fuels for our energy requirements. Additionally, with the increasing focus on the environment, an opportunity exists for thermochemical processes to be used to process waste feedstocks such as used tires and industrial wastes thast pose disposal problems, thereby providing an additional economic incentive for the commercialization of the technology.

The thermochemical area has undergone significant changes in R&D approach and expectation in Canada over the past ten years. There is now a renewed interest in these technologies as a result of the speciality and commodity chemicals potentially available in the pyrolysis oil and the current concerns about the environment. The European Community (EC) has had a bioenergy R&D program underway for the past fifteen years. The most promising route for the EC at present seems to be that of thermochemical with the production of pyrolytic or bio-crude oil. A high priority is being given to the commercialization of biomass pyrolysis as a means of obtaining liquid fuels from renewable resources, especially for electricity generation. Approximately half the EC R&D budget for bioenergy is in this technology area. The economic viability of these thermochemical process is promising in the near to mid term, and its integration into conventional energy systems presents no major problems.

With the great similarity between the R&D programs, both Canada and the EC have expressed an interest in establishing a collaborative R&D program in the thermochemical processing area. The objective of this callaboration is to share in the development of advanced technologies through the exchange of information and scientific personnel and the establishment of research activities in areas of common interest in order to stiumlate industry interest and encourage more rapid implementation of pyrolysis technologies at an industrial scale.

In this regard, the first joint meeting arising from this collaborative effort was held in Ottawa, October 23-25, 1990. Contractors from the Joint Opportunities for Unconventional or Long Term Energy (JOULE) Programme of the EC and the Bioenergy Development Program of Energy, Mines and Resources Canada gave technical updates on their research and development activities. The meeting was divided into four technical sessions: process updates; analytical methods; isolation and separation processes; and economics and market prospects. This forum allowed for the formal and informal exchange of information and ideas between research groups from Canada and the EC. This is expected to lead to further interaction and cooperation in the future, to the benefit of both groups and the development of thermochemical conversion.

E. HOGAN, T. BRIDGWATER, G. GRASSI, and J. ROBERT

AVANT-PROPOS

Les technologies de conversion thermochimique offrent la possibilité de convertir les biocharges en liquides et en charbon de bois qui serviraient de combustibles ou de charges d'alimentation, afin de réduire notre dépendance à l'égard des combustibles fossiles utilisés pour répondre à nos besoins énergétiques. En outre, étant donné l'intérêt accru manifesté à l'égard de l'environnement, on pourrait avoir recours aux processus thermochimiques pour traiter des déchets tels que les pneus usagés et les déchets industriels dont l'élimination pose des problèmes, ce qui constituerait une autre raison économique de commercialiser la technologie.

Au Canada, le domaine de la thermochimie a connu des changements considérables, tant du point de vue de la recherche et du développement que des attentes, au cours des dix dernières années. On s'intéresse davantage à ces technologies en raison de la disponibilité accrue de produits de base et de produits chimiques spéciaux qui proviennent de la pyrolyse de l'huile, et des préoccupations environnementales actuelles. Depuis quinze ans, la Communauté européenne (CE) mène un programme de R-D bioénergétiques. Actuellement, la voie thermochimique semble être la plus prometteuse pour la CE, avec la production d'huiles pyrolytiques ou biohuiles. On accorde la priorité à la commercialisation de la pyrolyse de la biomasse pour obtenir des combustibles liquides à partir de ressources renouvelables, notamment pour la production d'électricité. La CE consacre environ 50 % de son budget de R-D bioénergétiques à ce domaine technologique. La viabilité économique de ces processus thermochimiques est prometteuse à court et à moyen termes, et son intégration aux systèmes énergétiques classiques ne présente aucun problème majeur.

En raison de la ressemblance de leurs programmes de R-D, le Canada et la CE se sont montrés intéressés à établir un programme de R-D concertés en matière de traitement thermochimique. Cette collaboration a pour objet la participation conjointe à la mise au point de technologies de pointe, grâce à un échange d'information, de scientifiques et à l'établissement d'activités de recherche dans des domaines d'intérêt commun, afin d'encourager l'intérêt de l'industrie ainsi que la mise en oeuvre plus rapide, à l'échelle industrielle, des technologies de pyrolyse.

À cet égard, la première rencontre découlant de cet effort concerté a eu lieu à Ottawa, du 23 au 25 octobre 1990. Des entrepreneurs oeuvrant dans le cadre du programme *Joint Opportunities for Unconventional or Long Term Energy (JOULE),* de la CE, et du Programme de développement de la bioénergie, d'Énergie, Mines et Ressources Canada, ont présenté des mises à jour techniques sur leurs activités de recherche et de développement. La rencontre a comporté quatre séances techniques : mise à jour des processus; méthodes analytiques; processus d'isolement et de séparation; et perspectives économiques et d'utilisation. Le forum a donné lieu à un échange formel et informel d'information et d'idées par les groupes de chercheurs du Canada et de la CE. On prévoit qu'il mènera à une interaction et à une collaboration plus poussées, au grand profit des deux groupes et de la mise au point de techniques de conversion thermochimique.

E. HOGAN, T. BRIDGWATER, G. GRASSI, et J. ROBERT

Table of Contents/Table des matières

Preface i
Avant Propos ii
Introductory Remarks
A. Dolenko 1
Opening Overview
G. Grassi 3
The European Energy from Biomass Research Programme
G. Grassi, A.V. Bridgwater 6
Thermochemical Conversion R&D Activities in Canada
J.E. Robert, E.N. Hogan 16

Session I: Process Updates

The European Community R&D Programme on Biomass Pyrolysis and Related Activities
A.V. Bridgwater, G. Grassi 20
Upgrading of Pyrolysis Oils by Hydrotreating
E. Laurent, P. Grange, B. Delmon 32
Flash Pyrolysis of Biomass in an Entrained Bed Pilot Plant Reactor
K. Maniatis, J. Baeyens, H. Peeters, G. Roggeman 39
Energy from Biomass
C. Rossi 44
IEA Direct Liquefaction Project
D. Beckman 50
Rapid Thermal Processing (RTP): Biomass Fast Pyrolysis Overview
R.G. Graham, B.A. Freel, M.A. Bergougnou 52
New Applications of the Waterloo Fast Pyrolysis Process
J. Piskorz, D.S. Scott, D. Radlein, S. Czernik 64
Fast Pyrolysis Demonstration Plant
P.B. Fransham 74
Converting Sludge to Fuel - A Status Report
H.W. Campbell 78
Hydrolysis, Liquefaction, Solvolysis and Fractionation: an Overview
E. Chornet, R.P.Overend 86
Developments in the Steam/Water Liquefaction of Wood
S.G. Allen, D.G.B. Boocock, A.Z. Chowdhury 90
Biomass Liquefaction at SERI
J. Diebold, J. Scahill, R. Bain, H. Chum, S. Black, T. Milne, R. Evans, B. Rajai 101
Vacuum Pyrolysis of Used Tires, Petroleum Sludges and Forestry Wastes: Technological Development and Implementation Perspectives
C. Roy, B. de Caumia, H. Pakdel, P. Plante, D. Blanchette, B. Labrecque 109
Preliminary Mass Balance Testing of the Continuous Ablation Reactor
J.W. Black, D.B. Brown 123

Session II: Analytical

Biomass Liquefaction: Centralized Analysis
C.S. Alleyne, E.S. Skelton, J. McKinley 126
Molecular Weight Determination of Lignins and Celluloses Derived from Liquefaction/Fractionation Processes
P.F. Vidal, R.P. Overend, E. Chornet 129
Analytical Methodology for Aqueous/Steam Treatments
J. Bouchard, E. Chornet, R.P. Overend 137
Characterization of Vacuum Pyrolysis Oils of Diverse Origins
H. Pakdel, C. Roy 144

Session III: Isolation and Separation Processes and Products

Catalytic Upgrading of Pyrolytic Oils Over HZSM-5
R.K. Sharma, N.N. Bakhshi 157
Potential of Fast Pyrolysis for the Production of Chemicals
D.S. Scott, D. Radlein, J. Piskorz, P. Majerski 171
Levoglucosan from Pyrolysis Oils: Isolation and Applications
C.J. Longley, J. Howard, A.E. Morrison 179
Biomass Derived Alkaline Carboxylate Road Deicers
K.H. Oehr, G. Barrass 181
Upgrading of Biomass Oils to Cetane Enhancers
D. Soveran (Abstract Only) 184
La dépolymérisation d'une lignine de vapocraquage
M. Heitz, J. Lapointe, E. Chornet, R.P. Overend 185
The Conversion of Pentoses into Furfural Via a Novel Jet Reactor
C. Dubois, N. Abatzoglou, R.P. Overend, E. Chornet 198

Session IV: Economic Market Prospects

Costs of Biomass Pyrolysis Derived Liquid Fuels
A.V. Bridgwater 201
The Economics of Transport Fuels from Biomass
A.V. Bridgwater 209
Commercialization of Fast Pyrolysis Products
G. Underwood 226
Economics of Flash Pyrolysis of Peat
P.B. Fransham 229
Scale-up of the Ablative Fast Pyrolysis Process
D.A. Johnson, W.A. Ayres, G. Tomberlin 236
Some Problems of Technology Development for the Small Innovators
A. Wong 241

Workshop/Atelier 243

Author Index/Index des auteurs 255

TEXT OF INTRODUCTORY REMARKS

A. Dolenko
Alternte Energy Division
Canada Centre for Mineral and Technology
Energy, Mines and Resources, Canada

I have the pleasant task of welcoming you all to Ottawa. It is very gratifying to see such a large turnout for this meeting. I know a number of you have come from far away, and I would particularly like to welcome those of you from out of town, other parts of Canada and the United States, and as far away as Europe. I thank Dr. Giuliano Grassi for taking the time out of his busy schedule as head of the bioenergy program for the Commission of European Communities, to attend this meeting over the next few days. I also thank him for his efforts in helping to develop a very close collaboration of the programs between EEC and Canada; I think these programs are doing wonderful things for the development of the biomass liquefaction technologies.

We have also made a special effort at this meeting, to invite and have representatives from the private sector. A number of the technologies we have been supporting over the years are reaching the stage of commercialization and scale-up. It is very important to have the participation and help from the private sector in order to get these technologies into the marketplace. We are, therefore, very glad to have representatives from the petrochemical industry and from the chemical companies in Canada. I think we have some excellent opportunities in this area of biomass liquefaction and I am sure you are going to hear over the next couple of days about some of the exciting developments which have been made. Finally, I would like to welcome our colleagues from the Energy Research Laboratory of the Canada Centre for Mineral and Energy Technology, otherwise known as CANMET.

ERL, Energy Research Laboratory, is one of our sister divisions within CANMET and they have had a long and distinguished history of doing research and development in coal, co-liquefaction, heavy oil upgrading and natural gas conversion. They have also done some excellent work in biomass combustion, looking at combustion properties, safety, and emissions. We are very glad to have some representatives here from ERL, to allow them to get an idea of what has been going on in this program over the past several years.

I think the future for bioenergy in general is looking up, with events of the last few months in the Persian Gulf, and governments under increasing pressure to address a number of the environmental concerns that are being pushed by the general public. I also think those of you who have been involved in our programs over the years, particularly our contractors, know that we have come through some rather difficult times. Five or six years ago, the Federal Government in Canada was spending about $20-22 million on bioenergy research, development and demonstration; this year our budget is $3.5 million. So, you can see we have come a long way down, but I think we are on the way back up.

As I said, governments everywhere are under tremendous pressure to address environmental concerns, such as global warming due to CO_2, acid rain problems, urban pollution, and waste management in general. I think biomass really offers an opportunity to address a number of these concerns. Biomass in general, bioenergy is CO_2 neutral provided it is used on a sustainable basis and it does not have sulphur in it, which will contribute to the acid rain problems.

However, that is not to say bioenergy and biomass in general does not have its environmental problems. There are a lot of "nasties" which can be emitted from uncontrolled biomass combustion, and our friends at Environment Canada are looking over our shoulders and will be imposing, or are imposing, regulations which will ensure the best available technologies are being used. But, biomass/bioenergy does have opportunities in this area, and I think that the challenge is being put to us to develop environmentally clean conversion technologies.

Within the next couple of months - those of you who live in Canada have heard a lot about this in the press - the Federal Government will be announcing its "Green Plan", the so-called plan to address these environmental problems and provide a direction and agenda for over the next several years, to come up with solutions to the environmental problems faced by this country and those of the rest of the world. Hopefully this "Green Plan" will provide additional funds for research, development and demonstration of the technologies in alternative energy, and in particular in bioenergy.

So, in closing I just want to say, again, welcome to Ottawa; it is really very heart warming and gratifying to have such a large turnout. I would also like to take this opportunity to thank Ed Hogan for all of his hard work and diligence in organizing this meeting.

Opening Overview

G. Grassi
Commission of European Communities
DG XII, 200 rue de la Loi, B1049 Brussels, Belgium

Firstly, I would like to thank the Canadian Authorities on behalf of the Commission of the European Communities and the Director-General for Science, Research and Development, for inviting us to participate in this contractors' meeting in Canada. This is our first opportunity to have a mutual exchange of information for cooperation, and I hope that this will develop in the future. Our next contractors' meeting will take place in Florence in November, and we would like to extend an invitation to the Canadian Authorities and to anyone present here who would like to attend that meeting.

The European Commission started R&D on energy from biomass 15 years ago after the first oil crisis. We are now implementing R&D with plans for large scale integrated projects to link all the activities together. Three years ago, due to the very low cost of oil and energy in general, the biomass R&D Programme was no longer considered as a priority sector. Consequently, the biomass programme was heavily penalized with a budget reduction of nearly 40%. The situation, however, is now rapidly improving with plans for a significant expansion. Our efforts to overcome this difficult situation were intensified by two means:

1. Discussion with policy-makers and politicians, supported by our analyses and data, to convince them that bioenergy will become an important strategic area in the long term, not only for energy, but for other reasons such as rural development, environment and assistance to developing countries. More stringent contemporary environmental legislations relating for example to SO_2 and NO_x emissions make bio-fuels more attractive from an economic point of view. Other circumstances such as the crisis of the Common Agricultural Policy and the GATT agreements, give a new dimension to the biomass activity.

2. Developing the concept of integration of activities for both energy and industrial use. For the E.C's situation two promising industrial (non-energy) sectors were identified: pulp for paper production from biomass through innovative technologies (the E.C. has a very large import deficit of 8.2 million t/year) and the production of organic fertilizers. The basic idea was that, as the industrial market is able to absorb raw materials at higher prices, a part of the feedstock could be available at lower cost allowing competitive low-pollution bio-fuels to be produced. For example, at an average cost of biomass in the EC of 50 ECU (about 75 C$/t) pulp for the paper industry could be produced at 75 ECU/t (113 C$/t) and energy at 25 ECU/toe (38 C$/t).

The development of a new promising technology, implemented by the University of Hamburg and commercialised by Kraftanlagen Co., Heidelberg (Germany) was supported. At the laboratory stage, trials have been very satisfactory. In the next few months, verification of this process at an industrial level will be carried out in a 4 t/day pilot plant. A publication on this subject is being prepared, printed on paper produced by this installation from sweet sorghum bagasse.

The European Parliament in particular, is giving strong support and has recently asked for an immediate supplementary budget for non-nuclear energies, including biomass of 40 MECU in 1991 (60 million C$).

Beyond this, a new "Agriculture and Agro-industrial Research programme" (which also includes biomass R&D activity) has been recently approved in a first reading by the Council of Ministers with a total budget of 330 MECU, although the budget dedicated to bioenergy is not yet known. Cooperation with Third World Countries has also been proposed. We strongly believe that cooperation will be useful for everybody. The number of people involved with biomass R&D is limited so, by pooling the available knowledge and expertise, a faster and more cost effective impact will be obtained.

The structure of the present Biomass R&D programme covers six main sectors of activity:

(1) Production of biomass (short-rotation forestry, sweet-sorghum).

(2) Harvesting, transportation, storage of biomass.

(3) Biological conversion (acid and enzymatic hydrolysis).

(4) Thermochemical conversion (pyrolysis, catalytic processes).

(5) Integrated pilot projects.

(6) Assistance to large integrated biomass schemes - LEBEN projects.

The basic objective of the present and future R&D programme is the production of liquid 'bio-fuels':

- Liquid fuel or bio-crude-oil bio-oil from thermochemical processing by pyrolysis or lignocellulosic materials.

- Oxgenated fuels such as methanol and fuel alcohol produced through gasification processes.
- Bio-ethanol produced from sugar or starch for the transportation market.

For biomass conversion technologies, the most promising route at present seems to be that of thermochemical, i.e., the production of pyrolytic oil or bio-crude-oil. The economic viability of this process is promising in the medium term, and its integration into conventional energy systems presents no major problems. This liquid is of moderate heating value, is easily transported, can be burned directly in thermal power stations; can possibly be injected into the flow of a conventional petroleum refinery, burned in a gas turbine, or upgraded by hydrotreating or zeolite-based processes to obtain light hydrocarbons. The technologies for producing bio-oil are evolving rapidly with improving process performance, higher yields and better quality products. There is, therefore, considerable justification for a substantial and robust R&D programme in this area. Globally, nearly 50% of our budget is now dedicated to pyrolysis and upgrading activities.

The first and very modest contract for an exploratory study on pyrolysis conversion costing 7,000 ECU (C$ 10,000) was negotiated in 1982. Six years later, we received a commitment for a technological guarantee for the first European technology. We are now proceeding very quickly, especially through the help of medium and large industries.

Two basic C-4 biomass crops are envisaged: sweet-sorghum and miscanthus, the development of which have been supported for the past 10 years. Sweet-sorghum is a crop which could grow anywhere in Europe even with the northern European climate, although it may take perhaps 10 years to genetically adapt the crop to north European conditions. The average productivity of sugar is 10 tons per hectare; in the south of Europe 14 tons per hectre per year and in the north of Europe, between 6 and 8. Due to the favourable perspectives, efforts are now concentrating on C-4 annual crops, taking into account that in Europe there will be about 10 million or even 20 million hectares at the end of the century, which gives a future biomass potential of 300 - 600 tons (dry)/year. In addition, farmers prefer to grow annual crops with the possibility of deciding what to do the following year. The opportunities seem very promising.

In the case of pulp for paper production, it is possible to produce twice as much per year and per hectare from miscanthus or sweet-sorghum crops in comparison with what is obtained from S.R.F. For biological conversion, there is some activity in acid hydrolysis and enzymatic hydrolysis, but enzymatic conversion is now considered a second priority. It is anticipated that the cost of sugars obtained from sweet-feed-stocks will be twice the cost of sugars obtained from sweet-sorghum. The introduction of sweet-sorghum on a wide scale will make a large amount of sugar available. What can be done with this raw material while awaiting the development of the bio-ethanol market?

As the bagasse from sweet-sorghum can be up-graded by newly advanced technologies such as pulp for paper and electricity production via pyrolysis, very low cost sugars will be available which could therefore be utilized for competitive:

- bio-ethanol production and/or peak electricity production in ethanol filled gas turbines
- biogas and/or electricity production in bio-gas filled turbines, together with compost production to be recycled to the high yield plantations.

As far as integrated projects are concerned, several pilot-programmes are being implemented which represent the first modules of regional projects. Each project costs about I.5-3 million ECUs (2.5-4.5 C$ million); it is hoped next year we are able to considerably strengthen this kind of activity.

Another interesting product, which could be integrated in large biomass schemes, is desalinated water which is particularly interesting for Mediterranean regions. For coastal areas, biomass could offer a significant contribution for combined production of desalinated water and electricity. Water is one of the essential elements for high biomass productivity. In southern Europe, when water is available, the high solar irradiation and temperature can increase crop yields by 20-30% compared to northern Europe. This extra yield seems sufficient to recover the investment of water desalination by reverse osmosis and electricity production and the operating cost of 3.7 kWh/m^3 water at scales of 300 000-500 000 m^3/day.

Conclusion

1. Full exploitation of the high biomass potential in the E.C. of 300 Mtoe after the year 2000, could alleviate the problems of the European Common Agricultural Policy and also offer significant benefits from the point of view of rural development, environment and assistance to developing countries.

2. New biomass resources, advanced technologies and integration of activities seem able to open up very large markets such as for bio-fuels, electricity, chemicals, organic fertilizers, pulp for paper and desalinated water at competitive costs.

3. In particular, electricity production in decentralized systems of 1-50 MW, to satisfy peak-power demand, seems very promising. Equally, bio-crude-oil production by flash-pyrolysis and its upgrading is at

present the most promising conversion technology, because it is fast and low-cost; it is adapted to all types of biomass resources and is able to give access to many large markets such as heat, transport, electricity and chemicals.

4. International cooperation will constitute an important instrument for a faster and more cost effective progress of the bioenergy sector.

The European Energy from Biomass Research Programme

G. Grassi[1] and A.V. Bridgwater[2]

[1]Commission of the European Communities, DGXII, Rue de la Loi 200, B1049 Brussels, Belgium

[2]Chemical Engineering Department, Aston University, Aston Triangle, Birmingham, B4 7ET, UK

The European community Directorate for Research in Science and Technology has recently initiated a new four year framework programme which includes a substantial activity on non-nuclear energy known as JOULE - Joint Opportunities for Unconventional and Long term Energy in Europe. This programme includes a major activity in the area of Energy from Biomass with a financial allocation for the current two year programme of about US$ 12 million. The topics covered include biomass production, conversion to fuels by biochemical and thermochemical processes, utilisation of these fuels in a variety of applications and studies of integrated systems for implementation.

This paper describes the current European Community Energy from Biomass programme, the pattern of activities and their objectives, and future plans for development and expansion into related areas. These include pulp for paper, fertilisers and chemicals as well as liquid and gaseous fuels from bio-chemical and thermo-chemical conversion technologies.

La Direction générale de la recherche en science et en technologie de la Communauté européenne a récemment commencé un nouveau programme cadre d'une durée de quatre ans qui comporte un volet important portant sur l'énergie non nucléaire. Il s'agit du programme JOULE (Initiatives conjointes sur les énergies nouvelles et à long terme en Europe). Un des volets majeurs du programme s'intéresse à l'énergie dérivée de la biomasse; le programme de deux ans actuellement en cours reçoit un financement d'environ douze millions de dollars américains. On y étudie la production de la biomasse, la conversion en carburants grâce à des procédés biochimiques et thermochimiques, l'utilisation de ces carburants à diverses fins ainsi que des systèmes intégrés de mise en oeuvre.

L'article décrit le programme actuel d'énergie dérivée de la biomasse de la Communauté européene, le type d'activités et leurs objectifs ainsi que les prévisions de développement et d'expansion dans des domaines connexes, dont la pulpe pour le papier, les fertilisants, les produits chimiques ainsi que les carburants liquides et gazeux produits grâce aux techniques de conversion biochimiques et thermochimiques.

Introduction

The European Community Programme on Energy from Biomass was the first Community wide research programme to be initiated, and is now part of an annual RD&D programme totalling about $2000 M/y covering almost all industrial sectors. The current Energy from Biomass Programme has an annual budget of about $5 million, and plans are in hand to increase this budget to about $20 million/y and expand the programme to cover a wider range of agro-energy and agro-industry topics.

The overall emphasis is towards eventual commercial exploitation of the results of the programme, and the more specific objectives include:

- security of long term energy supplies for Europe,
- contributions to the development of industrial markets,
- improvement of the environment by utilising wastes and residues,
- making a positive contribution to the greenhouse effect,
- better management of surplus agricultural and marginal land,
- provision of opportunities for socio-economic development of the less developed regions of Europe particularly towards the south.

Resources and Infrastructure

Europe has considerable resources available and considerable potential for expansion by use of under-utilised land that will result from limiting food production in the Community. Land will be abandoned from traditional crops causing some areas to change from food crops to biomass plantations to produce raw materials for the energy market and industry. About 20 million hectares of agricultural land and 10 to 20 million hectares of marginal land are likely to become available for biomass production by the year 2000. Table 1 gives current and possible future arisings of a range of biomass products.

Production

Previous work in the European Community has already identified the most promising wood species as principaly poplar (and possibly willow) for temperate zones and eucalyptus for hot, humid zones. For each of these species, further research is needed on physiology, genetics, production techniques, including harvesting and integration with conversion technologies and studies of the environmental impact. For sugar producing species, the Jerusalem artichoke is more suitable for temperate zones, while sweet sorghum is preferred for warmer and more humid climates. The potential for biomass production is summarised in Table 2, and Table

3 shows the current and target productivities for different biomass forms.

Examination of new species is envisaged such as: miscanthus or other C4 plants for temperate zones, and robinia and cynara for arid zones. Methodologies to find the best path for the study of new species and production models will be developed that consider linkages with other parts of the programme. It will also be necessary to continue the development of harvesting techniques and methods, as well as techniques for the separation, drying and storage of the raw materials. Some early cost estimates of wood and sweet sorghum are given in Table 4, with projections of target improvements.

Conversion

There are several relevant technologies for conversion of biomass into higher value products, and these are classified as biological, thermal and physical, and shown in Figure 1.

Bio-conversion

Microbial and enzymic processing of biomass can yield sugar solutions from lignocellulose, for further conversion to alcohols and other solvents of interest to the fuel and chemical industries. Any sugar or starch crop can yield ethanol through conventional and commercially available yeast based fermentation, while anaerobic digestion in fabricated digestors or landfill produces methane from liquid or solid wastes and residues.

For lignocellulosics, hydrolysis is an essential first step in biological conversion. Acid hydrolysis is already well established at the laboratory level and now requires applications at the demonstration and industrial scale. Enzymatic hydrolysis, which appears to be very promising, is, however, at an early state of development and still needs substantial fundamental research effort such as genetic manipulation to improve yields.

The production of ethanol by fermentation is already well established commercially, but innovative studies for further improvements are still needed, such as improvement of reactors and separation systems.

The status of the various technologies is summarised in Table 5.

Thermochemical Conversion

The range of technologies available for thermochemical conversion of biomass is summarised in Table 6.

The most promising thermochemical conversion technology currently seems to be the production of pyrolytic oils or bio-oil. The economic viability of this technology is promising in the medium term, and its integration into conventional energy systems presents no major problems. This fluid is easily transported, it can be burnt directly in thermal power stations, injected into a conventional petroleum refinery, burnt in a gas turbine, or upgraded by hydrotreating or zeolite based processes to obtain light hydrocarbons for transport fuel. The technologies for producing bio-oil are evolving rapidly with improving process performance, larger yields and better quality products, and the basic process types are summarised in Table 7, with product options shown in Figure 2. The emphasis is on flash pyrolysis technologies that give high yields of bio-oil. Research in this area represents the largest single area of research in the current R&D programme.

Gasification has been researched extensively over the last 12 years, with the highlight being the construction of four major pilot plants to produce synthesis gas for methanol production in 1985-88. Gasification is considered to have reached the point of commercial viability, but the economics and short term prospects are inadequate for industry to justify demonstration and implementation. The current research programme, therefore, has been orientated away from gasification to the more innovative thermal processes, although gasification is seen to have substantial prospects and R&D will continue in the next programme.

Liquefaction is viewed as a longer term interesting possibility and a low level of support will continue to be maintained.

Combustion is generally viewed as a relatively mature technology for simple heat generation, with R&D needed in environment areas. Some recent developments have generated exciting possibilities for power production at lower cost and/or higher efficiency, and these are being supported.

Thermochemical conversion technologies could also be applied to the treatment of urban waste, using analogous technologies with minor modifications, and producing very similar products. This has the advantage of being a particularly low cost feedstock offering substantial environmental benefits from new methods of waste management.

Pulp for Paper

The manufacture of paper pulp is strategically important. A new process has been developed permitting the utilisation of conventional biomass instead of slow-growing traditional wood. This is a non-polluting technology which involves a closed water cycle and avoids all chlorine emissions. This activity would be integrated with the production of compost and organic fertilisers by aerobic and anaerobic digestion.

Transportation Fuels

The production of alternative transport fuels is still a major priority for Europe, and gasification processes for oxygenates and hydrocarbons will be supported in the next programme to compliment the current activity on pyrolysis technologies.

Utilisation

For energy applications, research must not only consider optimisation of existing installations to operate on the new fuel, but also devise and develop new applications and new products. One example is the decentralised production of electricity by the new technology of high temperature ceramic gas turbines which offers interesting cost and efficiency benefits. Another example is described below as an integrated approach to biomass utilisation in which a single feedstock is processed to meet market requirements in the most effective way.

Integration

A key feature of the European R&D programme is rapid implementation of the technologies that are being developed, with integration of biomass production with conversion and utilisation. The strategy is best described by reference to the example of sweet sorghum as depicted in Figure 3. The sorghum is an annual crop that makes it more financially attractive for farmers. When harvested the sugar can be extracted and fermented to ethanol. The bagasse can be used as a feedstock for either pulping, or for pyrolysis to fuels and chemicals, or for composting. The balance of products can be adjusted to local requirements and economically optimised.

Although this opportunity is not economically viable under present circumstances, implementation can be justified as a contribution to the socio-economic resuscitation of the less developed areas of Europe as in the LEBEN projects described below. This has the advantage of large scale demonstration and integration of technologies with the requirement of absolute economic justification, so that the technology will be available and proven when it is needed.

LEBEN Projects

To demonstrate the integration of biomass production, conversion and utilisation, several opportunities have been identified where biomass exploitation would offer substantial socio-economic benefits in less developed areas of Europe. These opportunities are known as LEBEN projects for Large European Bio-Energy Network, where infrastructural, economic and social developments will be accomplished by exploiting the biomass resources of the region. The objectives of these projects and those currently approved or possible are summarised in Table 8.

In addition to these European developments, attention is also devoted to both developed and developing countries where collaborative arrangements have been established as summarised in Table 9. The objectives are to provide opportunities for implementation of more advanced technologies in Europe, and to support the introduction of new technologies into developing countries.

RD&D Strategy

From consideration of the infrastructural resources available and projected demands over the full range of technical, economic, political, environmental and social aspects, a strategic plan for research, development, demonstration and implementation has been derived. This is depicted as target dates for both demonstration and commercialisation of the targeted technologies in Table 10.

The financial implications of meeting these targets in terms of an R&D budget are depicted in Figure 4, with the consequences of a "fast" spend and a "slow" spend rate on meeting these targets.

Conclusions

The European Community R&D programme has evolved considerably since it was founded as a result of the oil crises of the 1970s. The main focus of the current and future R&D programme is summarised in Table 11.

Bio-conversion is focussed on fundamental and applied research to improve productivity and yields of pretreatment methods for fermentation processes, through enzyme hydrolysis. Acid hydrolysis is already virtually commercialised, and fermentation systems are widely available commercially. There is interest in integrating bio-ethanol production with utilisation of what are traditionally viewed as byproducts, such as bagasse, for production of pulp, pyrolytic fuel products and compost.

In thermochemical processing, gasification has been developed to a point where commercialisation is feasible, and pyrolysis for production of liquid fuels is receiving substantial attention, with liquefaction viewed as a longer term opportunity. Upgrading to hydrocardons and chemicals is a potentially important area. Production and utilisation of less orthodox products is also being investigated.

Other products of interest in an integrated biomass exploitation facility include pulp for paper using newly developed processes that minimise environmental impacts; compost for soil conditioning; and recovery of chemical specialities from pyrolysis processes.

Application and integration of these technologies is a primary objective, and demonstration is most likely to be realised in the context of LEBEN projects.

Discussion:

K. Oehr (B.C. Research): Two questions

One is, what impact does the cost of hydrogen have on your economics? And two, what is the major source of hydrogen in Europe for hydrogenation? I assume, it is a nuclear derived via electrolysis or is there some other source?

A.V. Bridgwater (Speaker): In the case of both oxygen and hydrogen, we treated them as a utility, with a declared price. With something as ambitious as covering such a spread of technologies and fees and processes, we have taken all the utilities as an import or export, to a grid, as it were.

In a real situation you would, in fact, generate your utilities on-site, your fuel, your steam, your power and your hydrogen. But we have escribed costs to hydrogen which were shown on the utility costs here (refers to slide). We put in there, I think, 1,500 pounds a ton as a cost of hydrogen. That is shown here as a cost. This oxygen is escribed, a cost as a utility that you would buy in.

If you bear in mind, that we are talking, in the medium term even, of a minor contribution to the total transport fuel or liquid fuels production of Europe. One would in fact probably associate the upgrading with a refinery which has a hydrogen surplus. It is easy to cost the hydrogen as a utility rather than to think about a separate production. One can also do that, of course. In fact, one of the things is to again extend the program to derive a cost for producing hydrogen by gasification and you could then use that figure to plug into this.

K. Oehr: (Question off-mike)

Speaker: That approach is currently being looked at in Switzerland. In fact, they are looking at oxygen gasification and upgrading with the hydrogen from electrolysis using low-price tar.

Now, it can be expanded in lots of directions. The problem is to constrain it to get an output that was meaningful and interesting, but without putting too many limitations on how it would run. In fact, the computer program, in principle, is available for people who want it. We want to make sure it has been debugged adequately first.

J. Piskorz (University of Waterloo): You have put in a tremendous amount of work, and it is very difficult work. I only wish these numbers were right and we could accept them without any hesitance. Your assumptions were very often based on very experimental stage technologies, SERI, etc. How much bias is in those numbers? I mean how much uncertainty?

Speaker: There are different levels of uncertainty with different parts of the process. The front-end pre-treatment is based on established technology in the pulp and paper industry using conventional solids, handling, storage, processing, drying, screening, and so on. So, there is a low uncertainty there.

The synthesis of products, if you take methanol for example, the technology is fairly well developed and it is well optimized. You can buy it; you can licence it; and that, if you like it, is also very well established. The Shell S&DS process has been very well researched, but it is less certain because they have not built a plant yet. One can escribe in fact different levels of certainty to different parts of the Process.

On the other end, the biomass gasification processes have been less well developed and less well explored. The biggest facility, I believe, is the French one at 2.5 tons per hour. That has had very limited operation experience.

So, we know there is greater uncertainty there. The highest level of uncertainty, which in fact, was specifically mentioned in the last two slides concerns the flash pyrolysis with zeolites, in which there is almost no realistic data available. It is mostly based on what Jim Diebold has achieved at SERI. I am sure he would agree that it is highly uncertain, and it needs a lot of further work at resolving and optimizing the system.

That is one of the things the IA Group, in fact, are going to be looking at. They are going to look in much more detail at both the flash pyrolysis and zeolite upgrading and synthesis of gasoline, to try to resolve the uncertainties, and to use expertise they have accumulated over the last ten years or so (expertise you have heard about from Dave Beckman) in order to produce figures that are more plausible. The point about this is, to take the data as reported and presented, either from literature or from discussions, but to be consistent and as objective as possible. We have tried not to influence the results, but to take data as reported and model it.

However, you are quite right, there are highly different levels of uncertainty. One of the purposes of this, in fact, is to not just to compare technologies or process routes, but to identify where you ought to spend more effort at resolving uncertainties to get maximum benefit from it. It is a planning tool as well as an evaluation tool.

D. Wilson (Dow Chemical): I would like to know what computer hardware and/or software was used to develop this program?

Speaker: It is written on Pascal and it will run on an IBM computer, in a compiled form. We will obviously want to retain the source code, but we plan to make it available in a compiled form.

Table 1 European Biomass Resources		
	M dry t/y	
	Now	Future
Wood	5	75
Other energy crops	-	250
Wood wastes	50	70
Agricultural wastes	250	250
MSW/refuse	60	75
Industrial & commercial wastes	90	100

Table 2 Biomass Production in Europe			
		M dry t/y	
		Now	Future
Wood	short rotation forestry	1	150
	conventional forestry	85	170
	scrub woodland	5	100
Sweet sorghum		2	250
Agricultural wastes, e.g. straw		40	250
MSW / refuse		100	150

Table 3 Biomass Productivity			
	dry t/ha/y		
	Present	Potential	Location
Agricultural crops			
Sweet sorghum	25	35	SC Europe
Miscanthus etc	20	30	S Europe
J. artichokes	20	25	SC Europe
J. artichoke tubers	10	14	SC Europe
Trees			
Eucalyptus	10-15	17	SC Europe
Poplar	12	16	NWC Europe
Willow	10	15	N&W Europe
Robinia	5	8	S Europe
Conifers	5-6	8-10	NWC Europe

Table 4 Biomass Production Costs, ECU/dry t				
	Wood		Sorghum	
	now	future	now	future
Biomass	10	5	42	15
Harvesting	20	15	5	4
Storage	5	4	-	-
Transport	5	4	18	11
Total cost	40	28	65	30

Table 5 Biochemical Conversion Status			
		Status	Commercial
Hydrolysis	- acid	commercial	5- 10 years
	- enzyme	pilot	
Fermentation		commercial	
Distillation		commercial	

Table 6 Thermochemical Conversion Processes and Products	
Technology	Main primary products
Pyrolysis	bio-oil, charcoal
Gasification	fuel gas
Liquefaction	bio-oil
Combustion	heat

Table 7 Pyrolysis Processes	
Type of pyrolysis	Main product(s)
Slow	30 - 35% wt charcoal
Conventional	20 - 25% wt charcoal, and 10 - 15% wt bio-oil
Flash	40 - 60% wt bio-oil

Table 8 LEBEN - Large European Bio-Energy Network	
Objectives	
An instrument of regional redevelopment	
An economic solution for massive exploitation of renewable resources in anticipation of oil price increases	
Demonstration of integrated systems for materials and energy production	
Acquisition of operating experience	
Optimisation of integrated systems	
Projects planned	
Abruzzo	Italy
Campania	Italy
Catalonia	Spain
Coimbra	Portugal
Evritania	Greece
Galicia	Spain
Groningen	Netherlands
Munster	Eire
Saar	West Germany
Sicily	Italy
Strathmore	UK
Umbria	Italy

Table 9 Cooperation			
Objective			
Acceleration of biomass exploitation through cooperation with countries with more advanced technologies and those with opportunities for exploitation.			
Countries involved			
Brazil	Canada	Costa Rica	Finland
USA	USSR	Venezuela	

Topics for Cooperation	
Production of biomass	New potential crops
Thermal conversion	Process improvement and implementation
Hydrolysis	Process improvement and implementation
Compost	Process improvement and implementation
Environmental impacts	of biomass production, biomass conversion and product utilisation
System development utilisation	integration of production, conversion and

Table 10 Targets for Implementation	Year of implementation
Demonstration of:	
Conventional energy forestry	1992
Short rotation forestry	1993
Advanced combustion for power	1994
Advanced pyrolysis	1995
Pulp for paper from sweet sorghum	1995
Compost	1996
Sweet sorghum production	1996
Char slurry fired boilers, kilns etc	1996
Gas turbine fired on bio-oil and char slurry	1997
Advanced gasification	1998
Hydrocarbon synthesis	2003
Commercialisation of:	
Advanced combustion for power	1994
Conventional energy forestry	1998
Pulp for paper from sweet sorghum	1999
Short rotation forestry	2000
Sweet sorghum production	2000
Advanced pyrolysis	2000
Ethanol for fuel	2000
Compost	2002
Advanced gasification	2009
Methanol synthesis	2010
Hydrocarbon synthesis	2010
Hydrogen and ammonia production	2012

Table 11 R,D&D Strategy

- More intensive biomass production
- Fuels and electricity production through biomass conversion processes
- Materials and chemicals production through biomass conversion processes
- Greater emphasis on environmental aspects

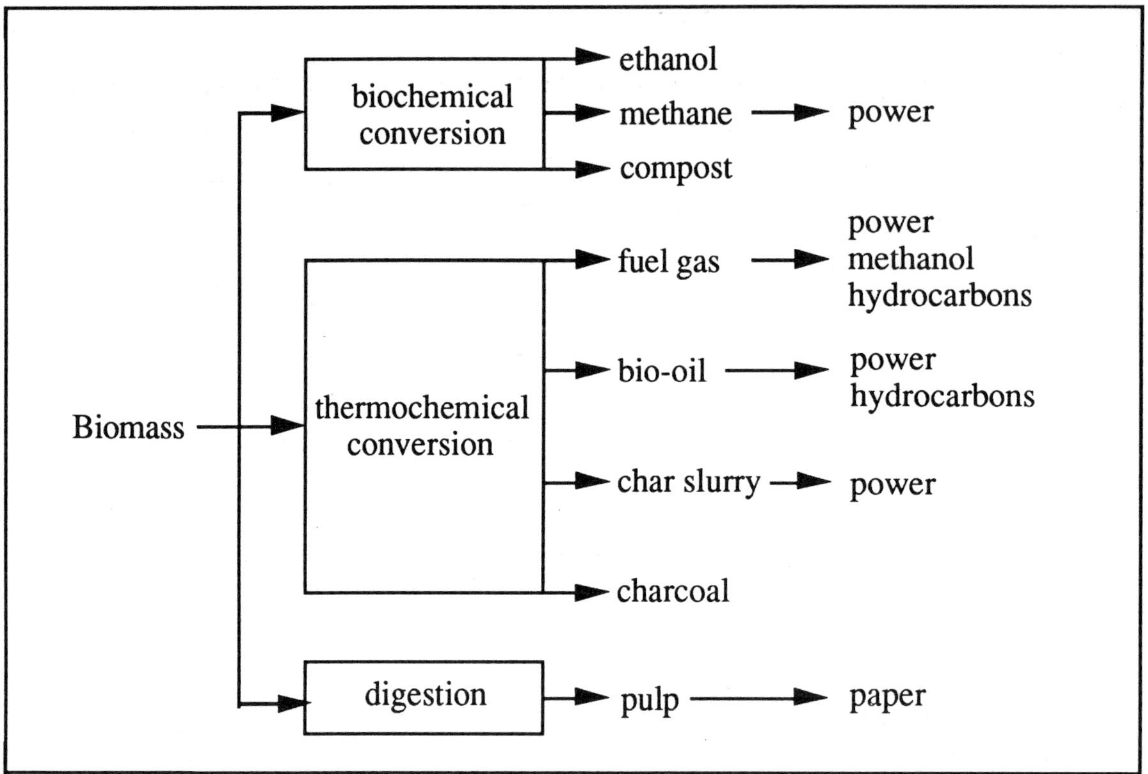

biochemical conversion
ethanol
methane
power
compost
Biomass
thermochemical conversion
fuel gas
power
methanol
hydrocarbons
bio-oil
power
hydrocarbons
char slurry
power
charcoal
digestion
pulp
paper

Figure 1 Basic Processes for Upgrading Biomass to More Valuable Products

bio-oil
Pyrolysis
char
Slurry
Slurry
Upgrade
gasoline
etc
Boiler
heat
power
& heat
Turbine

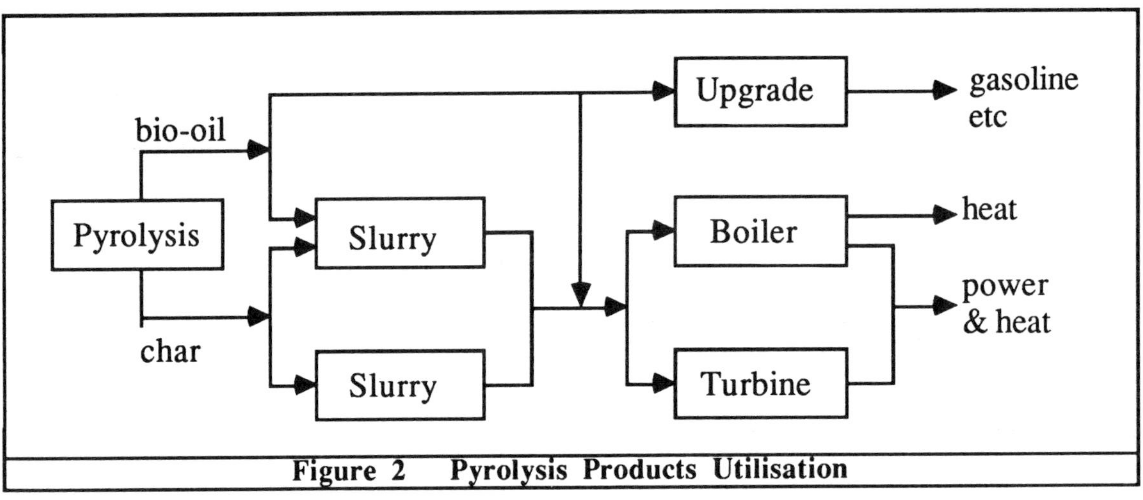

Figure 2 Pyrolysis Products Utilisation

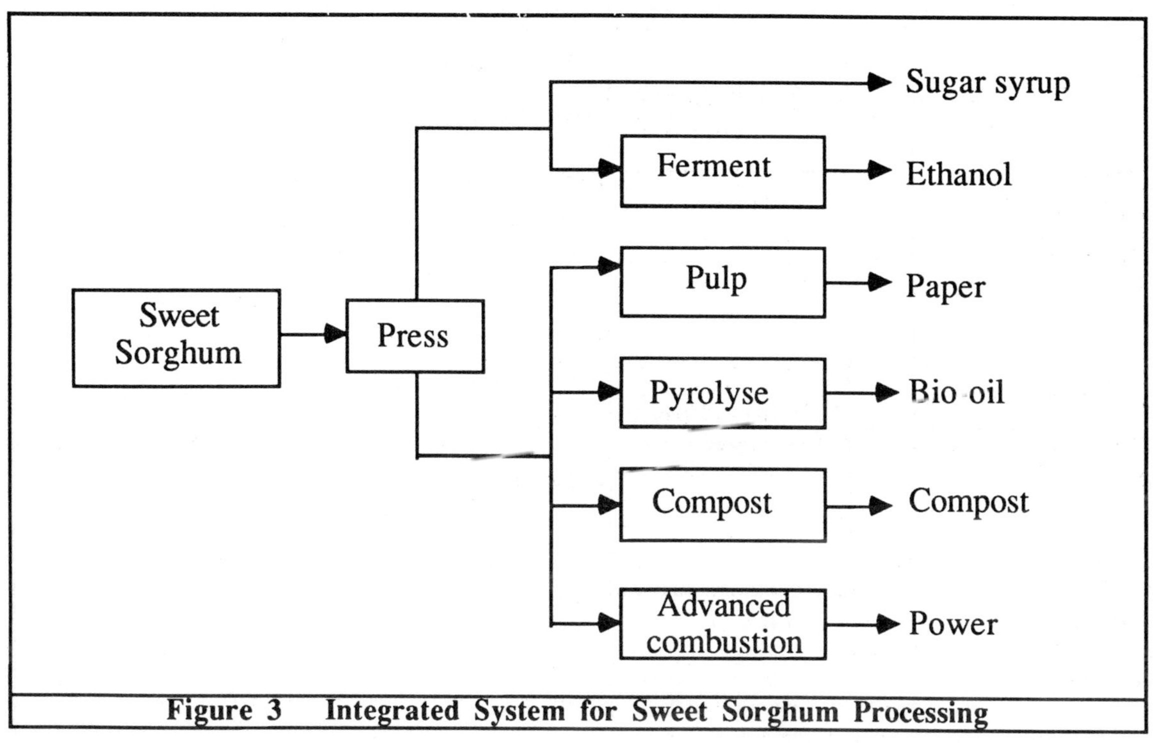

Figure 3 Integrated System for Sweet Sorghum Processing

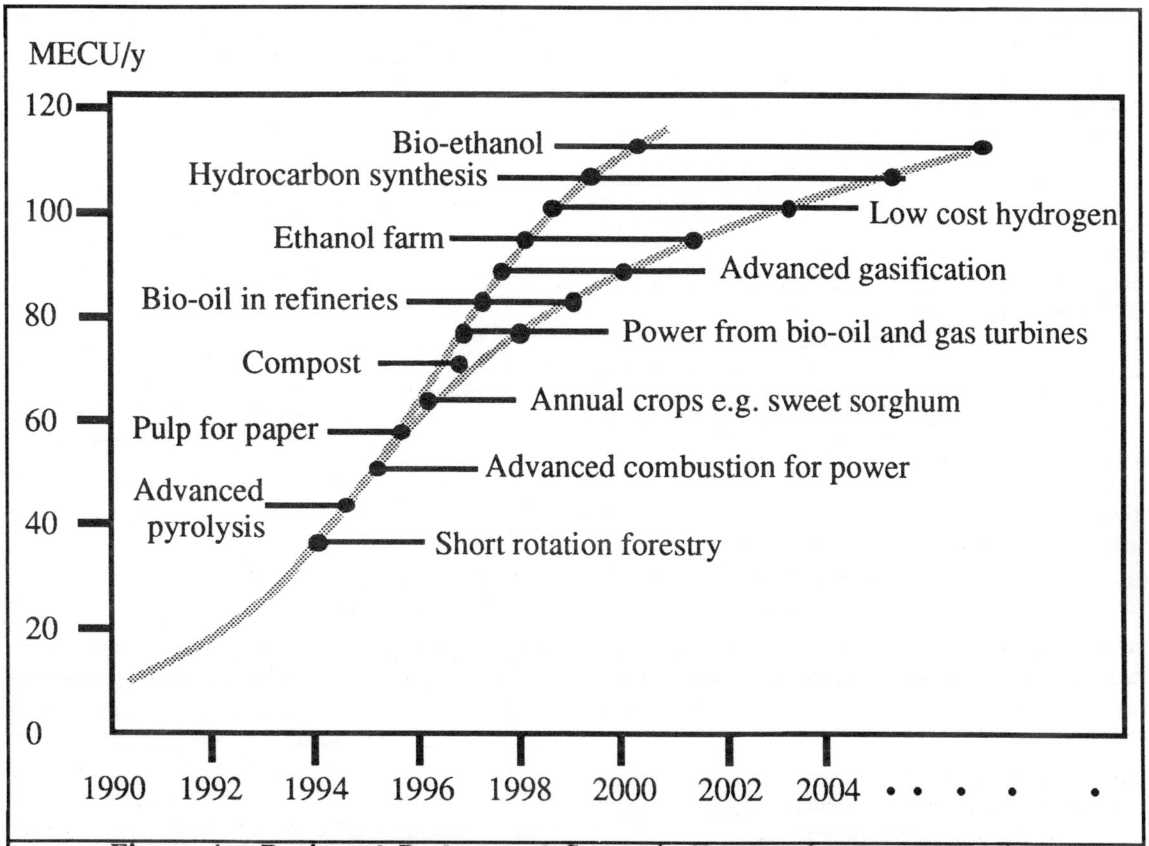

MECU/y
120
100
80
60
40
20
0
Bio-ethanol
Hydrocarbon synthesis
Low cost hydrogen
Ethanol farm
Advanced gasification
Bio-oil in refineries
Power from bio-oil and gas turbines
Compost
Annual crops e.g. sweet sorghum
Pulp for paper
Advanced combustion for power
Advanced pyrolysis
Short rotation forestry
1990 1992 1994 1996 1998 2000 2002 2004 • • • • •

Figure 4 Projected Budget and Strategic Targets for Demonstration

Thermochemical Conversion R&D Activities in Canada

E.N. Hogan and J.E. Robert
Bioenergy Supply Technology, CANMET, Energy, Mines and Resources Canada
Ottawa, Ontario, Canada K1A 0E4

This paper will present an overview of the R&D activities under way in Canada in the areas of gasification and liquefaction. Gasification received considerable support a decade ago but there is little R&D activity being supported now. A number of pyrolysis technologies have been developed in Canada and are at the pilot or precommercial stage. Research is focusing on the testing of new feedstocks, particularly those presenting disposal problems. Work is also focusing on the upgrading and/or isolation of value added chemicals and fuels from the pyrolysis oils.

Le document présente un aperçu général des activités de R-D actuellement en cours au Canada, dans les domaines de la liquéfaction et de la gazéification. Il y a une dizaine d'année la gazéifications à fait l'objet d'un grand intérêt qui a diminué depuis. Au Canada, la technologie de la pyrolyse à fait quelques percées et en est à l'étape pilote ou précommerciale. La recherche s'intéresse principalement à de nouvelles sources d'alimentation pour la pyrolyse surtout dans les domaines de déchets difficilement recyclables. Des travaux sont en cours visant l'amélioration ou la production de produits chimiques et carburants utiles à l'aide de la pyrolyse.

Introduction

Biomass is an energy source that provides Canada with a cost effective, sustainable and environmentally acceptable alternative to fossil fuels. Bioenergy currently contributes approximately 7% (540 PJ/yr) of Canada's primary energy supply. Almost all of this bioenergy is derived from the combustion of wood waste and pulping liquors to provide process heat in the forest products industry and from the combustion of wood for space heating in the residential sector.

Ample opportunity exists for bioenergy to substantially increase its contribution in the future, especially in the light of the growing concern with environmental issues. The potential supply of waste biomass has been estimated at 2,700 PJ with mill residues accounting for 700 PJ and forest wastes the remaining 2,000 PJ. The potential additional demand for bioenergy is expected to be between 500 - 700 PJ. Part of this anticipated contribution will come from the pyrolysis/gasification of biomass and industrial and municipal waste. Continued research will be required to ensure that these thermochemical conversion technologies reach the market place. Canada's biomass thermochemical conversions R&D is funded primarily through the bioenergy Development Program (BDP) of Energy, Mines and Resources Canada. The objective of the BDP is to promote the development of technologies which can convert any type of biomass into energy, prepared fuels or chemicals that can reduce traditional fossil fuel consumption patterns.

The technology areas that have been identified as requiring continued R&D are: fuel preparation and materials handling; biomass combustion; biochemical conversion and thermochemical conversion.

This paper will focus on the technology area of thermochemical conversion and examine the status, projects under way and future research directions.

Thermochemical conversion covers the processes of gasification and pyrolysis/liquefaction, in which biomass feedstocks undergo irreversible chemical changes to produce a number of end products. These products include gases and oils of varying quality, sugar solutions, chemicals (olefins, phenolics, speciality products), gasoline or diesel fuel and higher value carbon products such as activated charcoal. Feedstocks for this process are diverse and include whole biomass, fractionated biomass components such as lignin, wood waste, peat, tires, waste sludge and municipal solid waste.

Gasification

Gasification R&D received considerable support in the 1970's and early 1980's because it offered several advantages over the direct combustion of biomass including higher combustion temperatures, more efficient and controlled combustion and lower emissions. The gas produced could be used to provide energy for space heating, it could be used in turbines, boilers, engines or as a synthesis gas for methanol or ammonia production. Omnifuel Ltd. designed and installed the world's first large-scale industrial fluidized bed gasifier at a forest products complex in Hearst, Ontario. Based on this technology, Biosyn (a joint federal-provincial initiative) was established to develop and demonstrate the production of methanol from wood. Since only the initial gasification stage was considered to be non-proven, the project involved the construction and long term testing of a 10 tonne/h pressurized, oxygen-blown fluidized-bed gasifier at Ste. Juste-de-Bretanieres, Quebec. The results from the project proved technically successful - Biosyn scaled up pressurized gasification of biomass in Canada by a factor of 100 and achieved throughputs 7 to 10 times greater than other pressurized biomass units in the world at that time. The closure of the Biosyn demonstration was due to the economic disadvantage that syngas from biomass faces in comparison to that from natural gas in

Canada. For example, the predicted methanol production cost for Biosyn was projected to have been 18¢/litre in comparison to a price of natural gas methanol of 13¢/litre. A spin off project (BIODEV) operated the Biosyn gasifier in a pressurized air blown mode to establish its suitability to supply a low energy gas to an internal combustion engine to generate electricity.

The current availability of low cost natural gas and petrofuels has resulted in an almost complete cessation of gasification R&D work in Canada. While there is little interest on the part of industry to take up gasification technology at this time, it is still felt that opportunities do exist for these technologies in the future. The introduction of gasification systems into gas fired turbines for electrical generation, could result in a potential peak generation of 1000 MW. In the forest products industry in Canada, the lime kiln is the only remaining operation which still has a requirement for fossil fuels. The use of gasification in lime kilns is a commercially viable process in Europe. While economics do not favour the introduction of this technology into Canada at this time, a combination of environmental regulations and an increase in gas prices could interest a forest products company in introducing the technology into Canada through a demonstration project in the near to mid term.

In order to facilitate the use of gasifiers in electricity generation, EMR is supporting research at the University of Sherbrooke, in conjunction with a waste transfer company and the Centre Québécois de Valorisation de la Biomasse, to examine the potential of using the Biosyn technology to gasify processed municipal solid waste. The major emphasis is on the hot gas clean up to generate a fuel gas suitable for gas turbine electricity generation. Researchers at the University of Waterloo, in an extension of their work on pyrolysis, have a research project under way in which they replace the sand in their Fluidized bed reactor with a nickel based catalyst and hydrogen atmosphere. At operating conditions similar to the Waterloo Fast Pyrolysis Process, conversions of 85% of the carbon fed to methane were achieved, and later tests with different catalysts achieved conversions of 40% or more of carbon fed to C_{2+} hydrocarbons, with a high percentage of C_{4+} compounds. Current research is under way to determine whether this catalytic gasification process can provide a syngas rich in carbon monoxide and hydrogen, which can be used to partially upgrade heavy oil emulsions.

Liquefaction

Since much of Canada's energy demand is tied to liquid fuels, considerable interest has been shown in another thermochemical process - liquefaction, which converts solid biomass feedstocks into liquids either for fuels or as chemical feedstocks. This area has also undergone a significant change in R&D approach and expectations over the last ten years. Initial work was targeted on the development of technologies/reactors and products that could substitute for natural gas and fuel oil. With the government cutbacks in R&D in the mid 80's, the plan then was to conclude much of the work on the development of reactors and revert to low cost basic research, focusing on the production of chemicals from the pyrolytic oils. A number of research developments quickly occurred however that changed this strategy.

It was realized that the oils produced by biomass pyrolysis could be fractionated and refined in a fashion analogous to petroleum. These oils were produced in good yields and found to contain a variety of high value speciality and commodity chemicals. These include those in the aqueous carbohydrate fraction such as acetic acid, food flavourings, sugars, aldehydes and fine chemicals and those in the aromatic lignin derived fraction. Also, pyrolysis processes began to be applied to low quality feedstocks such as refuse, waste tires and industrial organic wastes that posed environmental disposal problems, providing an additional economic incentive to the technologies.

The objective of the research in the pyrolysis area now is to develop technologies that can thermochemically convert biomass into refinery feedstocks, as well as into liquid or gaseous fuels for the generation of electricity or process heat. Priorities include the optimization and scale-up of processes proven at the pilot stage, the improvement and industrial demonstration of fractionation and isolation processes and the development of new methods to upgrade the oils, gases and chars to industrial chemicals and products.

Reactor Technologies

The BDP has supported the development of four fast pyrolysis technologies which are at the pilot or semi-commercial stage. These processes characteristically have a very rapid heating rate, a short reaction time and quick product cooling which preserves the initial biomass decomposition products as liquids in high yields of 60 - 80%. This is achieved with minimal production of char and gases. The following is a short description of these reactors and their research status.

The University of Waterloo has developed an atmospheric pressurized flash pyrolysis unit based on a blow-through fluidized bed reactor. The process is being demonstrated in a 100 kg/h mobile pilot unit developed by Encon Enterprises to produce an activated absorbent under the tradename "Berthinate" from peat and in another 100 kg/h plant being built by Union Fenosa to produce fuel oil for electricity generation in Spain.

Ensyn Engineering has developed a transport reactor employing a solid particulate heat carrier that they have called Rapid Thermal Processing (RTP). RTP evolved from an initial concept of extremely rapid heat transfer and short contact times first investigated at the University of Western Ontario. The first commercial biomass fast pyrolysis unit (100 kg/h) for the production of speciality chemicals has been built by Ensyn Engineering for a U.S. specialty chemical company. Ensyn is in the process of finalizing designs for a scale-up to a 1000 kg/h plant and expect to have completed the construction in 1992.

A multi hearth vacuum reactor has been developed at the University of Laval. The process has been successfully demonstrated in a 250 kg/h plant in St. Amable, Quebec for the production of carbon black and fuel oil from used tires. The reactor also has potential for the production of fine chemicals and in the environmental area for the treatment of contaminated soils and waste sludge.

BBC Engineering has recently developed a system involving a simple continuous ablative transport reactor and are testing the process on shredded rubber tires. The system provides a continuous steady force for ablation, eliminates inert solids recycle yet maintains very high heat transfer rates. The reactor has been successfully scaled up to 50 kg/h and tested on a range of feedstocks including oil shale.

In this area of reactor development, the focus has been on the extension of the pyrolysis processes, which up until now have been based largely on wood feedstocks, to a wide range of low value organic wastes, many of which are causing disposal problems.

These include low quality plant oils, sewage sludge, tall oil from the kraft pulping process, newsprint recycling sludge, and urban and various industrial wastes.

The BDP has also supported the development of hydroliquefaction technologies that are still at the lab scale.

A non catalytic approach to hydro liquefaction, called rapid aqueous pyrolysis has been developed at the University of Toronto. Wood is heated to 350°C for 3-10 minutes resulting in oil yields between 45 and 55% and high phenol proportions.

The low pressure, low temperature liquefaction of wood using aqueous hydrogen iodine (HI) is being studied at McGill University. A major technical and economic problem has been the loss of the HI in the process and this has now been addressed with a new reactor recycle configuration.

Pyrolysis Products

With the current low petroleum and natural gas prices the use of pyrolysis oils only as a fuel source is presently uneconomic. The recent rekindling of interest in thermochemical conversion technologies has been partly due to the variety of commodity and specialty chemicals such as acetic acid, sugars, aldehydes and lignin based products that can be recovered from the product. As such, the second major focus of the BDP research programme in the thermochemical field has been on the development of cost effective methods for fractionation, identification, isolation, upgrading and utilization of the pyrolytic oils to produce higher value added products.

A number of processes have been developed that show good technical and economic promise. These include:

- work at B.C. Research to develop novel recovery processes to produce calcium magnesium acetate (CMA) a road deicer, and to isolate pure levoglucosans and convert these reactive sugars to polymers and catalysts.
- research in cooperation with the University of Waterloo and B.C. Research to extract chemicals, such as aldehydes and ketones, from pyrolysis oils and convert these oils to higher value products.
- research at the Saskatchewan Research Council has developed a process that can convert biomass oils - vegetable or wood oils, into a specified diesel fuel product using conventional petroleum hydro-treating refining technology. Arbokem Ltd., a Vancouver, B.C. based company involved in tall oil production has licensed this process from EMR.
- studies at the University of Laval on the production and upgrading of carbon black, activated charcoal and fine chemicals from the vacuum pyrolysis process of various feedstocks.
- work under way at the University of Saskatchewan to upgrade pyrolytic oils using HZSM-5 catalyst to gasoline derivatives.
- tests with the European Community and Ensyn Engineering to determine the suitability of using pyrolytic oils in gas turbines and to determine the emissions from the combustion of the bio-oils.
- work by Ensyn Engineering and the University of Sherbrooke to examine the feasibility of pyrolysing lignin to produce products such as transportation fuel additives, petrochemical derivatives and other chemicals, i.e., adhesives.

Further work will be required to optimize these isolation/separation processes and prove them on an industrial scale. Additionally, the current research has identified a number of other chemical products that are present in the pyrolysis oil and which could be valuable if new technologies for separating them are developed. This project area will have an increase R&D in activity over the next two to three years since the production of value added chemicals is considered crucial to the commercialization of pyrolysis technologies.

A final, but very important, aspect of the Canadian liquefaction program has been the centralized analysis program coordinated by B.C. Research (BCR). In the early stages of liquefaction research in Canada, there was little uniformity between research groups carrying out oil analyses and specifications. Consequently, it was impossible to compare the various liquefaction processes being developed. Recognizing this, EMR funded a number of efforts to develop standard characterization techniques for biomass derived pyrolysis oils. BCR performs various tests such as elemental analysis, moisture and ash content, GC-MS, etc., on the various oils submitted by the pyrolysis groups. Additionally, they are developing new or improved analytical methodologies in conjunction with the Canadian pyrolysis community as the needs arise.

Other Activities

As is shown by this meeting, there is great interest in establishing a collaborative research effort between the EEC and Canada on thermochemical conversion. A number of joint research projects such as combustion emission tests, gas turbine testing of the oils and stability tests are under way or being planned. It is also intended that EC and Canadian contractors participate more actively in each others meetings and conferences, and that the exchange of specific personnel be encouraged in order to further strengthen the cooperative effort.

Conclusion

As in other bioenergy areas, the major problem facing the introduction of thermochemical technologies is the low oil and natural gas prices. Despite this, significant progress is being made in the implementation of these technologies where specific circumstances warrant, such as high disposal costs for excess biomass residues, or for the production of high value speciality chemicals.

The new impetus now receiving R&D attention is in two areas; the identification of feedstocks such as clarifier and newspaper recycling sludge from the forest industry, or industrial wastes such as auto fluff and demolition wood which present environmental disposal problems, and; the upgrading to fuels or the isolation of high value chemicals from the pyrolytic oils.

Most of the processes have been proven at the lab scale, and the future objective will be to get them to industrial scale applications. The next stage will be to demonstrate product separation, recovery and upgrading processes at the process development unit stage.

The European Community R&D Programme on Biomass Pyrolysis and Related Activities

A.V. Bridgwater[1] and G. Grassi[2]

[1]Coordinator of EC R&D Pyrolysis Activity, Energy Research Group, Chemical Engineering Department
Aston University, Birmingham B4 7ET UK

[2]Manager of EC R&D Pyrolysis Activity, Commission of the European Communities
DGXII, 200 rue de la Loi, B1049 Brussels, Belgium

The European Comunity is now in the middle of a three year programme on Energy from Biomass in which the dominant activity is thermochemical conversion, upgrading and utilisation to produce fuels and chemicals. This paper describes the structure of this programme, and concludes by summarising the opportunities for developing an EEC - Canada Scientific and Technical Collaboration in this area.

La Communauté européenne est maintenant à mi-chemin d'un programme de trois ans sur l'énergie dérivée de la biomasse, dont l'activité principale est la conversion thermochimique, son amélioration et son utilisation pour produire des carburants et des produits chimiques. Le document décrit la structure du programme et se termine par un résumé des possibilités de création d'une collaboration entre la CEE et le Canada dans ce domaine au plan scientifique et technique.

Introduction

The Energy from Biomass Programme is one of the Renewable Energy sectors in the EEC JOULE Programme. The total budget of this programme is about II.5 million ECUs, about CAN$ 17 million, of which about 43% is allocated to the pyrolysis and thermochemical area. The whole Energy from Biomass programme has been divided into five main areas of which one is the thermochemical conversion as shown in Figure 1.

The thermochemical activity has been further subdivided into five main areas as shown in Figure 2, which covers advanced pyrolysis technologies, upgrading and characterisation, testing products and environmental considerations, direct liquefaction technologies, and techno-economics and modelling. In addition there is a strong link to the LEBEN projects **(1)** in which these technologies will be implemented and demonstrated when ready. Each of these sectors is described below.

The interactions of these six sectors is depicted in Figure 3. There is movement of information and materials between these sectors as shown in this figure. Table 1 lists the projects currently approved, and nearly all of these have now commenced. Table 2 reproduces the General Technical Annex for each contractor in the programme which applies to all the projects in this area and clearly establishes the respnsibility for each project in providing data, materials and support to other projects as depicted in Figure 3.

Pyrolysis Technologies

An essential feature of a liquid fuels research and development programme is the production of liquids in high yield and low cost. Several routes are being followed in the current programme: pyrolysis to bio-oil or bio-crude-oil; gasification to syngas for synthesis of fuels; and pressure liquefaction to bio-oil. This sector covers advanced pyrolysis processes and gasification.

There has been extensive demonstration of flash or fast pyrolysis processes in Canada and the USA that give high yields of bio-oil or bio-crude-oil of up to 80% wt on feed. There are five projects in the current programme that aim to provide similar experience and expertise in Europe.

1) The only advanced gasification project in Europe at ASCAB is being supported to gain valuable data on fluid bed pressurised oxygen gasification of biomass with secondary reforming in the 2.5t/h Stein Industrie pilot plant at Clamecy.

2) A laboratory scale air blown circulating fluid bed is planned by CRES at around 5-10 kg/h. This offers the advantage of generating heat for pyrolysis inside the reactor and avoids the need for external recycle of gas and gas heat exchange systems. Thus a more efficient and lower cost pyrolysis system is anticipated.

3) A 200 kg/h pilot plant is being built by Egemin based on entrained flow flash pyrolysis. Ground wood particles of -3 mm are contacted with hot gas in a down flowing entrained reactor with rapid quenching of pyrolysis products. Feed rate of 60 - 400 kg/h, residence time of 0.2 to 0.8 seconds and gas temperature up to 1000°C are proposed.

4) Fluid bed pyrolysis will be investigated by LNETI at two scales of operation using hot gas to effect heat transfer. Initial work will be carried out on a 10 cm fluid bed with subsequent research on a 30 cm square fluid bed with top feeding. Several parameters will be investigated including pressure, temperature and bed additives or secondary

reactants including zeolites, zinc chloride, carbonates and alkali.

(5&6) A significantly different approach to solving heat transfer problems in flash pyrolysis is effected by ablative pyrolysis in which biomass is pressed against a hot moving surface. Heat transfer is effected through a thin liquid film between the hot surface and particle which is then vapourised. The rate of reaction is a function of particle pressure and relative motion. One approach being investigated is a tumbling rotating cone at the University of Twente and another is a force fed variable geometry pinch roll or 'vee' at the University of Aston.

Upgrading and Characterisation

The liquid product from flash pyrolysis can either be used directly to substitute for fuel oil, it can be converted into a higher value energy product such as electricity, or it can be upgraded into a higher quality hydrocarbon product. All these aspects are being investigated in this sector. A key element of this group of projects is the analysis and characterisation of both the crude liquids and the upgraded products.

1) There are many individual chemical components of pyrolysis liquids that may be recovered either individually or as families. CPERI is examining methods of recovering certain fractions of chemicals and converting these to high value chemicals such as ethers to use as motor fuel additives.

2) Upgrading by hydrotreating is being studied at the University of Louvain using both orthodox and modified hydrotreating catalysts. The product contains little or no oxygen and requires reforming and/or blending to give a conventional gasoline or diesel fuel product. Previous experience on batch reactions will be used to build a continuous unit. The work will be related to the investigations at Veba into bio-oil incorporation into a conventional refinery.

3) Analysis and characterisation of the primary products from pyrolysis and the upgraded products is an essential part of the programme and this activity at the Université de Paris P&M Curie will play a key role in developing methods of analysis and characterisation as well as providing consistent analyses of products from many of the pyrolysis and upgrading projects.

4) An alternative to hydrotreating bio-oil to remove oxygen and produce hydrocarbons is through zeolite catalysts. These can either be close coupled to the pyrolysis system in which the vapours of bio-oil are formed, or can be separated with revapourisation of the bio-oil although this latter method is not preferred. Small scale tests with proprietary catalysts have produced interesting results with high yields of aromatic hydrocarbons and olefins which could, for example, give high quality gasoline. More data is needed on different catalyst systems including multifunctional catalysts to obtain design data for a pilot plant, and this activity is being led by the University of Aston in collaboration with other European and Canadian institutions.

5) An alternative approach to dedicated hydrotreatment is incorporation of the bio-oil into an oil refinery to utilise the excess hydrogen and existing process units. This is being studied by Veba Oel who will examine the optimum feed location and the effect of conventional hydrotreatment and related operations. Included will be studies on the physical and chemical properties of both bio-oil from different sources and mixtures of crude oil and bio-oil.

Testing, Slurries and Environment

A significant aspect of the programme is the development and demonstration of practical opportunities to utilise the products from pyrolysis and/or upgrading processes. There is considerable interest in the concept of bio-fuels but without demonstrating that they can be successfully and economically used in a variety of applications, they only have an academic future. This group of projects is, therefore, crucial to the success of the overall programme by showing that bio-oils and upgraded products can be successfully used by industry as well as identifying further R&D in utilisation studies.

1) A circulating fluid bed gasifier for refuse derived fuel is being constructed by Aerimpianti near Florence, Italy which will be the subject of an environmental monitoring programme. The gas generated will be burned to raise steam by conventional steam turbines. As well as RDF, it is also planned to gasify wood and other forms of biomass, with similar environmental monitoring.

2) A novel approach to char-water slurries has been the invention, by CADET, of a ternary mixture of charcoal, mineral oil and water known as Bio-UCF. This project is to further develop the product including optimisation of the production process and utilisation.

3) A process for successfully treating waste-waters from the olive oil industry in Italy has been developed by IRIN based on membrane technology. A mobile test system will be used to assess the effectiveness of this process for treating

pyrolysis waste-waters from those processes that produce water such as the KTI/Italenergie process at Raiano in Italy and the Bio-Alternative process when operated in the mode where water is produced.

4) Char-water slurries have been demonstrated as effective liquid fuels, but only small scale batch production has been carried out so far with no optimisation on particle sizes or additives. ENEL and KTI will continue to investigate the fundamental aspects of char-water slurry production and characteristics and also develop a continuous unit for char-water slurry production.

5) Conversion into electricity with a conventional small scale gas turbine directly fired with bio-oil is being examined by Noel Penny, UK. The first phase is to assess the suitability of existing turbine combustors to identify combustion problems and find solutions through modification of the combustor, modification of operation of the system, or modification of the bio-oil by additives or upgrading. The results will determine the course of the second phase of the work. This second phase is planned to directly fire the gas turbine with bio-oil coupled to a brake or generator to demonstrate the feasibility, obtain operational experience and design data for commercialisation.

6) One of the LEBEN Projects is in Galicia, Spain where a mobile flash pyrolysis unit of 250 kg/h wood feed is being constructed based on the University of Waterloo Process by Union Electrica Fenosa. One of the main problems currently in implementing bio-fuel technologies is the lack of experience and expertise in utilisation of the liquid product, so this project is aimed at exploring the combustion of bio-oil and char slurries in a variety of applications to both gain operational experience and identify problem areas that require further R&D to make implementation successful.

Liquefaction

Pyrolysis is the more developed of the direct liquefaction technologies due to the high initial and operating costs of pressure liquefaction processes. These latter processes do, however, have longer term potential through the possibility of a more controlled degradation of biomass or wastes in a system that is inherently more compact and more efficient from employing mostly liquid phase components. Several possibilities are being studied:

1) Continued development of pressurised liquefaction of biomass in solvents to produce oils and/or aromatic feedstocks is being carried out by IWC, including the development of a continuous system. Continuous hydrotreating of the oils is included in the work which will be in association with the projects at Veba and UCL. In addition work is continuing on the analytical characterisation of product oils. Post reaction separation by distillation has given interesting yields of phenols such as phenol and para-cresol.

2) A two step process for wood liquefaction by solvolysis and upgrading is being studied at the University of Compiegne. Phenol is used as the solvent at moderate temperatures and pressures followed by separation by distillation and upgrading by hydrotreating. The objectives of the work are to define the optimum conditions for the dissolution of wood, for the separation and recovery of the solvent and for the hydrotreating.

3) Supercritical extraction is being employed at the University of L'Aquila for the liquefaction of biomass and the recovery of valuable components from bio-oil. The solubility and phase equilibria of biomass and bio-oil in several supercritical fluids is being studied to identify the optimum conditions. A thermodynamic model will be used as a basic for simulating the behaviour of a multicomponent system.

4) Pressurised liquefaction of black liquor for fuel production is being studied by VTT in collaboration with IWC. Experimental work is being undertaken on both the continuous reactor at VTT and the reactor at IWC for comparison of performance. Different raw materials and upgraded products will be compared on the conversion and upgrading plants and analytical methods are being developed in parallel.

Technoeconomics and Modelling

The justification for an extensive R&D programme on biomass conversion and utilisation is that the resultant bio-fuels will be economically attractive for industry and commerce. In order to establish the most cost effective technologies, and identify the most attractive opportunities, it is necessary to have a robust technical and economic database available for selection, optimisation and planning.

1) In order to evaluate and compare pyrolysis, liquefaction and upgrading technologies, a computer model is being developed by the University of Aston. This will include performance and economic models of each relevant activity in the current JOULE programme as well as from LEBEN projects, Canada, USA, Australia, Finland, and Sweden. Sensitivity analyses on feedstock, feed cost, scale of operation, product quality and or-

thodox financial criteria will be included. This work will be carried out in collaboration with VTT, Finland who are leading the IEA Bioenergy Agreement liquefaction group - ALPS. Significant advantages of biomass derived fuels include their low sulphur content which enhances their value and indigenous production which also enhances value through import savings, provision of employment and investment and infrastructural development. These effects will also be quantified in the project.

2) A parallel activity at the University of Naples will generate robust models of pyrolysis processes, that will utilise data generated from practical experimentation in the Energy from Biomass Programme to improve and validate theoretical models.

Collaboration

A framework has now been established in which the EEC and Canada can collaborate in the development and implementation of technologies for the production of fuels and chemicals from biomass. R&D programmes in both locations will be examined at a series of workshops, seminars and meetings over the next year or so from which common objectives and aspirations can be established to more clearly define a cooperative programme. Some of the initial possibilities are summarised in Table 3 to illustrate the opportunities for collaboration at all levels of research, development demonstration and commercialisation of biomass pyrolysis, upgrading and utilisation.

Discussion:

N. Bakhshi (University of Saskatchewan): What sort of cooperating work are you doing with zeolite?

A.V. Bridgwater (Speaker): This activity is not doing work on zeolite. It is to establish a program on zeolite.

One of the activities I wanted to achieve here, is to establish contacts with workers like yourself on zeolite in order to agree with some common objectives and discuss ways of working closely together. Zeolites appear to offer some interesting possibilities for production of very valuable fuels and chemicals from biomass. I think it is an area we would like to get more involved in.

There is no current work in the European programme on zeolites but I would like to see some.

Participant: What is the concept behind the work by Vega Oel as opposed to a conventional refinery?

Speaker: They are going to look at two aspects. One is the incorporation of a bio-crude oil or bio-oil into a refinery stream and an assessment of where it might be fed in and what the possible consequences are; either to put it into the first crude oil stream or into a subsequent partially refined stream of refinery. They also have extensive hydro-treating facilities which is a pilot scale. The intention is, that they will work more closely with the University of Louvain to look at larger scale processing in a more conventional petroleum refinery, hydro-treated operation, and how that might work with bio-oils - either alone, or in combination.

So, it is really a very practical applied look at how a conventional refinery might handle these liquids - both to see if they can and how they might.

Participant: What type of work are you doing on the impact to the environment?

Speaker: At the moment not a great deal of work. There is a particular contract in Aerimpianti who are building an RDF, refused-derived-fuel gasifier in Italy, and we are providing some support to the analysis of the emissions and the impact of those emissions on the environment. The EEC is also a member of the International Energy Agency Waste Water Activity, again, to identify what the problems are in order to commit resources in the future. It seems to be an area of considerable importance where you are going to have to pay a lot more attention. At the moment there is not very much activity beyond a recognition of its importance.

TABLE 1
Summary of Thermochemical Conversion Contracts in the ENERGY FROM BIOMASS Programme of JOULE

Contractor	Project
PYROLYSIS TECHNOLOGIES	
ASCAB, France	Operation of a 2500 kg/h pressurised oxygen fluid bed gasifier, with secondary gasification
CRES, Greece	Construction and testing of a 3 kg/h flash fluid bed pyrolyser
Egemin, Belgium	Construction and operation of an 200 kg/h entrained flow pyrolyser
LNETI, Portugal	Construction and testing a fluid bed pyrolyser
University of Aston, UK	Construction and testing an ablative pyrolyser
University of Twente, Netherlands	Construction and testing an ablative pyrolyser
UPGRADING AND CHARACTERISATION	
CPERI, Greece	Production of chemicals and fuel additives from bio-oils
University of Louvain, Belgium	Hydrotreating of pyrolysis oils
University P&M Curie, France	Characterisation and analytical procedures for bio-oils
University of Aston, UK	Design and development of processes for upgrading bio-oils
Veba Oel, Germany	Evaluation of upgrading bio-oils in a conventional refinery.
TESTING, SLURRIES AND ENVIRONMENT	
Aerimpianti, Italy	Environmental monitoring of a refuse derived fuel gasifier
CADET, France	Testing char-water slurries
Fac-Consult, Belgium	Testing bio-oil in a turbine
IRIN, Italy	Waste-water processing
ENEL and KTI, Italy	Development of a continuous process for char slurries
Noel Penny, UK	Combustion testing for gas turbines
Union Electrica Fenosa, Spain	Testing bio-oils in a variety of applications
LIQUEFACTION	
IWC Hamburg, Germany	Continued development of pressurised liquefaction of biomass in solvents
University of Compiegne, France	Supercritical liquefaction of biomass
University of L'Aquila, Italy	Solvolysis for liquefaction of biomass
VTT, Finland	Pressurised liquefaction of black liquor for fuel production
TECHNOECONOMICS AND MODELLING	
University of Aston, UK	Technical and economic assessments of pyrolysis and upgrading processes
University of Naples, Italy	Development of models of pyrolysis processes
CONTRACTS NOT YET FINALISED	
Bio-oil standards,	Environmental problems
Mobile pyrolysis network	Electricity production network

TABLE 2
GENERAL TECHNICAL ANNEX
for all Energy From Biomass Contracts In Thermochemical Conversion

These projects are part of a co-ordinated CEC activity in the field of thermochemical conversion of biomass. The contractors will, therefore, participate in a Coordinated Network dealing with various R & D topics to facilitate interaction and effect collaboration as follows:

1 Pyrolysis plants will be built and operated by the following organisations: **CRES, Egemin, Aston University, Twente University and LNETI.** Samples of bio-oils from each process will be made available for analysis by **P&M Curie University** in collaboration with **UCL** and **CTFT**, for upgrading by **UCL** and **CPERI**, and for testing on a variety of applications by **UEF** and **Noel Penny** as detailed in the relevant sections below. Data will be made available to the **University of Aston** for technical and economic assessment and the **University of Naples** for modelling pyrolysis processes.

2 The liquid products from each of the pyrolysis plants detailed above in section 1 will be characterised by physical and chemical properties and by detailed analysis by **P&M Curie University** in collaboration with **UCL** and **CTFT**. The products will be compared with samples from other processes from Canada and the USA.

3 Samples of the liquid products from some or all of the pyrolysis processes detailed above in section 1 will be upgraded to hydrocarbons through hydrotreating by **UCL** in collaboration with **VEBA**. The products from hydrotreating will be characterised as detailed in section 2 by **P&M Curie University** in collaboration with **UCL** and **CTFT**.

4 Zeolite upgrading will be coordinated by **Aston University** on bio-oils from **Egemin, LNETI, CRES** and/or Bio Alternative in the context of the LEBEN project and/or from other sources. Tests will be commissioned and small units designed for installation on one or more of the above pyrolysis units. The products will be characterised as detailed in section 2 by **P&M Curie University** in collaboration with **UCL** and **CTFT**.

5 High pressure liquefaction of biomass will be carried out by the **University of Compiegne, University of L'Aquila** and **Institute of Wood Chemistry** in collaboration with **VTT**. The liquid products will be characterised as detailed in section 2 by **P&M Curie University** in collaboration with **UCL** and **CTFT**.

6 Char-water and char-oil slurries will be manufactured by **ENEL and KTI** and **CADET** and tested by them and **UEF** for combustion and power generation.

7 The opportunities for incorporation of pyrolysis oil into a conventional refinery for upgrading and blending will be assessed by **Veba**. Relevant data on product characteristics will be supplied by **P&M Curie University** in collaboration with **UCL** and **CTFT**. Relevant data on pyrolysis and liquefaction processes will be supplied by **Egemin, Aston University, Twente University, LNETI, CRES, University of Compiegne, University of L'Aquila** and **Institute of Wood Chemistry** in collaboration with **VTT**. Relevant data on process performance and economics will be shared with **Aston University** in the context of technoeconomic evaluation.

TABLE 2
GENERAL TECHNICAL ANNEX continued

8 High added value chemicals will be manufactured by **CPERI** from bio-oils supplied by one or more of: **Egemin, Aston University, Twente University, LNETI, CRES, University of Compiegne, Institute of Wood Chemistry,** and **University of L'Aquila** as they become available. Comparisons will be made with oils from other sources such as from Canada and the USA in the context of the agreed collaboration.

9 Combustion trials of bio-oil and char slurries will be carried by **UEF**, and by **Noel Penny** on a gas turbine. Oils from one or more of: **Egemin, Aston University, Twente University, LNETI, CRES, University of Compiegne, Institute of Wood Chemistry,** and **University of L'Aquila** will be used as they become available. Comparisons will be made with bio-oils from other sources such as Bio-Alternative, KTI/Italenergie, and from Canada and the USA in the context of the agreed collaboration.

10 **IRIN** will apply their experience of integrated waste-water treatment from the olive oil and similar industries to.pyrolysis waste-waters from one or more of the processes referred to in section 1 and/or other relevant processes as material is available. This will assess the feasibility of different techniques and produce outline treatment plant designs and costs. Relevant data will be supplied to **Aston University** for evaluation.

11 **TÜV** will develop suitable standards for assessment and utilisation of bio-oils in a variety of applications using relevant information on characteristics from **P&M Curie University, UCL** and **CTFT**; performance data from **CADET, UEF,** and **Noel Penny**; and upgraded product data from **UCL, VEBA** and **CPERI.** Data will be utilised as soon as it becomes available.

12 **Aerimpianti** will monitor emissions and other environmental aspects resulting from the operation of their refuse derived fuel gasification plant. The consequences for environmental control measures will be assessed.

13 On gasification, **ASCAB** will derive results from the pressurized fluid bed oxygen gasifier at Clamecy on eucalyptus feeds to establish performance criteria and viability assessment.

14 Other projects that are initiated during the course of this programme will contribute to this Coordination in a similar way as the projects listed above.

15 Other activities in any of the above areas may also be carried out in the context of agreed collaboration with other countries and organisations including Canada, Brazil and the International Energy Agency Bioenergy Agreement, and these activities will also form part of this coordinated programme.

16 All contractors will provide confidential information to the Commission, and performance and cost data to **Aston University** for techno-economic assessment.

TABLE 2
GENERAL TECHNICAL ANNEX continued

Coordination

All the above contractors will fully cooperate in the Coordination Activity to effect interaction and collaboration between all activities, which will be managed by the Commission with the assistance of.the coordinators as indicated in the attached list.

Coordination will be effected through regular workshops, sectoral seminars and working groups in addition to Contractors Meetings. These thermochemical coordination meetings will review progress on each project, discuss required interactions and collaboration, and set target dates for deliveries of samples, materials and and information to each contractor. Coordinators will be responsible for reporting progress to the Commission, and negotiating implementation of necessary collaborative interactions

TABLE 3

COLLABORATIVE RESEARCH PROGRAMME IN BIOMASS PYROLYSIS between THE COMMISSION OF EUROPEAN COMMUNITIES and CANADA

INTRODUCTION

A major new biomass pyrolysis programme has recently commenced in Europe supported by the EEC R&D Directorate General XII as described in Tables 1 and 2 above. This is valued at about CAN$11 million, and includes advanced pyrolysis processes, upgrading, and utilisation studies.

OBJECTIVES

The main objectives of this collaboration include:

- to obtain an overview of the main activities in Canada in the field of liquid fuel and chemicals production from biomass,
- to provide an overview of the corresponding EEC programme
- to assess the orientation and future development plans of the current range of research activities in both the EEC and Canada,
- to identify the opportunities for collaboration between the EEC and Canada in the area of liquid fuels and chemicals production from biomass by thermochemical processing
- to establish an framework to proceed with implementing the collaboration agreement.

POSSIBLE AREAS OF COLLABORATION

The following areas of collaboration have already been identified for research and development leading to implementation of pyrolysis technologies in Europe.

- Joint Canada-EEC workshops on pyrolysis technology achievements and problems
- Characterisation of liquid products
- Vacuum pyrolysis
- Catalytic upgrading
- Waste disposal
- Toxicity assessment and measurement
- Liquid stabilisation
- Liquid compatibility
- Chemicals in co-production with fuel
- Combustion of pyrolytic products
- Process optimisation
- Slurry production, handling and combustion
- Charcoal activation
- Feed pretreatment, both physical and chemical

TABLE 3

COLLABORATIVE RESEARCH PROGRAMME IN BIOMASS PYROLYSIS continued

IMPLEMENTATION OF COLLABORATION

The first step is to agree to have a joint programme of research and development that will lead to academic and industrial cooperation followed by implementation of pyrolysis technologies in both countries. This can be followed by a regular series of workshops to agree objectives, programmes, activities and priorities, in which industrialists would be expected to play a key role.

Subsequent activities are likely to include exchange of scientific personnel, exchange of information on agreed activities, establishment of an agreed scientific programme in areas of common interest, stimulation of industrial cooperation, and implementation of pyrolysis technologies at an industrial scale.

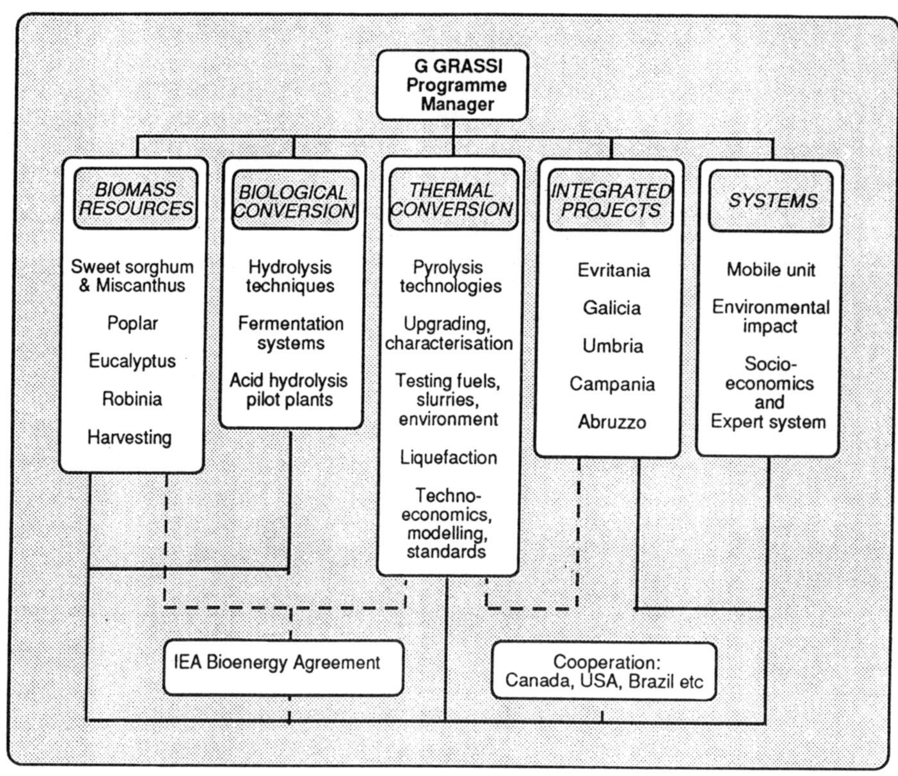

Figure 1
Structure of the EEC Energy from Biomass Programme

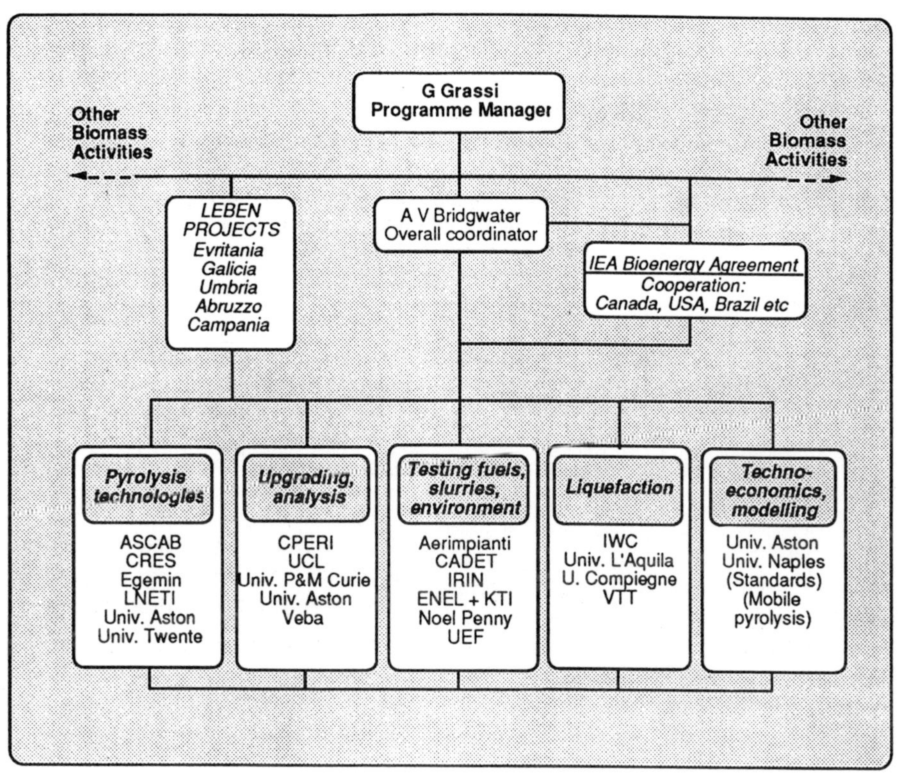

Figure 2
The Thermochemical Area of the EEC Energy from Biomass Programme

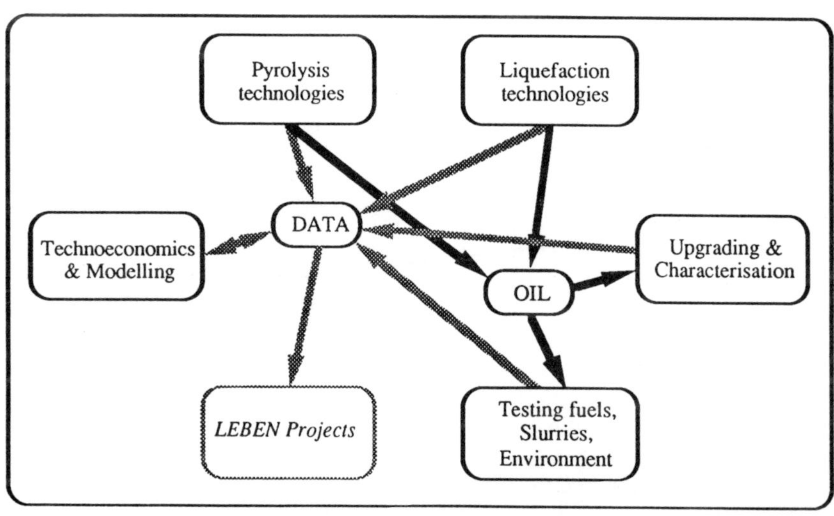

Figure 3
Interactions Between the Major Areas of Activity in the EC Pyrolysis Programme

Upgrading of Pyrolysis Oils by Hydrotreating

E. Laurent, P. Grange and B. Delmon
Unité de Catalyse et Chimie des Matériaux Divisés, Université Catholique de Louvain
Place Croix du Sud 2, Boite 17, B-1348 Louvain-la-Neuve, Belgium

Several aspects of the upgrading of pyrolysis oils by hydrotreating studied at the Université Catholique de Louvain are summarized. The necessity of a stabilization before or during the treatment is outlined. Practical solutions are envisaged. A pretreatment at lower temperature or the coprocessing with a fraction of hydrotreated oil permit a hydrotreating at common temperature. Influences of operating conditions are briefly discussed. New perspectives in research are presented.

On résume plusieurs des aspects du procédé d'amélioration, par hydrotraitement, des huiles de pyrolyse, qui a été étudié à l'Université Catholique de Louvain. On explique brièvement pourquoi la stabilisation est nécessaire avant et pendant le traitement. On examine des solutions pratiques. L'application d'un prétraitement à une température plus faible ou le cotraitement avec une fraction d'huile hydrotraitée permet de réaliser l'hydrotraitement à une même température. On examine brièvement l'effet des conditions de traitement. On présente les nouvelles perspectives en matière de recherche.

Introduction

Oils can be produced from biomass through several processes with high yield. Among these processes, pyrolysis has been directed these last years towards fast degradation of polymeric structures and rapid elimination of primary products. These last developments of pyrolysis are call fast **(1)**, flash **(2)** or vacuum pyrolysis **(3)**. Yields as high as 50-70% of liquids can be obtained. Much interest has been given to these oils as renewable energetic fuel. They offer the advantages to be more energetic than wood or charcoal on a volume basis and to be liquid, important properties considering the handling and uses.

The bio-oils are usually used in direct combustion even though they rarely meet the standards required for fuels. They are very viscous, corrosive, not completely volatile and present high oxygen contents which are indirectly responsible for their general properties.

In order to increase the possibilities of use of pyrolysis oils as clean fuel, they need to be processed to reduce viscosity, corrosivity, instability and heteroatoms contents. Up to now, two catalytic processes have been investigated. One is a dehydration-decarboxylation over zeolitic catalysts. The base of the process is the aim of zeolite ZSM-5 to convert small oxygenated organic molecules to a gasoline type highly aromatic fuel. In the case of Methanol, low deactivation rates are achieved which can not be obtained with complex crude pyrolysis oils. Nevertheless, yields as high as 30% of dry feed oil in C5-C10 fraction have been reported by Kaliaguine **(4)** with high small organic content vacuum pyrolysis oils. The system included a precoker and catalyst life time was between 250 and 500 seconds.

The second approach is to treat pyrolysis oils under hydrogen pressure in presence of traditional sulfided Cobalt-Molybdenum or Nickel-Molybdenum catalysts. The treatment aims at the removal of oxygen and nitrogen as water and ammonia and, to some extent, at the hydrogenation and hydrocracking of large molecules.

Different searchers have shown the applicability of such treatment to pyrolysis oils in spite of the instability of the oils at a temperature higher than 200°C. Elliot and Baker **(5)** performed hydrotreatment of pyrolysis oils under upflow conditions. They observed that a stabilization step at a temperature between 250°C and 280°C is needed before upgrading in more severe conditions. Yields of 31% in gasoline type products have been achieved on a wet oil basis, thus, yields of nearly 45% on a dry oil basis. Elliot, moreover, reports a one-step hydrotreatment using a non isothermal reactor **(6)**.

Sheu et al **(7)** also studied the hydrotreatment of diluted pyrolysis oils using a non-isothermal reactor. The influences of several reaction parameters are pointed out and kinetic aspects are developed. Gagnon **(8)** studied the possibility to pretreat vacuum pyrolysis oils in presence of Ruthenium catalyst at low temperature before the upgrading itself. The instability of the oils is evidenced at a temperature as low as 100°C and the influence of the pretreatment temperature on the upgrading efficiency is demonstrated.

At the Université Catholique de Louvain, Churin studied the upgrading of pyrolysis oils by hydrotreating. The present paper summarizes the information obtained about reaction procedure, reaction conditions and catalytic aspects. The conclusions are written as perspectives for the future research in this field.

Experimentation

The oils used in this study have been produced in the biomass pyrolysis demonstration unit of Raiano, Italy (oil I) and in the experimental carbonization unit of S.A. Bio-Alternative, Zwitzerland (oil II). General characteristics and composition of these oils are presented in Table I.

Hydrotreating experiments were conducted in a one-liter batch autoclave with the catalysts free in the liquid phase. Reaction temperature ranged between 300°C and 400°C and hydrogen pressure between 7 MPa and 12 MPa. Industrial Ni-Mo and Co-Mo catalysts were used in their sulfide state.

Results

Necessity for a stabilization step (9).

A rapid heating of pyrolysis oils to the reaction temperature induces thermal degradation. The polymerization rate of the molecules constituting the oils is higher than the rate of the deoxygenation reaction. A stabilization step at lower temperature is a possibility that has been tested successfully by several groups.

To determine the optimum pretreatment temperature, the experiment reported on (Fig. 1) was performed. The pressure was allowed to evolve freely in the batch reactor in function of the temperature. The straight line is the evolution of pressure without hydrogen consumption and the curve represents the pressure evolution on the pyrolysis oil and catalyst system.

One can observe that hydrogen consumption starts at a temperature as low as about 120°C. The hydrogen consumption begins to be significant at a temperature 220°C. The upgrading experiments were thus performed with a pretreatment at the temperature of 220°C with traditional sulfided catalysts.

Beneficial influence of dilution with hydrogen donor solvent (10).

The dilution of oxygenated molecules constituting the oils is theoretically advantageous because polymerization is hindered. The use of a hydrogen donor solvent presents, moreover, the possibility to stabilize free radicals. The effects of the co-processing of pyrolysis oils with tetralin are presented in Table II.

It can be seen that tetralin has a positive influence on the HDO and HDN. Its effects on the percentage of volatile fraction is more pronounced.

The results can be explained by the role of tetralin which limits the formation of high molecular weight compounds. Deactivation of catalyst is limited because heavy compounds which are coke precursor are not formed or are formed at a lower rate.

On a larger commercial scale, tetralin has to be replaced by a fraction of treated pyrolysis oil or coprocessed with a petroleum fraction. Tests have been done where tetralin was replaced by hydrotreated pyrolysis oil. The histograms presented in Figure 2 compare virgin pyrolysis oil with treated pyrolysis oil (Ni-Mo) without solvent, with tetralin and with hydrotreated oil as solvent. The five fractions are obtained by liquid chromatography (SESC) on a silica column (11).

It appears that a fraction of hydrotreated pyrolysis oil can replace tetralin with only a minor loss of efficiency. Recirculation of the heavy fraction of treated pyrolysis oil in some kind of extinction hydrotreating is a valuable solution to the instability problems.

Control of the treatment via reaction conditions and catalyst type

It is planned to roughly outline the influence of several reaction parameters on the hydrotreatment of pyrolysis oils.

Catalyst Type (12)

The two main hydrotreating catalysts are Ni-Mo and Co-Mo supported on g-alumina used principally and respectively in hydrodenitrification, hydrogenation reactions and in HDS reactions.

In the case of the hydrodeoxygenation of pyrolytic oils, these two catalysts show similar activity. A difference is visible considering the hydrogenation and hydrocracking activity. The Nickel-Molybdenum consumes more hydrogen to obtain the same oxygen removal level (440 l/kg versus 380 l/Kg for CoMo). As a consequence, it shows a better ability to convert heavy fractions but also produces a larger quantity of gas.

Temperature (13)

The effect of temperature of treatment between 300°C and 400°C does not correspond to temperature influence on the HDO kinetic of model compounds. This fact has to be related to the low stability of pyrolysis oils. A higher temperature in this range 300°C - 400°C, in spite of a stabilization pretreatment, causes a higher polymerization during heat-up period and beginning of reaction. Heavy compounds present more resistance to deoxygenation and produce more coke deactivating the catalyst.

Pressure (13)

The hydrogen pressure, in the range 7 - 12 MPa, has a very low influence on the deoxygenation reaction. High pressure is not required for HDO of model compounds. The effect of increase in pressure is mainly an increase in the production of secondary gaseous products. Nevertheless, the necessity of high hydrogen pressure to avoid high coking rate has not been investigated.

Conclusions and Perspectives

The necessity of a pretreatment step is evident before hydrotreating pyrolysis oils. Using traditional sulfided catalysts, a temperature of 220°C is a minimum.

The possibility to dilute pyrolysis oil with a fraction of treated oil or a fraction of petroleum has been assessed. A higher stability during the treatment is obtained and "extinction hydrotreating" has been postulated as a good solution.

Taking account of these results and information of other groups, it is intended to direct research towards:

1) hydrotreating of pyrolysis oils under flow conditions which are representative of larger scale conditions. Use of non-isothermal reactor temperature is envisaged.

2) optimization of pretreatment step. This represents mainly the testing of new catalysts (cheap and disposable) and search for optimum conditions (temperature, pressure, contact time).

3) optimization of hydrotreating conditions and specially, search of minimum reaction conditions severity needed for a product usable in power generator engines.

4) precise quantification of hydrogen consumption. The objective is to reduce hydrogen consumption to the minimum level. This consumption will be related to severity of the treatment, catalyst type and oil characteristics.

5) once standard reaction conditions are set up, comparison of several pyrolysis oils in hydrotreating will be performed. Influence of composition, related to pyrolysis conditions, pyrolysis technique or raw material, will be investigated.

The discussion of all the results of pyrolysis oil hydrotreating needs fundamental evidence about the HDO mechanism of oxygenated compounds. The literature, with respect to this subject, shows discrepancies. It is intended to develop HDO reaction on model compounds and focus on reactivity of oxygenated molecules, catalyst activity and selectivity. These results will serve as a comparison base for peculiarities of pyrolysis oil discussion.

References

1. Knight, J., Gordon, C.W., Kovac, R.J., and Newmann, C.J., in Proceedings of the 1985 Biomass Thermochemical Contractors Meeting, Minneapolis, p. 99.
2. Scott, D., and Pirskorz, J., Can. J. Chem. Eng. 62 (1982), 404.
3. Roy, C., de Caumia, D., Blanchette, D., Lenrieux, R., and Kaliaguine, S., in Energy from Biomass and Wastes IX, D.L. Klass, ed. IGT, Chicago, 1985, p. 1085.
4. Renaud, M., Grandmaison, J.L., Roy, C., and Kaliaguine, S., Amer. Chem. Soc., Div. Petr. Chem. prepts. 32, 2 (1987), 276.
5. Elliot, D.C., and Baker, E.G., in Energy from Biomass and Wastes X, D.L. Klass, ed., Chicago, 1987, p. 765.
6. Baker, E.G., and Elliot, D.C., in Research in Thermochemical Biomass Conversion, A.V. Bridgwater and J.L. Kuester, eds., Elsevier, 1988, p. 883.
7. Sheu, Y-H.E, Anthony, R.G. and Soltes, E.J., Fuel Proc. Tech., 19 (1988), 31.
8. Gagnon, J., and Kaliaguine, S., Ind. Eng. Chem. Res., 27, 10 (1988), 1783.
9. Churin, E., Grange, P., and Delmon, B., in Biomass for Energy and Industry, 5th Conference, ed., Grassi, Gosse and do Santos, Lisbon, 1989, p. 2.616.
10. Churin, E., in "Euroforum New Energies", ed. Stephens et al, Saarbrücken, 1988, p. 638.
11. Churin, E., Maggi, R., and Delmon, B., in Biomass for Energy and Industry, 5th Conference, ed., Grassi, Gosse and do Santos, Lisbon, 1989, p. 2.621.
12. Churin, E., Maggi, R., Grange, P., and Delmon, B., in Research in Thermochemical Biomass Conversion, ed. Bridgwater and Kuester, Phoenix, 1988, p. 896.
13. Churin, E., Grange, P., and Delmon, B., EEC contract EN3B-0097-B.

Discussion:

N. Bakhshi (University of Saskatchewan): Do you have any idea of the full size of your catalysts?

E. Laurent (Speaker): Yes, the average size is around 100 for hydro-treating catalysts, this is cobalt molybdenum. I think maybe it could be possible to develop a catalyst with a combination of hydrotreating catalyst and zeolite catalyst as two combinations, and maybe have more selective hydrocracking of every fraction contained in pyrolysis oil without having too much gas formation - having a hydro-treating component in this same catalyst. These are also some of our projects for developing catalysts.

N. Bakhshi: Is your reaction an isothermo operation or is it a non-isothermo operation?

Speaker: It is a batch reactor. There is a hydro-treatment step at the lower temperature, but you have to increase the temperature until we reach a reaction temperature of about 350 degrees centigrade. It is not really an isothermo reaction.

A. Uppal (Participant): How would you classify the end product? Would you classify it as a heating oil or kerosene?

Speaker: We have not yet a really good characterization method for the product of the treatment, but we would like to have a simulated distillation on techniques like "GBC" to characterize the quantity of light fraction produced by the treatment and also the conversion of every fraction of the oil.

A. Uppal: But actually, what is the boiling range of the product?

Speaker: It is possible when you prepare for a severe treatment where all the oil is in the boiling range, 0 to 350 degrees centigrade.

E. Chornet (University of Sherbrooke): What is the percentage of water in this oil at the beginning? And what do you have to do, if anything, to remove the water?

Speaker: In fact the oil from this carbonization unit, contains a very low quantity of water, about 5%. But before we do the test, we dry the oil on the vacuum in order to try to remove the water.

We are also performing experiments with some other compounds at this time in order to study the inference of water or ammonia on the catalyst activity and selectivity. But we avoid the presence of water at the beginning of the reaction.

E. Chornet: So, there is the utilization step prior to processing?

Speaker: Yes.

Table I
Characteristics of pyrolysis oils studied

	OIL 1	OIL 2
Viscosity Cp (60°C)	480	240
Density (g/CC)	1.19	1.17
Water content	10.0	4.5
Elemental composition		
C%	71.4	58.8
H%	7.2	6.9
0%	16.9	33.7
N%	1.7	0.6

Table II
Influence of tretralin on the processing of pyrolysis oils

	Without tetralin	Tetralin/oil ratio 0.5	Tetralin/oil ratio 1.0
% O	4.7	3	2.2
% HDO	70	80	85
% HDN	58	75	85
Volatile fraction, Wt %	50	75	95

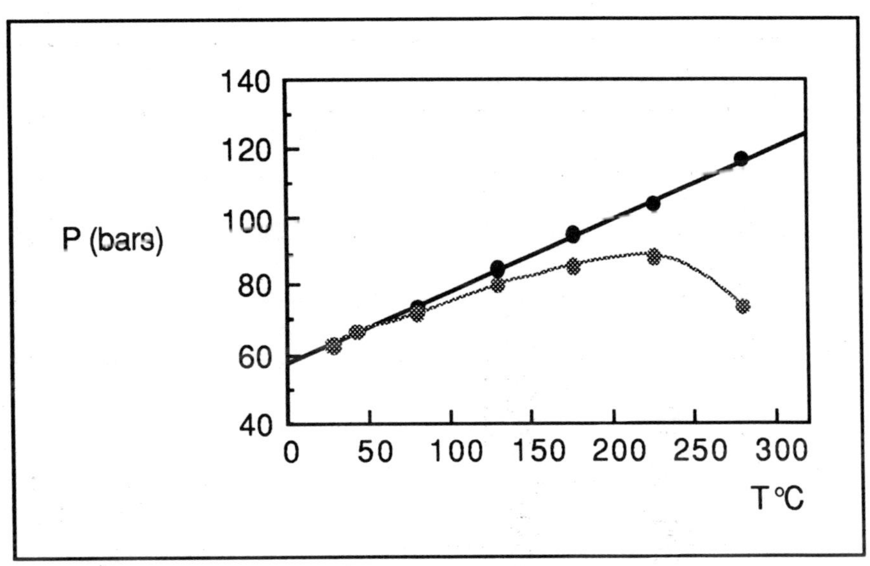

Figure 1
Hydrogen pressure evolution in batch reactor with and without consumption.

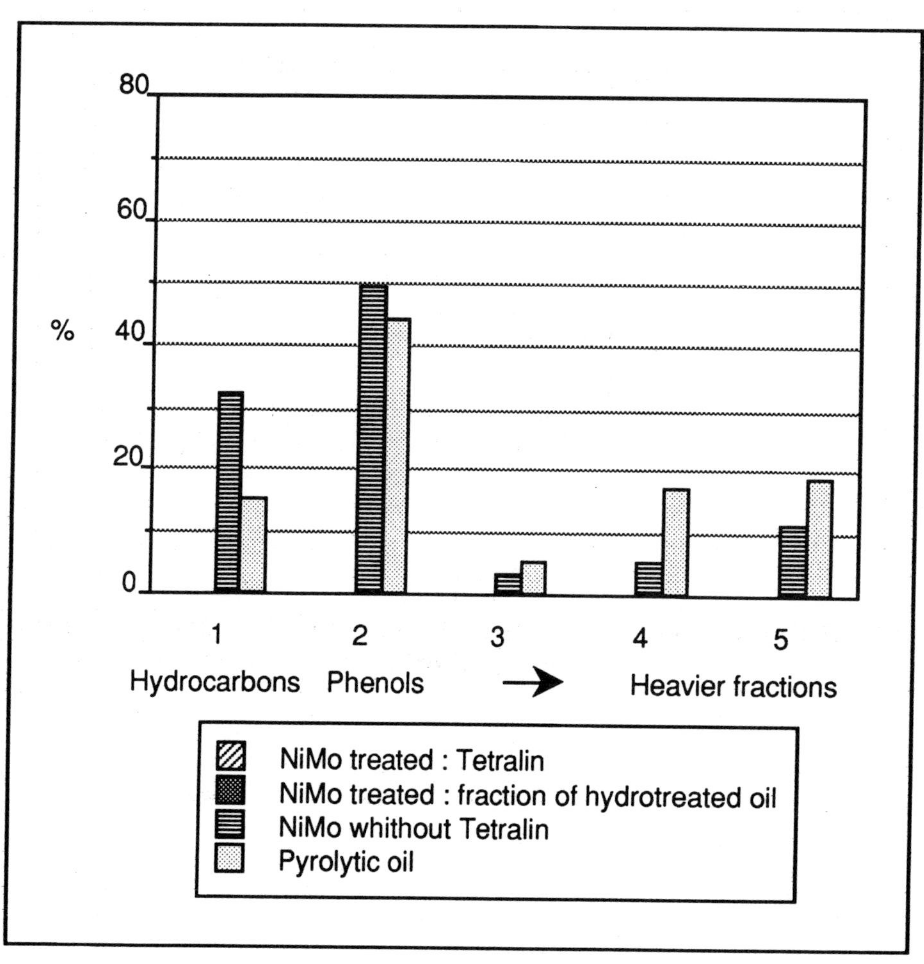

Figure 2
Fractions of pyrolysis oil obtained by SESC and treated differently

Flash Pyrolysis of Biomass in an Entrained Bed Pilot Plant Reactor

K. Maniatis[1], J. Baeyens[2], H. Peeters[3] and G. Roggeman[3]
[1]Dept. of Chemical Eng., Free University of Brussels, Pleinlaan 2, Brussels 1050, Belgium
[2]Dept. of Chemical Eng., Catholic University of Leuven, (KUL), Kardinaal Mercierlaan 92, 3030 Leuven-Heverlee, Belgium
[3]Egemin NV, Bredabaan1201, 2120 Schoten, Belgium

Biomass pyrolysis technology has been extensively studied for the production of bio-oil and several processes have been developed in US, Canada and Europe. In the framework of the JOULE programme of the Directorate General for Research and Development of the Commission of the European Communities EGEMIN has developed an entrained bed flash pyrolysis process and the construction of the plant started in October 1990. The capacity of the plant is 200 kg/h and it is expected that it will be fully operational by June 1991. The conceptual design as well as the operating parameters of the pilot plant are presented in this paper. The expected bio-oil yield is 60 wt% at an average operating temperature of 550° C and mean residence time of 0.6 s.

La pyrolyse de la biomasse a fait l'objet de nombreuses études pour la production de bio-pétrole; plusieurs procédés ont été mis au point aux États-Unis, au Canada et en Europe. Dans le cadre du programme JOULE, de la Direction générale pour la recherche et le développement de la Commission des communautés européenne, EGEMIN a mis au point un procédé de pyrolyse-éclair à lit entraîné. La construction de l'usine a débuté en octobre 1990. L'usine devrait avoir atteint sa capacité de production de 200 kg/h d'ici juin 1991. Le document présente le design conceptuel ainsi que les paramètres de fonctionnement de l'usine pilote. Le rendement prévu de bio-pétrole est de 60 pourcent de la masse à une température moyenne de 550 °C et un temps de séjour moyen de 0,6 s.

Introduction

The potential application of thermochemical biomass conversion systems for fuel and chemicals production is generally considered technically feasible. There are three main thermochemical processes by which bioimass can be upgraded to a higher value product and these are: Pyrolysis, Gasification and Liquefaction. From these the pyrolysis processes are attracting increasing attention due to the production of liquid fuels (liquefaction processes are still considered unproven for industrial scale plants). In particular, flash pyrolysis processes have resulted in very high yields (up to 70 wt%) of bio-oil. Their ease of handling, storage and transportation as well as their utilisation in several heat and power applications with little or no modification to existing equipment has made liquid fuels the desirable products for near future alternative energy scenarios. Moreover a variety of feedstocks (such as wood, energy crops, agricultural residues et al.) can be considered for flash pyrolysis processes as long as they meet the criteria of low ash content, low moisture content and small particle size.

EGEMIN NV is a dynamic engineering company involved in several areas such as Process Automation and Integration, Project Management, Sensors Development et al. with licensees in the EEC, US and Japan. Investments in new products and processes in the R&D department have resulted in several successful projects in Bio-Thermodynamics, Environmental Systems and Thermochemical Biomass Conversion. Since the mid-eighties EGEMIN had identified the potential of biomass resources as an alternative renewable source of energy and several studies were carried out in association with Belgian Universities in order to identify the most optimum conversion routes. The European Commission's call for tenders in the framework of the JOULE programme offered the possibility to implement this project and accelerated the efforts of EGEMIN to bring the flash pyrolysis process to commercialisation.

Experimental

Process Specifications

A thorough literature review indicated that although entrained bed reactors are well suited for flash pyrolysis due to the ease in residence time control as well as the particle size characteristics of biomass, no fully proven and reliable process existed. It was therefore decided to develop an entrained bed process that would operate under flash pyrolysis conditions. A market search also identified several wood working industries in Belgium which produce large quantities of wood waste and are unable to utilise it for their own benefit. A factory was selected which produces a relatively well defined wood waste with mean particle diameter of about 1mm which is ideal for flash pyrolysis. This factory will supply the feedstock for the initial experiments of the pilot plant.

The process design was based on published results as well as practical experience from the team of experts of EGEMIN and the process specifications are summarised in Table 1. Table 2 shows the expected yields of the various products. The performance of the entrained bed as presented by the data of Table 2 is rather conservative and it is expected that during the optimisation stage the yield of the bio-oil will exceed 60 wt%. The process has been designed in a rather flexible way so that for most parameters of importance there is a turn down ratio of 1:2.

Table 1.
EGEMIN Flash Pyrolysis Process Specifications

Parameter	Units
Specific throughput	1700 kg/h.m^2
Capacity	200 kg/h
Operating Temperature	550°C
Operating Pressure	1 bar
Average Residense Time	0.6 s
Mean particle size	1 mm

Table 2.
Expected Yields of Pyrolysis Products

Product	Yield %
Bio-oil	60
Gas	18
Char	12

The Conceptual Design

The conceptual process flowsheet is shown in **Figure 1** and this will form the basis of the EGEMIN flash pyrolysis process.

The pretreatment section consists of a wood chopper, a sieving machine for particle size separation and a drier. This part of the plant is used to produce the feedstock from green wood with the strict specifications for the process requirements.

The biomass is fed into hot recycled pyrolysis gas at the top of the reactor where it undergoes flash pyrolysis with an overall residence time in the reactor of less than a second. The products pass through a cyclone where the char is separated from the gas stream and is collected in a container. The bio-oil is quenched in a condensor and the residual gas is cleaned in a mist eliminator. The bio-oil from the previous two process steps is collected in separate vessels. Then the pyrolysis gas is separated into two streams, one which is used as carrier gas for the entrained bed reactor and the other which is used as fuel in a fluidized bed combustor. In order to raise the temperature of the recycled gas to an appropriate level, the char collected from the cyclone is also used as fuel in the fluidized bed combustor. If needed, it is possible to supply propane to preheat the recycled gas in the fluidized bed. A flare is provided for burning off the unused pyrolysis gas and for emergency as well as a fan in order to overcome the pressure drop over the condensor and mist eliminator.

The Actual Design

Due to budgetary limitations it has not been possible to implement the conceptual design and it was necessary to consider minor modifications. These include the elimination of the fuel pretreatment section and the fluidized bed combustor and the replacement of the mist eliminator by two parallel filters. A sub-stoichiometric propane burner is used to provide the heat and if necessary the carrier gas (the carrier gas is the flue gases from the propane burner) for the entrained bed reactor. The temperature of the flue gases from the propane burner is controlled by nitrogen addition upstream the reactor. The actual process flow sheet is given in **Figure 2.**

Process Description

The plant consists of the fuel reception and storage system, the feeding system, the entrained bed reactor, the propane burner, the condensation system and the recycle loop.

The Fuel Reception and Storage System

The biomass is delivered ready for use to the pilot plant location. It is transported there by truck and the truck empties the load pneumatically into a main storage silo of about 10 m^3 which can provide sufficient material for several days operation. The silo is equipped with a bag filter for the pneumatic feeding and with 4 extraction screws at its bottom for feeding the biomass to the buffer silo. The 4 screws as well as the geometry of the storage silo ensure that bridges will not be formed. The storage silo is provided with max and min level indicators.

The Feeding System

The feeding system consists of the buffer silo, a gas tight screw extractor, an inclined transport screw and the high-speed feeding screw.

The Reactor

The reactor is cylindrical and vertically placed. The inlet for the carrier gas (which is accomplished with a tangential flow) is very close to the inlet for the biomass from the feeding screw. This envisages the creation of a swirl for improved mixing and heat transfer.

The Condensation System

This consists of the cyclone, the condenser and the filter. The condenser is a venturi scrubber, with direct cooling by recycled oil. The bio-oil is collected into a large vessel below the scrubber while the pyrolysis gases leaving the vessel pass through a woven-fibre filter.

The Propane Burner

A substoichiometric propane burner is used to produce the carrier gas (or to supplement the recycled carrier gas). The control of the temperature of the carrier gas - which should be about 800°C at the reactor inlet - is accomplished by cooling with carrier gas and/or nitrogen.

Aknowlegements

EGEMIN NV as well as the authors are grateful for the financial support provided by the Directorate General for Research, Development and Technology of the Commission of the European Communities JOULE Programme for this project.

Discussion:

E. Chornet (Université de Sherbrooke): A key point of this technology is a particle size of the feed. What would be the particle size of your feed?

A.V. Maniatis (Speaker): Around one millimeter. The average particle size is one millimeter in diameter. It is very, very fine.

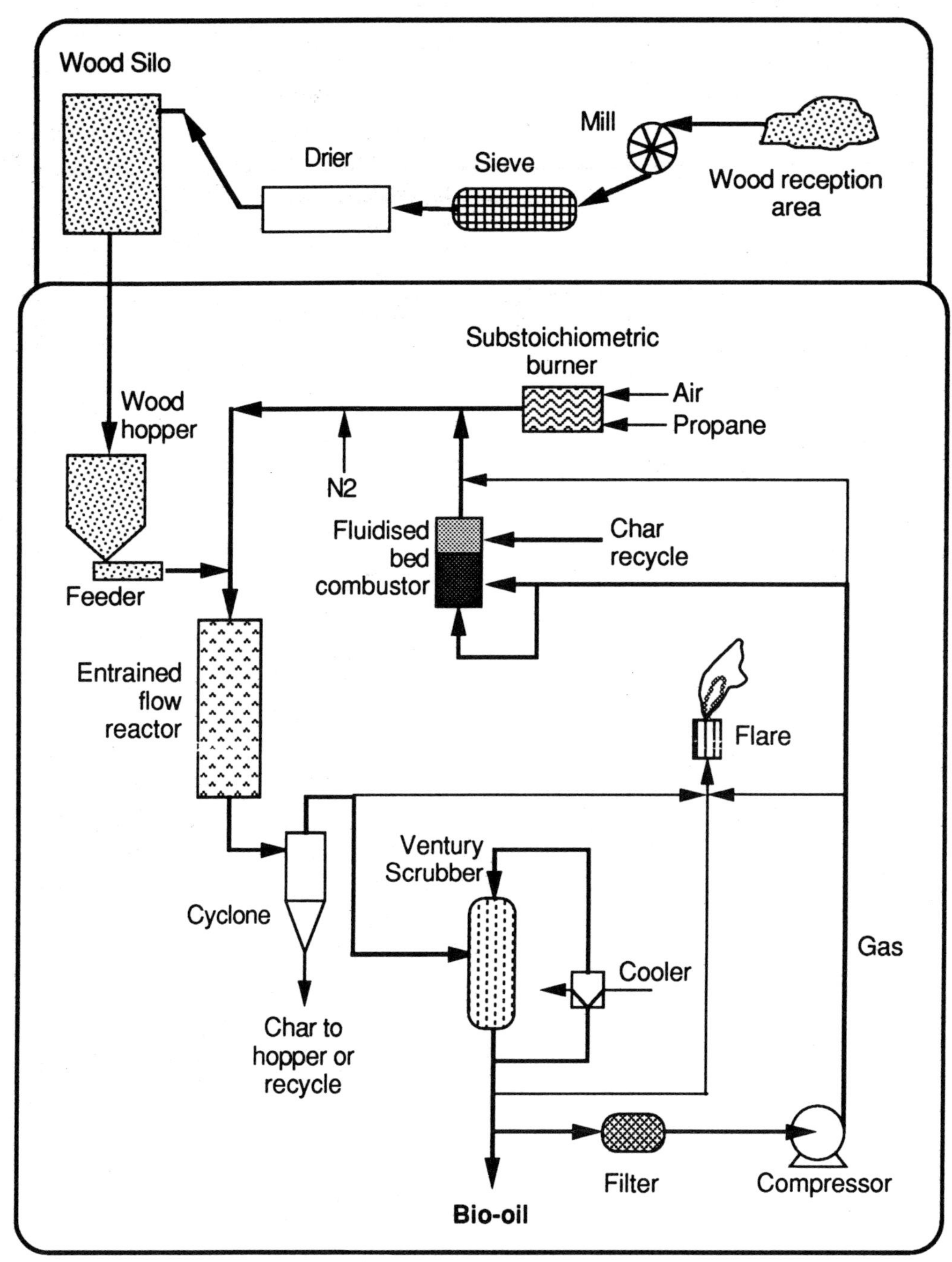

FIGURE 1
The EGEMIN Flash Pyrolysis Process; Conceptual Design.
(Thin lines indicate optional operation)

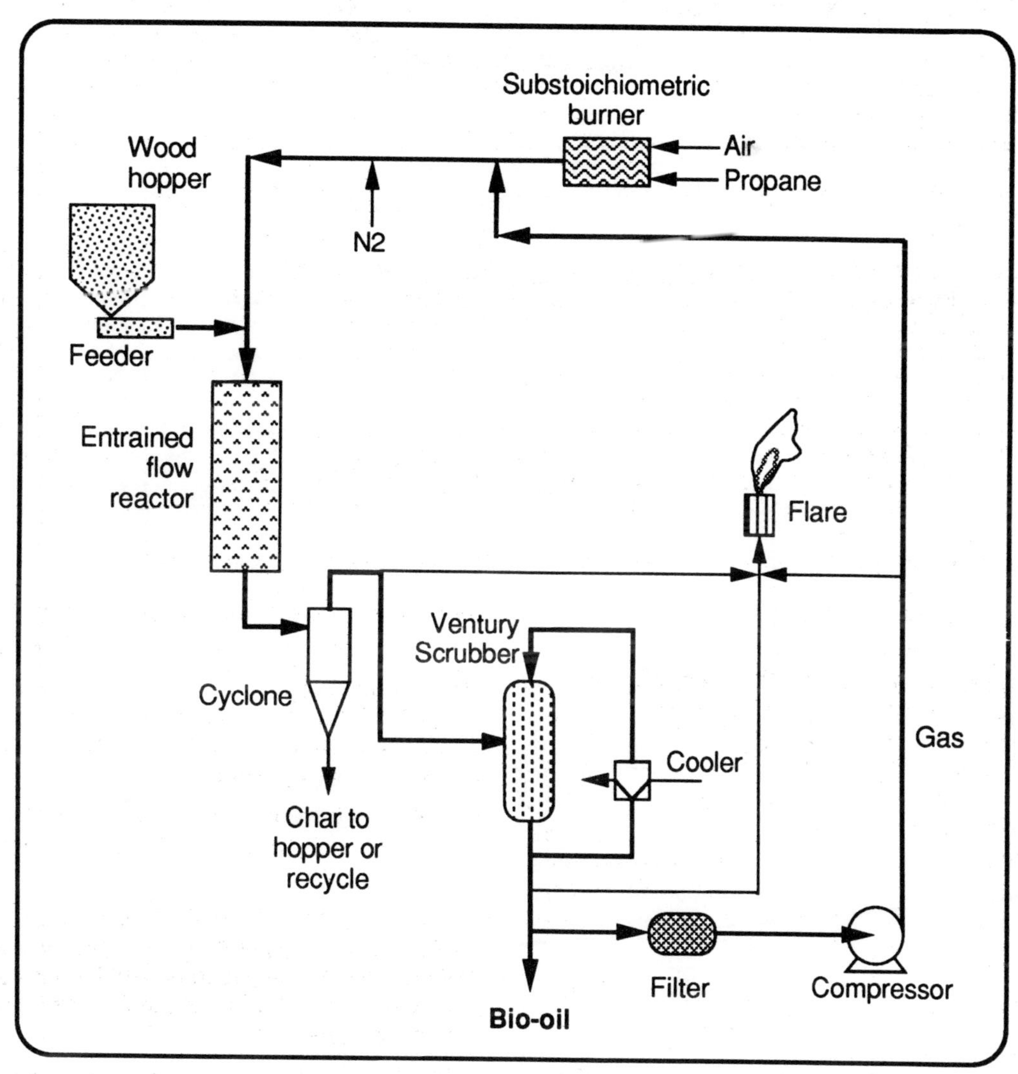

FIGURE 2
The EGEMIN Flash Pyrolysis Actual Design

Energy from Biomass[1]

C. Rossi

ENEL - Thermal and Nuclear Research Center, Via A. Pisano, 120 - 56100 Pisa, Italy

The paper presents the research and development activity planned in Italy on the biomass energy utilization for electricity production. This research will be carried out also with the financial support by the Commission of the European Communities, within the framework of coordinated activity on Biomass, R&D program, Directorate General XII.
In the first part of the paper a study system and two Leben integrated projects (Umbria and Abruzzo) are discussed, in particular as regards the role of ENEL, the National Electricity Board of Italy, that will operate in close collaboration with regional local authorities.
In the second part, the author, as sector coordinator, presents an overview of the research and development activities that will be carried out within the framework of the Joule program sponsored by EEC, Energy from Biomass Subprogram, Thermochemical Conversion, coordinated activity on pyrolysis, in the specific sector "Testing fuels, slurries and environment".

L'article fait état d'un project de recherche-développement prévu en Italie sur l'utilisation de l'énergie de la biomasse pour la production d'électricité. Dans le cadre des activités coordonnées sur la biomasse, le programme de R.-D. et la Direction générale XII, le project reçoit une aide financière de la Commission des communautés européennes.
La première partie de l'article est consacrée à un système d'étude et à deux projets intégrés Leben (Ombrie et Abruzzes), plus particulièrement au rôle d'ENEL, la commission nationale d'électricité de l'Italie, qui collaborera étroitement avec les pouvoirs locaux.
Dans la seconde partie, l'auteur, en sa qualité de coordonnateur de secteur, présente un aperçu des activités de recherche-développement qui seront menées dans le cadre du programme JOULE financé par la CEE, les sous-programme énergie dérivée de la biomasse, la conversion thermochimique, l'"activité coordonnée sur la pyrolyse dans le secteur particulier de la vérification des carburants, des combustibles en suspension et de l'environnement.

[1]Research activity sponsored by the Commission of the European Communities - Joule program

Energy from Biomass the Enel Research Activity

Considering the potential availability of biomass in Italy, estimated around 8.106 t/year, corresponding to an energy production of 2.5 Mtep/year, ENEL, according to the objectives indicated by the National Energy Plan has arranged a comprehensive research program, that considers the utilization of renewable resources as vegetable biomass, for electricity production, in order to diversify the energy sources, to increase the utilization of national resources, and with the objective of the environment safeguard.

Some activities, as the direct firing of rice husk and straw, in collaboration with a private firm, to cogenerate heat and electricity, are already in progress, while several others, also with the possibility of financial support of the Commission of the European Communities, DG-XII, Bio Energy and Biomass R&D, have been just planned and are going to be submitted to the final authorization of the ENEL Board of Directors. The quoted EEC financial support is made within the framework of the Joule program.

Electricity production by an advanced turbo generator fired with biomass

This project is one of the study systems considered by EEC coordinated activity on biomass and takes into account decentralized production of electric energy by means of new crops and innovative technology.

Electricity production is based on powder production from C4 plants (e.g., sorghum) fibers and their use as a fuel for a special ceramic fast rotating gas turbine, coupled with an electricity generator.

The final objective is to build up a plant with a 100 kWe capacity for demonstration of the concept and performance evaluation.

ENEL will collaborate with Daimler Benz, Ferruzzi Group and other university partners, as regards feasibility study, market analysis, flue gas treatment for particulate abatement and state of the art of medium-high temperatures heat exchangers.

ENEL will also carry out a study to evaluate the optimal sharing factor between electricity and heat generation, taking into account the electricity price and the costs to deliver heat.

All the results obtained with the above activities will be evaluated in a final assessment, to be completed prior to the start-up the second phase of the project consisting in developing the new combustion chamber, performed by Daimler Benz, and in designing a new heat exchanger, requested by the heavy operating conditions of the cycles.

The study results will be useful for scale-up to plants with a 300400 kW capacity.

The Leben Integrated projects of Umbria and Abruzzo

The Italian regions Umbria and Abruzzo are involved in the Leben Project (Large European Bio Energy Network) sponsored by the Commission of the European Communities and aimed at developing demonstration integrated biomass-based systems for materials and energy production.

Within the framework of this project, ENEL has planned an experimental activity in the field of thermal conversion of biomass by means of the pyrolysis process, in close collaboration with the Italian regional authorities of Umbria and Abruzzo.

As a matter of fact, ENEL, in virtue of its own great experience in the field of large energy conversion plants, is requested to provide regions with the necessary and suitable technical contribution for the assessment of new advanced technologies. Their possible implementation in different agro-industrial situations within the regional area are very important, especially as regards the expected social-economic benefits.

The pyrolytic process, in opposition to the direct firing of biomass, is considered as a viable system to convert vegetable biomass to suitable liquid, solid and gaseous fuel (bio-oil, charcoal, gas) that, by virtue of their characteristics and the higher specific energy content, present advantages in transportation, storage, combustion and flexibility in production and marketing.

These biofuels can be utilized in industrial plants for heat and electricity production, as well as in thermoelectric power plants, as a supplemental fuel.

The input feedstock to pyrolysis reactor will consist of residual coppice forestry and of energy crops in marginal and set-aside lands available for the culture of woody species suitable for energy and industry (Short Rotation Intensive Culture - SRIC).

The regions are requested to supply the biomass necessary for the experiments, while ENEL will provide the implementation of the energy conversion process:

- build up of pyrolysis plants,
- operation and optimization of biofuels production,
- handling, storage and combustion tests with bio-oil and charcoal

The regions, with the help of their organizations, will cover all the agricultural aspects as seeding, production, harvesting, chipping and transportation of biomass to the utilization plant.

Leben Umbria project

The objectives of the research, carried out by Region Umbria in collaboration with ENEL, are:

- study of the integrated agro-energy cycle;
- production, processing and transportation of biomass to an ENEL thermoelectric power plant located in Umbria;
- technical, economic, energy and environmental evaluations of pyrolysis process and biofuels utilization in the above mentioned plant.

Umbria region will supply biomass of suitable quantity and quality, namely coppice and minor amounts of sorghum and robinia.

The project requires the contruction of a conventional pyrolysis plant for production of charcoal and liquids, to be located close the ENEL thermoelectric unit.

The pyrolysis plant capacity will be chosen according to the actual biomass availability, but however it will not exceed 22.5 t/hr.

Exhaustive tests on pyrolysis reactor and on bio-fuels utilization at the ENEL power plant will be conducted according to a proper research program to be discussed with the Commission of the European Communities.

Leben Abruzzo project

The objective of this project is to demonstrate the feasibility of industrial exploitation of renewable energy resources in the Abruzzo Region.

To that purpose, the construction of a small mobile conventional pyrolysis pilot plant is planned, with a maximum 0.5 t/hr capacity, in order to investigate the technical and economic aspects related to the plant mobility, namely size, capacity, electricity and water supply and so on.

The Abruzzo Region, will provide the biomass, within the framework of a program that also includes production of pulp for paper and other solid and liquid fuels.

ENEL will be especially interested in the production of bio-oil and in the collection of technical and economic data related to the mobile plant operation.

Analytical methods and procedures for control of environmental impact will be investigated and transferred to other laboratories of EEC.

Exhaustive trials on the reliability of a 200 kWe gas turbine electricity generator group and on the pollutants emission will be performed for a suitable period of time.

The know-how gained, also as regards the environment impact of pyrolysis process, could be transferred not only to other similar local situations, but to a national level also.

JOULE Programme

Energy from Biomass Subprogram Thermochemical Conversion Coordinate activity on pyrolysis Sector: "Testing fuels, slurries and environment"

Introduction

The structure of the EEC-coordinated activity on pyrolysis is shown on the diagram of Fig. 1, which includes the Programme Manager, the overall coordinator and five sector coordinators, together with their allocation of individual projects.

The "Testing fuels, slurries and environment" sector, coordinated by the author, includes the contractors shown in Table 1, where it indicates the status as at October 1990 and the kECUs offered by the EEC.

The European Commission is trying to start the overall program as soon as possible, in order to avoid delays in the research development.

The major part the of contracts has been individually negotiated and one of the coordinator tasks, in this first stage, was to avoid overlap of activities. Once the contracts are finalized and the program is underway, he must follow the progress, also promoting close connection and relationship among contractors, when necessary.

Meetings will usually be held on the contractors premises, on a rotational basis, with the aim of concluding the contracts according to the schedule.

In the following part of the report a review and a concise description of the planned research activities are presented.

Union Electrica Fenosa

The activity is a part of an overall comprehensive, coordinated R&D work, in the sector of thermochemical conversion (pyrolysis).

The object of the project is to study the viability of a number of bio-fuels, char-water and char-oil slurries as fuel for gas turbines and other thermal machinery.

The work program consists of three tasks:

- fuels characterization;
- evaluation of compatibility of the fuel with parts and elements of the power equipment (handling, storage etc.);
- tests on burners (atmospheric and pressurized), gas turbines and diesel engines.

The duration of the research is estimated at two years.

Samples of fuels will be supplied by some of the teams participating in the collaborative project.

Cadet International

This company will manufacture the Bio-UCF (ultracarbofluid) fuel, a ternary mixture made of charcoal-oils-water and additives, after encouraging results from the previous DGXII sponsored program activity. The reference biomass will be birchtree; the main benefits expected are: low cost light fuel substitute; production of this fuel also in remote areas (islands) in case of biomass availability; utilization in internal combustion engines and in gas turbines. Also burner design and burning tests will be carried out.

The program activities will be performed with the collaboration of several units, like FAC - Consult, included in the Sector, and others not comprised.

The economic viability of Bio-UCF will be assessed, with regard to typical locations in Europe and in developing countries.

Aerimpianti

The first step of the research program concerns the pollution monitoring of a large scale RDF (refuse derived fuel) gassification plant, capable of producing around 7 MWe. The plant to be monitored is located in Greve in Chianti (Tuscany) near Florence. It is under construction and is scheduled to enter into operation within Spring 1991.

Information will be supplied as regards the treatment system installed downstream of the gas combustion system and recovery boiler. Comparison with other flue gas treatment and justification of the technical choice operated are considered too. Information will be furnished also on the pollutants monitoring criteria.

The combustible raw gas coming from the RDF gasification plant is directly fired in the boiler without any clean-up treatment before combustion; the pollutants emission is therefore transferred exclusively to the flue gas treatment system downstream of the combustion process.

All information collected will be useful for evaluation of environmental impact of similar plants.

ENEL + KTI

ENEL and KTI, Kinetics Technology International S.p.A., are going to put together their experiences and to share experimental facilities for the production of slurries, in order to develop and optimize a process to produce advanced ternary mixtures made of high percentage of charcoal, fossil coal and water.

Characterization of fuel and burning tests will be carried out, with a complete assessment of the results.

The charcoal necessary for experimental activity will be provided by different producers in Europe; also the char produced at the pyrolysis plant planned for the Leben Project Umbria could be utilized in the future.

Production tests will be carried out to finalize the production scheme with the target of finding the condition to obtain stable slurries with the highest solid concentration, to be transferred to industrial plants.

IFRF

The International Flame Research Foundation carried out in the past combustion tests on bio-oils utilizing a 1 MW pilot furnace. The activity has been temporarily suspended because all the available types of bio-oils have been tested.

FAC - Consult

Fac Consult is completing the research activity related to a previous contract; it is concerned with the feasibility study of utilizing bio-oils produced by pyrolytic processes in gas turbine.

The tests are carried out on a 300 kW kerosene fuelled gas turbine. The final target is a 100% of bio-oil as fuel.

The future investigations will be focused on the problems posed by polymerization of pyrolytic oils.

I.RI.N.

Irin Company is carrying out a study contract on the treatment of pyrolytic effluents from biomass pyrolysis process, accomplished by means of separation methods based on semipermeable membranes.

The duration of contract is six months, so the results of this study will be available by the end of the year.

The problem of disposing of residues of biomass pyrolysis process is very important but not completely solved. Previous research activity has been sponsored by EEC, also with experiments on preparation of slurries with charcoal and acid pyrolytic water, that unfortunately did not give good results.

Noel Penny

The contract, that is under finalization, concerns some experimental trials to be carried out on small scale gas turbine, available in the market, fed with bio-oils from different pyrolysis process (conventional and fast pyrolysis).

TABLE 1. Joule Program Coordinated R&D activity in the sector of thermochemical conversion (pyrolysis)
Sector "Testing fuels, slurries and environment"
Status at October 1990

NO.	PROPOSER	kECUs	SUBJECT	STATUS Proposed /Contract
1.	Union Electrica Fenosa (Spain)	260	Combustion characteristics of bio-oil, ready for char water and char-oil slurries. Combustion tests on gas turbines and power generation equipment.	Contract with EC
2.	CADET INTERNATIONAL (France)	200	Testing char-water slurries for oil ready for substitution in boilers, Diesel engine and gas turbine applications.	Contract with EC
3.	Aerimpianti (Italy)	50	Environmental monitoring of a refuse ready for derived fuel (RDF) gasifier.	Contract with EC
4.	ENEL + KTI (Italy)	130	Advanced char-coal-water under definition slurries preparation, characterization and burning tests.	Among contractors
5.	IFRF (The Netherlands)	148	Testing the combustion characteristics of bio-oil in a 1 MW furnace.	First part completed
6.	Fac-consult (Belgium)	130	Testing bio-oil in a gas turbine.	With EC (finalized)
7.	Irin (Italy)	30	Study on the treatment and clean-up of waste water from pyrolysis plants, by means of semipermeable membranes.	With EC (finalized)
8.	Noel Penny (UK)	30	Testing different bio-oils in small size gas turbines.	In hand

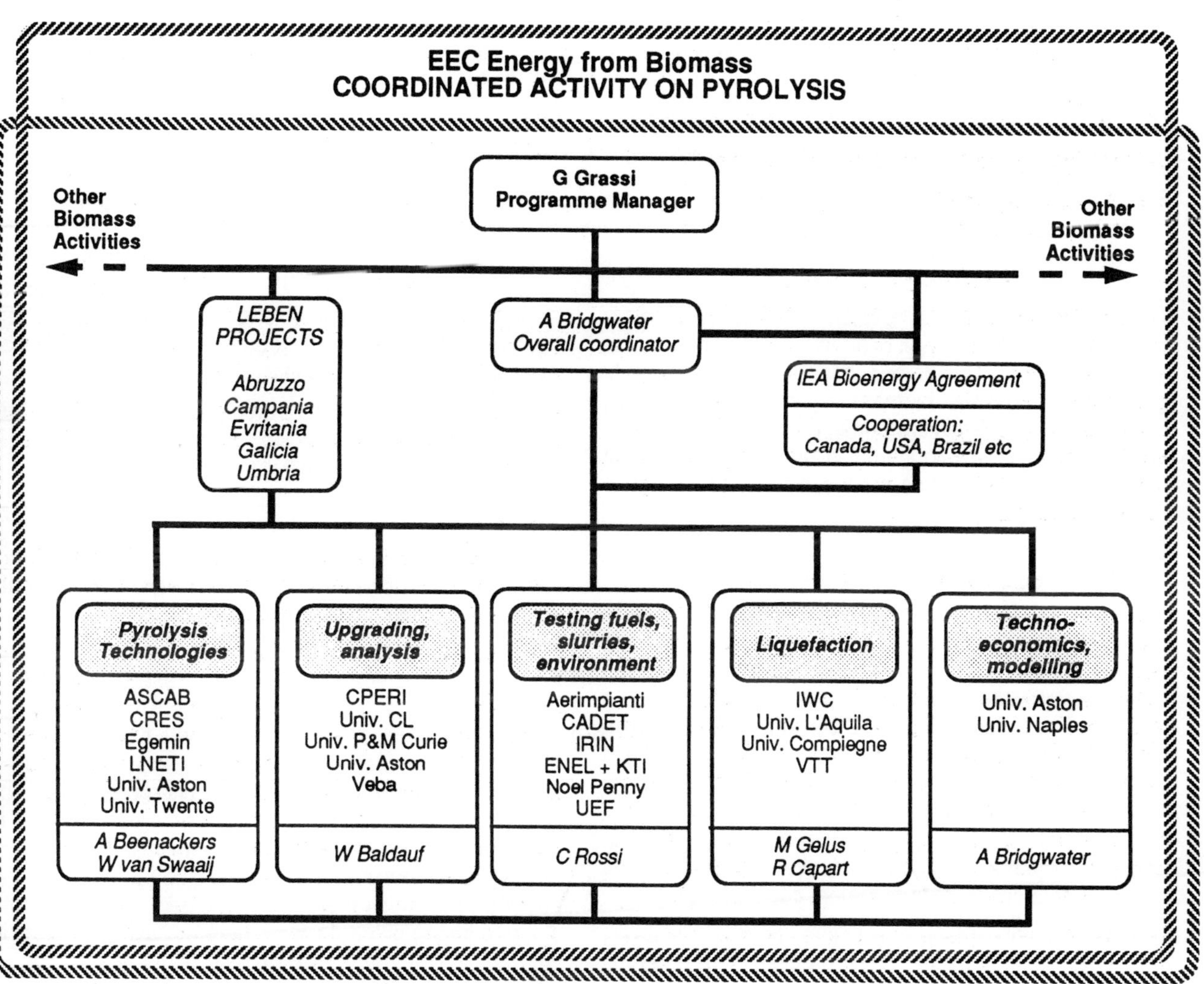

Fig. 1 - Commission of the European Communities
- Coordinated activity on pyrolysis -

IEA DIRECT LIQUEFACTION PROJECT

(Assessment of Liquefaction and Pyrolysis Systems, Update October 23, 1990)

D. Beckman, P. Eng.
Zeton Inc., Burlington, Ontario

This project began in the spring of 1989 with the overall objective of assessing direct thermal liquefaction and pyrolysis processes. The following countries/organizations are participating in this phase of the project which will be completed by the end of 1991: Canada, EEC, Finland, Italy, UK, and USA. The following summarizes the work that is underway in the project.

Le projet a débuté au printemps de 1989, l'objectif global était l'évaluation des procédés de liquéfaction thermique directe et de pyrolyse. Les pays et organismes suivants ont participé à cette phase du projet, qui devrait être terminée d'ici la fin de 1991 : le Canada, la CEE, la Finlande, l'Italie, la Grande-Bretagne et les États-Unis. L'article résume les travaux en cours dans le cadre de ce projet.

State-of-the-Art Review

The working group has compiled a review of the developments in the state-of-the-art of direct biomass liquefaction from 1984-1989. This review covers high pressure liquefaction, pyrolysis, biomass liquids upgrading and other biomass liquefaction research efforts. The review, which is entitled "Developments in Direct Liquefaction of Biomass: 1984-1989", has been submitted to Energy & Fuels for publication.

Technoeconomic Assessments

Assessments of the Solar Energy Research Institute (SERI) ablative pyrolysis process using zeolite upgrading to high octane gasoline, and the Manoil process (in collaboration with the University of Manchester Institute of Science and Technology) are currently underway. An updated methodology for determining the production costs of biomass conversion processes has been formulated. It will include procedures for determining the physical properties of biomass liquid and vapour streams required for mass and energy balances and procedures for equipment sizing and equipment cost correlations.

Two process concepts, present case and potential case, have been defined for the SERI process. The potential case process will utilize a silo dryer for drying of wood chips. This dryer allows the use of low temperature waste energy from the process. Silo dryers are commonly used for grain drying.

A preliminary process flowsheet for the Manoil process was prepared, however the material and energy balance data received was not sufficient to complete a process assessment at this time.

The lack of processing data on wastewater streams from thermochemical biomass conversion has been identified as a problem in assessing these processes. The working group plans to collaborate with the IEA wastewater activity in order to provide a technical and economic basis for processing these streams.

Chemical from Biomass

A preliminary database of chemicals present in direct liquefaction oils has been prepared. This database will be expanded as information is available.

An assessment of an integrated direct thermal liquefaction conversion process to chemicals and fuels is planned.

Discussion:

N. Bakhshi (University of Saskatchewan): You had there chemicals from biomass; could you elaborate on that? What sort of chemicals are you looking at?

Speaker: Well, we are looking at a number of different types of chemicals, whatever researchers have recovered. We don't do any development work ourselves; we just take a look at what other people have recovered. It would include commodity chemicals and in some cases specialty chemicals as well. I don't have any specific ones on my fingertips right now, but what we are trying to do is put together a list of the type of chemicals that different people have found in the different countries doing research. Yes, I think there are a number of these that are close to taking place. In fact, there are some projects in Canada, which I think you will hear about later on today, which our people are going to be talking about going ahead at your scale.

I think right now some of these are in the design phase or in the planning phase and will be going ahead. So, I don't really have that information on me, but I think you will hear a bit more about that today.

C. Roy (Université Laval): Just a brief question on this assessment study on equipment costing and sizing; when will that be available?

Speaker: Well, the plan is to have a final report available; it would be included in our final report for the project by the end of 1991.

We may be able to have a sort of working draft ready sooner which, if you are interested, I could try and get something to you; but I'm not sure when that would be ready - hopefully (maybe) by the mid part of next year.

Rapid Thermal Processing (RTP): Biomass Fast Pyrolysis Overview

R.G. Graham[1], B.A. Freel[1] and M.A. Bergougnou[2]
[1]Ensyn Engineering Associates, Inc., RR #5, 2610 Del Zotto Avenue
Gloucester, Ontario K1G 3N3, Canada
[2]Faculty of Engineering Science, The University of Western Ontario
London, Ontario N6A 5B9, Canada

Rapid Thermal Processing (RTP) employs any one of several proprietary reactor configurations to convert a variety of carbonaceous feedstocks to high yields of chemical and fuel products via fast pyrolysis or rapid cracking. The common denominator of each system is the ability to accomplish rapid heat transfer to the feedstocks with precise control of short contact times. Valuable intermediates which are formed but do not normally persist, can be preserved in optimum yields for chemical and fuel applications. Biomass and fossil fuel feedstocks have been processed in five RTP plants of input capacities increasing from 0.3 to 100 kg/h. The largest is now in production at a specialty chemical company in the U.S.A., and is believed to be the first and only commercial fast pyrolysis system in the world. Three larger "turnkey plants", with capacities of 5, 7 and 25 tonnes per day have been designed and costed in detail, and are scheduled for construction in 1991. Ranges of reactor residence times and temperatures are .03 to 1.5 s and 400-950° C, respectively. The maximum total liquid yield is about 80% from wood and 90% from cellulose. These yields are about three times those reported for liquid products from conventional thermal processing (i.e., slow pyrolysis). The liquids from RTP have the consistency of a light engine oil and contain various primary chemicals. Liquids from slow pyrolysis are typically viscous tars consisting of heavy secondary compounds. Maximum yields of total gas and gaseous hydrocarbons from biomass are 90 and 18 percent, respectively. Specific products of commercial interest derived from biomass are liquid fuels (fuel oil and transportation fuels), specialty and commodity chemicals, petrochemicals, and polymers, from wood, wood residues and other biomass feedstocks. In addition, a considerable amount of RTP research is being conducted in cooperation with European and American oil and petrochemical companies using a variety of fossil fuel feedstocks (crudes, heavy oil, resids, etc.) to produce transportation fuels, chemicals and petrochemicals. Because of a number of patents pending on the reactor systems and products-by-process (production, extraction and isolation), only selected, general results are reported in this paper.

Au cours du traitement thermique rapide, on fait appel à plusieurs configurations de réacteur brevetées pour convertir par pyrolyse rapide ou par craquage rapide, avec un rendement élevé, une variété de charges d'alimentation carbonées en produits chimiques et en combustibles. La capacité de transférer rapidement de la chaleur aux charges d'alimentation tout en réglant avec précision les courtes durées de contact est un facteur commun à chaque système. On peut ainsi préserver, avec un rendement optimal, les précieux intermédiaires qui sont formés mais qui normalement ne persistent pas, en vue de les utiliser comme produits chimiques ou comme combustibles. On a traité des charges d'alimentation constituées de biomasse et de combustibles fossiles dans cinq installations de traitement thermique rapide dont la capacité variait de 0,3 à 100 kg/h. La plus grosse de ces installations est présentement exploitée par une société de produits chimiques spéciaux aux États-Unis, et serait, estime-t-on, la première et la seule installation commerciale de pyrolyse rapide au monde. On a conçu et évalué en détail les coûts de production de trois "usines clés en main" d'une plus grande capacité, soit 5, 7 et 25 tonnes par jour; ces usines seront construites, prévoit-on, en 1991. Les temps de séjour et les températures varient de 0,03 à 1,5 secondes et de 400 à 950 °C respectivement. Le rendement total maximal en liquide est d'environ 80 % à partir du bois et 90 % à partir de la cellulose, soit à peu près trois fois plus élevé que les valeurs signalées pour les produits liquides obtenus par traitement thermique classique (c.-à-d. pyrolyse lente). Les liquides obtenus par traitement thermique rapide possède la consistance d'une huile moteur légère et contiennent divers produits chimiques primaires. Les liquides obtenus par pyrolyse lente sont des goudrons visqueux constitués de composés secondaires lourds. Le rendement maximal en gaz total et en hydrocarbures gazeux à partir de biomasse est de 90 % et de 18 % respectivement. Parmi les produits spécifiques d'intérêt commercial qui sont tirés de la biomasse, on compte les combustibles liquides (mazout et carburants), les produits chimiques spéciaux et les produits chimiques de base, les produits pétrochimiques et les polymères, qui sont obtenus à partir de bois, de résidus de bois et d'autres charges d'alimentation provenant de la biomasse. De plus, on effectue, en collaboration avec des sociétés pétrolières et pétrochimiques d'Europe et des États-Unis, de nombreuses recherches sur le traitement thermique rapide, à l'aide de diverses charges d'alimentation constituées de combustibles fossiles (bruts, pétroles lourds, etc.), en vue de fabriquer des carburants, des produits chimiques et des produits pétrochimiques. Vu le nombre de demandes de brevets visant ces systèmes et ces procédés (production, extraction et séparation), seuls des résultats de nature générale sont publiés dans la présente communication.

Executive Summary

The Process

Fundamental research has clearly shown that extremely rapid thermal processing (i.e., cracking or fast pyrolysis), characterized by quick feedstock heating and short processing time, takes advantage of chemical pathways which are not predictable nor possible using conventional thermal techniques. Rapid Thermal Processing (RTP) is a practical, fast heat transfer process operating at residence times between 30 ms and 1.5 s, and at temperatures between 400 and 950° C. Heat is rapidly transferred to petroleum or biomass feedstocks by the action of hot particulate, catalytic or non-catalytic heat carriers. The complex structures of these feed materials are broken into useful fuel and chemical "building blocks" by intense thermal action achieved by very rapid heating and a short (fraction of a second) processing time. This chemical process consists primarily of depolymerization reactions, and is better known as "fast pyrolysis" in the biomass community or "rapid cracking" in the petroleum industry. The initial decomposition products are preserved by rapid cooling which stops the chemical reactions before the valuable components can degrade to nonreactive or low-value residues. The residence time, which has a dramatic effect on the yields of individual chemical components, can be precisely controlled. The RTP process currently includes three different reactor system configurations, each of which is suited to a particular feedstock, product and type of heat carrier.

RTP Reactor Configurations

There are three basic types of proprietary RTP reactor systems capable of achieving fast pyrolysis conditions. All are recirculating, transported-bed designs which effect rapid heat transfer in a turbulent mixing zone. One of these, termed the RTP-1 design, is a downflow configuration in which the feedstock and heat carrier flow downwards in the reactor, and the heat carrier flows upwards in the transfer line. The remaining two, known as RTP-2 and RTP-3 designs, are upflow systems where the feed and heat carrier flow upwards in the reactor, and the heat carrier is recirculated downwards in the transfer line.

RTP Hardware History

Five RTP Systems have been constructed and tested from 1985 to 1990. Two of these are RTP-1 downflow designs and are rated at 5 and 10 kg/h respectively. Two of these are RTP-2 upflow designs rated at 30 and 100 kg/h, respectively. The 100 kg/h unit is currently in production at a U.S. chemical company producing specialty chemicals and fuel oil from wood. The fifth system, with a capacity of 50 kg/h, is an RTP-3 upflow design which is currently being tested at Ensyn's Gloucester (Ottawa) facility. With the exception of the fourth RTP plant, all of the RTP systems were constructed with private capital derived exclusively from three sources. These sources include the initial investment of the principals of the firm, the routine reinvestment of Ensyn's profits from consulting income, and joint-development projects with chemical companies. One plant was funded on a cost-share basis, with the Ministry of Energy contributing 40% of the capital cost.

Feedstocks

Feed materials which have been processed in RTP equipment include:

- wood and wood residues
- cellulose and lignin
- agricultural residues and other biomass
- conventional petroleum fuels
- non-conventional petroleum fuels (heavy oils)
- petroleum industry byproducts (gas oils, resids, etc.)
- other solid wastes (tires)

Products

Products which have been investigated include:

- liquid fuel oils (boiler fuel substitute)
- specialty chemicals (low market volume, high value)
- commodity chemicals
- polymers/copolymers/resins
- petrochemicals (ethylene, propylene, butadiene)
- liquid transportation fuels

RTP Business Opportunities

Ensyn and its investors have identified five areas in which there are sound business opportunities for RTP applications:

- specialty and commodity chemicals from wood (and other biomass)
- resins and Copolymers from pulping residues
- petrochemicals (olefine, plastics) from heavy petroleum feedstocks
- transportation fuels (primarily gasoline) from heavy petroleum feedstocks.
- conversion of non-hazardous organic solid wastes (newsprint, RDF, tires) to
- boiler fuels and chemicals.

This latter opportunity had been considered solely in the context of an environmental business (solid waste management). However, with the recent escalation of oil prices, it may now be viewed as the coupling of an energy conversion/recovery enterprise with an environmental business.

Ensyn's role would vary from licensor to manufacturer depending on the scope and magnitude of the particular business.

RTP Economics

A thorough economic analysis has been carried out on various potential commercial applications of the RTP process, by Ensyn's investors and business advisors. The economics look very favourable on the short-term for four specific applications which are now targeted for commercialization.Longer-term applications will be the subject of future R&D. The economics of specific commercial applications are beyond the scope of this paper.

Rapid Thermal Processing (RTP): Biomass Fast Pyrolysis Overview

Introduction

Rapid Thermal Processing (RTP) is an evolving technology for converting a variety of carbonaceous raw materials into useful products such as liquid fuels, fuel gases and valuable chemicals. Feed materials which have been processed in RTP equipment include:

- wood and wood residues
- cellulose and lignin
- agricultural residues and other biomass
- conventional petroleum fuels (crude oil)
- non-conventional fossil fuels (heavy oils)
- petroleum industry byproducts (asphalt and heavy distillates)

The complex structures of these feed materials are broken into useful fuel and chemical "building blocks" by intense thermal action achieved by very rapid heating and a short (fraction of a second) processing time. This chemical process consists characteristically of depolymerization reactions, and are better known as "fast pyrolysis" in the biomass community or "rapid cracking" in the petroleum industry. The initial decomposition products are preserved by rapid cooling which stops the chemical reactions before the valuable components can degrade to nonreactive or low-value residues.

Rapid Thermal Processing must be clearly distinguished from conventional "slow" pyrolysis in which the products are secondary, low-value chemicals. For example, the fast pyrolysis of wood or other biomass results in the production of a liquid whose consistency is that of a light engine oil and whose yield approaches 80%. On the other hand, slow pyrolysis of these materials produces a 30-35% yield of heavy (viscous) "tar". *Rapid Thermal Processing is fundamentally different from slow pyrolysis in terms of chemistry, overall yields and quality of products.*

Products which have been investigated and optimized from biomass and petroleum feedstocks via RTP include:

- petrochemicals (ethylene, propylene, butadiene)
- liquid transportation fuels
- liquid fuel oils (boiler fuel substitute)
- specialty chemicals (low market volume, high value) polymers/copolymers

Rapid Thermal Processing is currently carried out in one of three configurations of proprietary reactor systems which are well suited for scale-up and for stable, steady-state operation. Each system is designed for a particular range of temperatures and residence times, and are optimized for specific products from specific feedstocks. Together, they cover the broad range of temperature, residence time and other important parameters which characterize fast pyrolysis and rapid cracking.

To date, five RTP plants ranging from 0.3 to 100 kg/hour (feedstock input) have been designed and operated. The largest unit is currently operating "on location" at a U.S. specialty chemical company, producing boiler fuel and a slate of about six specialty chemicals (this may be expanded shortly) from biomass. Ensyn believes that this unit is the first (and only) commercial biomass fast pyrolysis plant in the world. Three larger RTP "turnkey plants", with capacities of 5, 7 and 25 tonnes per day have been designed and costed in detail, and are scheduled for construction in 1991.

Following almost a decade of systematic research and development and in response to the counsel of Ensyn's business partners, a five-year plan was drawn up to fully realize the potential of RTP. Among the many feedstock and product combinations which been investigated during the course of RTP development, five combinations have been highlighted in the business plan:

- chemicals and fuel oil from biomass
- polymers/copolymers from biomass
- petrochemicals from petroleum
- transportation fuels from heavy petroleum
- fuels and chemicals from non-hazardous solid wastes (solid waste management)

In each of these areas, Ensyn has established cooperative initiatives with key players including two of the principal petroleum companies in the world, two chemical companies which are the largest in North America in their respective fields of specialty chemicals and polymers/copolymers, a large biomass resource company, and one of the leading petrochemical companies in Canada. In addition, a sound business strategy has been developed and a secure financial base has been secured to achieve the goals of the business plan. This paper reviews the evolution of the development of the Rapid Thermal Process for biomass feedstock applications (specialty chemicals, boiler fuel and polymers/copolymers from biomass), and the hardware which was designed and tested over the course of this development. Selected general experimental results are given to illustrate that product yield and quality was maintained during each level of scale-up. Proprietary results are not reported. Practical lessons learned and resolved scale-up problems are briefly discussed.

A summary of the business strategy is also given to illustrate how the technology has been carried forward towards commercialization in a manner which has satisfactorily protected the interests of each participant. Since all of Ensyn's short-term, market-oriented research and development has been privately funded within the context of confidentiality, specific specialty chemicals and polymers/copolymers are not directly identified.

Background

Fundamental laboratory-scale research in the early 1970's clearly demonstrated a general phenomenon that very high yields of valuable chemicals, chemical intermediates (i.e., the building blocks in the petrochemical industry) and fuels could be produced from a variety of biomass and fossil fuel feedstocks under conditions of extremely rapid thermal and thermo-catalytic processing. This fundamental research indicated that both rapid heating of the feed material and precise control over a very short processing time are essential to achieve maximum yields of the desired products.

Conventional thermal processing equipment (chemical reactors) used in the petroleum, petrochemical and chemical industries cannot achieve the required rapid heating while still maintaining precise control over the necessary short processing time. Although many valuable petroleum, petrochemical and specialty chemical products are produced commercially in conventional equipment, a serious technical limitation exists which has prevented the optimization of yields, as well as the production of desirable chemicals/fuels from lower-grade feed materials.

The challenge in taking the promise of the fundamental research to a commercial concept was to engineer an innovative reactor system which could accomplish rapid heating of the carbonaceous feed while still having precise control over the process reaction time. In the late 1970's, a group of chemical engineers at the University of Western Ontario initiated the development of a rapid pyrolysis/cracking process which they termed "Ultrapyrolysis". This laboratory-scale "mini-plant" (Figure 1) incorporated a steady-state, semi-continuous reactor system rated at 0.3 kg/hour. The reactor was based on the principle of using heat carrier streams of inert, particulate solids to transfer heat by direct contact with a variety of solid, liquid and gaseous carbonaceous feedstocks. A barrel and cone-shaped thermal mixer was designed to rapidly heat the feed in fractions of a second. This mixer was coupled to a tubular reactor which enabled the precise control of processing time. The operating temperature range was about 500 to 1000° C, and the residence times were from 30 to 1000 milliseconds.

The experimental results were very promising, and the Ultrapyrolysis unit was proven to be an extremely valuable research unit. Nevertheless, severe limitations existed in the Ultrapyrolysis reactor design with respect to the potential commercialization. Excessive wear (erosion) of the reactor walls and "dead spots" of incomplete mixing had to be overcome, and provision made for scale-up to a commercial size. Two of the original process development engineers joined Ensyn when the firm was incorporated in 1984, in order to carry on process development at a larger scale. In the meantime, process development work has continued independently at the University of Western Ontario.

RTP Process Development and Hardware

The generic RTP process currently consists of several types of reactor systems which are simple yet innovative, and are suitable for scale-up in many commercial applications. The common denominator of each system is the rapid mixing (and resultant rapid heat transfer) and precise control of relatively short residence times. To date, three basic reactor system configurations have been developed for commercial applications. One is a downflow transported bed, while two are upflow transported beds. The optimum process residence time is the critical parameter for selecting the reactor configuration. As is illustrated in the next section (Experimental Results), residence time is one of the most important process parameters.

The RTP-1 downflow reactor system, described below, was developed to investigate a broad range of residence times. The systematic R&D carried out in this system gave a clear indication of optimum residence times (at a given temperature) for a given chemical or fuel product which can be derived from a given feedstock. For some valuable chemical products and fuel additives which are very reactive, the optimum yields occur well below 250 milliseconds. In this residence time range, the downflow design must be used for a scaled-up commercial application. This configuration must also be used in those applications where a particulate catalyst is used as the heat carrier, and catalyst backmixing must be avoided.

When residence time constraints are such that the optimum yield of the desired product occurs between 600 and 1200 milliseconds (at the optimum operating temperature), the RTP-2 upflow reactor system is suitable for commercial applications. This has the advantage over the RTP-1 design in terms of economics and process control.

A third system, the RTP-3 upflow reactor has been designed to allow for fast pyrolysis or rapid cracking processes whose optimum residence time is in the range of 250 to 600 milliseconds. This system involves a simple but significant modification to the RTP-2 reactor.

RTP-1 Reactor System

The term "Rapid Thermal Processing" (RTP) was formally adopted in 1985 when development work was undertaken by Ensyn to engineer reactor systems which would be durable and suitable for scale-up while still providing the two critical conditions necessary for fast thermal processing; rapid thorough mixing (heat transfer) and short contact times. As a result, a Rapid Thermal Processing pilot plant, rated nominally at 10 kg/h and representing a scale-up of more than 30 times the Ultrapyrolysis plant capacity, was designed, built and tested at Ensyn's Ottawa facility. At the heart of the plant is a simple yet innovative downflow mixer-reactor system which uses hot solids to rapidly heat the carbonaceous feedstock materials while ensuring minimal erosion, maximum heat transfer and precise control of the process time. This unit is capable of operating within a residence time range of 50 to 1500 milliseconds (i.e., from heat-up to quench), and from 400 to 900°C. Because of its flexible operating range of temperature and residence time and ability to feed a variety of inert (sand) or catalytic heat carriers, this unit (along with a similar 5 kg/h plant) remains as the "workhorse" of systematic research, process development and product optimization studies at Ensyn. Experiments conducted in this unit were translated into a viable business plan as a result of a clearly demonstrated *general* increase in yields of desired products over conventional processes using identical feedstocks. The most dramatic result was a twelve-fold increase in the yield of one specialty chemical from biomass (clearly a short residence time phenomenon). This increase in yield of the chemical has led to the development of a novel (patent pending) recovery technique, and to economical isolation and successful marketing. Another example includes the selectivity of desirable polymer/copolymer characteristics from biomass using a simple inexpensive catalyst and optimum temperature and residence time. Similar results were evident in the investigation of petrochemicals and high value transportation fuels from heavy petroleum feedstocks. These latter applications are beyond the scope and interest of the current workshop.

Figure 2 illustrates the principal components of the 10 kg/h RTP-1a unit. Hot solids (sand or catalyst) flow from a single or multiple heat carrier feeders to the mixer where they are injected towards the centre of the vessel. The carbonaceous feed material (solid, liquid, or gaseous) is delivered from one of several interchangeable feeders to the top of the reactor where it is injected into the cloud of turbulent hot solids in a proprietary mixer. Extremely rapid heating of the feed material is achieved as the feed and hot sand particles are quickly and thoroughly mixed. After the fast, intimate mixing is complete, the feed and solid heat carrier pass through a tubular reactor whose length is adjusted to control the processing time.

The products are rapidly cooled in a quencher and the solids are removed in a drop-out vessel (solids catchpot). Two condensers, an electrostatic precipitator and filter system are then used to separate the liquids from the gaseous products. The liquid and gaseous products are sampled and identified by standard analytical procedures. A second RTP-1 plant (RTP-1b), rated at 5 kg/h was commissioned in 1990 to reduce the ever-increasing scheduling demands placed on the RTP-1a unit.

RTP-2 Reactor Systems

In response to a specific industrial application for producing specialty chemicals from biomass, the basic features of RTP were incorporated into a second rapid thermal process unit which was assembled in 1988 and tested early in 1989. This unit, rated at 30 kg/h, was operated for 75 hours until sufficient data had been collected to ensure a successful design of the scaled-up commercial demonstration plant. More than two barrels of liquid product were collected for further analysis, extraction and testing. The reactor was an upflow configuration with the mixing section at the base. Residence times could be varied between 850 milliseconds and about 1.5 seconds. Temperatures were in the range of 450 to 650°C. No catalyst was used.

Successful operation of the 30 kg/h upflow system, gave rise to the basic design parameters for a similar 100 kg/h (nominal) upflow RTP plant. This system was constructed in mid-1989 and thoroughly tested before delivery to an American specialty chemical company in the fall of 1989. It incorporated complete recirculation of the solid heat carrier streams in a reactor system capable of operating between 450 and 600°C and at residence times in the range of 600 to 1100 milliseconds.

The plant, as depicted in Figure 3, consists of a lock-hoppered surge bin designed for integration into the pneumatic transport system which conveys particulate hardwood feedstocks around the existing specialty chemical plant. The lock is achieved using a pinch valve. The feed hopper, located directly under the surge bin, is also equipped with a pinch valve between the two vessels. Wood is conveyed via a metering screw to a mechanical feed screw system. The operation of the two screws is integrated to ensure no plugging or binding of the feed system. Wood is injected near the base of the reactor into the proprietary mixing section in which streams of heat carrier (particulate sand) are conveyed pneumatically by a portion of recycled product gas. The particulate heat carrier is stripped from the product vapours in a primary cyclone, reheated and then recirculated to carry heat to the fresh wood feed. A secondary cyclone removes the char and inorganic "fines" (ash and attrited heat carrier). This cyclone is coupled to the in-plant recovery system which transports the char to a combustor for process heat generation. A bank of condensers, filters and demisters is used to condense and recover the product, and to clean up the product gas prior to recirculation. The operating temperatures, method of contact, and sequence of the various components of the recovery train are critical parameters for efficient cleaning of the gas and complete recovery of the liquid product. Because of the complexity of biomass fast pyrolysis liquors, the menu for efficient recovery was a learned art and was not predictable from first principles nor from empirical design correlations. The crude liquid product is delivered to existing extraction processes where valuable specialty chemicals are isolated and concentrated. The first stage of the extraction process is based on water addition and phase separation, a technique which has been used by this chemical company on wood-derived pyrolysis oils for almost 30 years. A portion of the crude liquid product is diverted, without treatment or upgrading, to fuel a commercial boiler as needed, to provide process heat. Considerable RTP bio-oil combustion experience is being accumulated.

During the four levels of scale-up, there were never any serious problems with the reactor systems. They consistently operated in a stable, predictable and relatively isothermal manner. Minor feed system problems were rapidly overcome. The most serious problems were associated with efficient removal and recovery of the liquid product. By "educated" trial and error, in consultation with technical experts, the correct sequence of unit operations (at optimum operating conditions of temperature and loading) was quickly established while preserving the very favourable economics of the process. Effective removal and recovery of the liquid product also eliminated problems associated with the recirculation blower, which often failed in early trials when exposed to a high aerosol or condensibles loading in the gas stream.

RTP-3 Reactor System

The experience associated with the design and operation of the RTP-2 system gave rise to a third reactor system which has been termed RTP-3. This system is also an upflow design, but has incorporated a significant yet relatively simple modification to allow processing as low as 300 milliseconds while maintaining the heat transfer associated with the other two RTP reactor designs. A 50 kg/h unit has been constructed and thorough testing and characterization is in progress (Figure 4).

Experimental Results

One clear indication of successful scale-up is whether product yields are reproducible as the equipment gets larger. In the Ultrapyrolysis equipment, more than 400 fast pyrolysis experiments using wood and cellulose feedstocks were carried out in the temperature range of 600 to 900°C and at residence times between 60 and 1500 milliseconds. The overall yields of total gas, char and liquid were measured and were found to fit very well with a first order model **(1)**. These experiments have provided a thorough data base for comparison with subsequent experiments in other fast pyrolysis equipment. For cellulose trials, maximum liquid yields approached 90 percent, by mass, at 100 milliseconds and 650°C, while gas yields were about 10 percent and char yields were negligible at these conditions. Maximum total gas yields from cellulose exceeded 90 percent, by mass, at 900°C and residence times in the range of 300 to 700 milliseconds, while maximum ethylene, total olefinic gas, and total hydrocarbon gas yields were 7, 11 and 18 percent, respectively, at these reactor conditions.

For hardwood feedstock experiments in the Ultrapyrolysis system, maximum liquid yields approached 75 percent, by mass, at 200 milliseconds and 650°C, while char and gas yields were approximately 7 and 18 percent, respectively, at these conditions. No experiments were conducted at temperatures less than 650°C in the Ultrapyrolysis system. However, trends in the data suggested that liquid yields would increase further at reduced temperatures and residence times. This was verified later in the RTP-1 reactor system. Maximum total gas yields from hardwood approached 75 percent by mass at 850°C and residence times in the range of 100 to 500 milliseconds, while maximum ethylene, total olefinic gas, and total hydrocarbon gas yields were 5, 7 and 15 percent, respectively.

More than 300 fast pyrolysis experiments have been carried out in the RTP-1 reactor system using hardwood and cellulose over the temperature range of 450 to 900°C and at residence times between 90 and 1500 milliseconds. This provided a broad overlap of experiments conducted in the two systems at similar conditions of temperature and residence time. Data from the RTP-1 experiments was also fitted to a first order model and exhibited excellent agreement with the Ultrapyrolysis data **(2)**. For hardwood feedstock experiments conducted using the RTP-1 reactor, maximum liquid yields approached 77 percent, by mass, at 250 milliseconds and 550°C, while char and gas yields were approximately 11 and 12 percent, respectively, at these conditions. Maximum total gas yields from hardwood approached 80 percent by mass at 900°C and residence times in the range of 250 to 400 milliseconds, while maximum ethylene, total olefinic gas, and total hydrocarbon gas yields were 5, 8 and 15 percent, respectively.

The two RTP-2 reactor systems were not flexible research tools, but were "dedicated" to operate over a very small range of temperatures and residence times. As a result, the data from the 30 kg/h RTP-2 unit was largely gleaned from a steady-state run conducted at 500°C and 1400 milliseconds. Data from the operation of the 100 kg/h RTP-2 unit was recorded during a steady-state experiment carried out at 525°C and 800 milliseconds. Table 1 illustrates the effectiveness of the scale-up from 0.3 to 100 kg/h and summarizes the good agreement of the overall liquid product yields in different fast pyrolysis equipment at similar reactor temperatures and residence times.

Table 1. Total Liquid Product Yields from Hardwood for Fast Pyrolysis Experiments Conducted in Four Reactor Systems

Reactor Conditions		Total Liquid Product Yields for a Given Reactor (% by mass)			
Temp. (C)	Res.Time (ms)	Ultrapyr.	RTP-1a	RTP-2a (30 kg/h)	RTP-2b (100 kg/h)
700	100	71	70	-	-
650	200	67	66	-	-
525	800	-	72	-	74
500	1400	-	64	63	-

The importance of residence time is clearly indicated when the yields of specific valuable specialty chemicals (resident in the total liquid product) are measured as a function of residence time at a fixed temperature. For example, the yield of one particular specialty chemical (i.e., that which is of principal interest to our U.S. client) can be considered to be 1 on a dimensionless index using their current pyrolysis technology. At the same temperature, a reduction in the residence time by a factor of nine, gives a yield of 6 on the scale. A reduction in the residence time by a factor of 40, gives a yield of 12 (i.e., a twelvefold increase!). This remarkable increase in yield dictated that fast pyrolysis would be an essential, central part of future production strategies for this particular industry.

Business Strategy

In the early years of the development of RTP, there was only one strategy available to Ensyn to ensure that ownership and control of the technology was retained. The development of hardware and short-term market-oriented R&D had to be funded by Ensyn. Accordingly, Ensyn developed a profitable consulting division which invested profits into process development. In addition, the principals of the firm deferred a portion of their salaries until proprietary technology was well established. Initial external contracts were only entered into if they involved medium or long term market applications. The rationale was that a portion of the longer term interests would be shared proportionately with public and private sector investors, and profits earned by Ensyn would then be reinvested into short term interests.

As the technology was developed, Ensyn then identified medium-scale markets which could be exploited in a joint-venture fashion such that control of the technology was still maintained. These ventures have proven to be critical for the development of RTP since the joint-venture partners are directly linked to the market place, and once convinced that the technology will serve them well, they have a vested interest in seeing the technology implemented. Assuredly, financial benefits must be shared in a joint-venture arrangement, but it is our opinion that implementation of the technology could not advance as fast or as far without such an arrangement. In addition, our partners have contributed considerable "downstream" technology with respect to the extraction, isolation and upgrading of raw biomass-derived fuels and chemicals.

The first two phases of Ensyn's development, which have included internal funding (reinvesting profits from consulting and contract R&D) and joint-ventures in medium-scale applications, have led to current opportunities. With a limited but significant proven track record, Ensyn has been able to access reputable business expertise and assemble a five-year business plan for further commercial process development. This will take the corporation off the contract "treadmill" and provide for a period of time in which a greater portion of the corporations talent and expertise can be devoted to RTP process development. We look forward to the future with great anticipation.

Acknowledgements

In addition to a number of private corporations which have funded contract or joint-venture initiatives, Ensyn wishes to publically acknowledge the assistance of the National Research Council and Energy Mines and Resources in the Ultrapyrolysis work at the University of Western Ontario. The Technology Branch staff of Energy Mines and Resources Canada, particularly Ed Hogan, have showed considerable interest in fostering cooperation between Ensyn, various private companies in Canada and the European Economic Community. They have also shared directly in the funding of medium and long-term R&D projects at Ensyn; their assistance was critical for our survival in the early years. Ensyn would like to express deep appreciation to Ms. Gabriela Teodosiu and the staff of the Ontario Ministry of Energy for their support in the development of the RTP-2 reactor system. Finally, Dr. Stuart Smith of RockCliffe Research and Technology has played an essential role in focusing Ensyn's business opportunities. RTP implementation would not have progressed to its current level had it not been for his interest, support and expertise.

References

1. Bergougnou, M.A., Graham, R.G., Freel B.A. and Vogiatzis, L., An Investigation of the Fast Pyrolysis of Cellulose and Wood, Ottawa: National Research Council of Canada, 1983.

2. Freel, B.A., and Graham, R.G., Rapid Pyrolysis of Wood and Wood-Derived Liquids to Produce Olefins and High Quality Fuels, Ottawa: Energy Mines and Resources Canada, 1988.

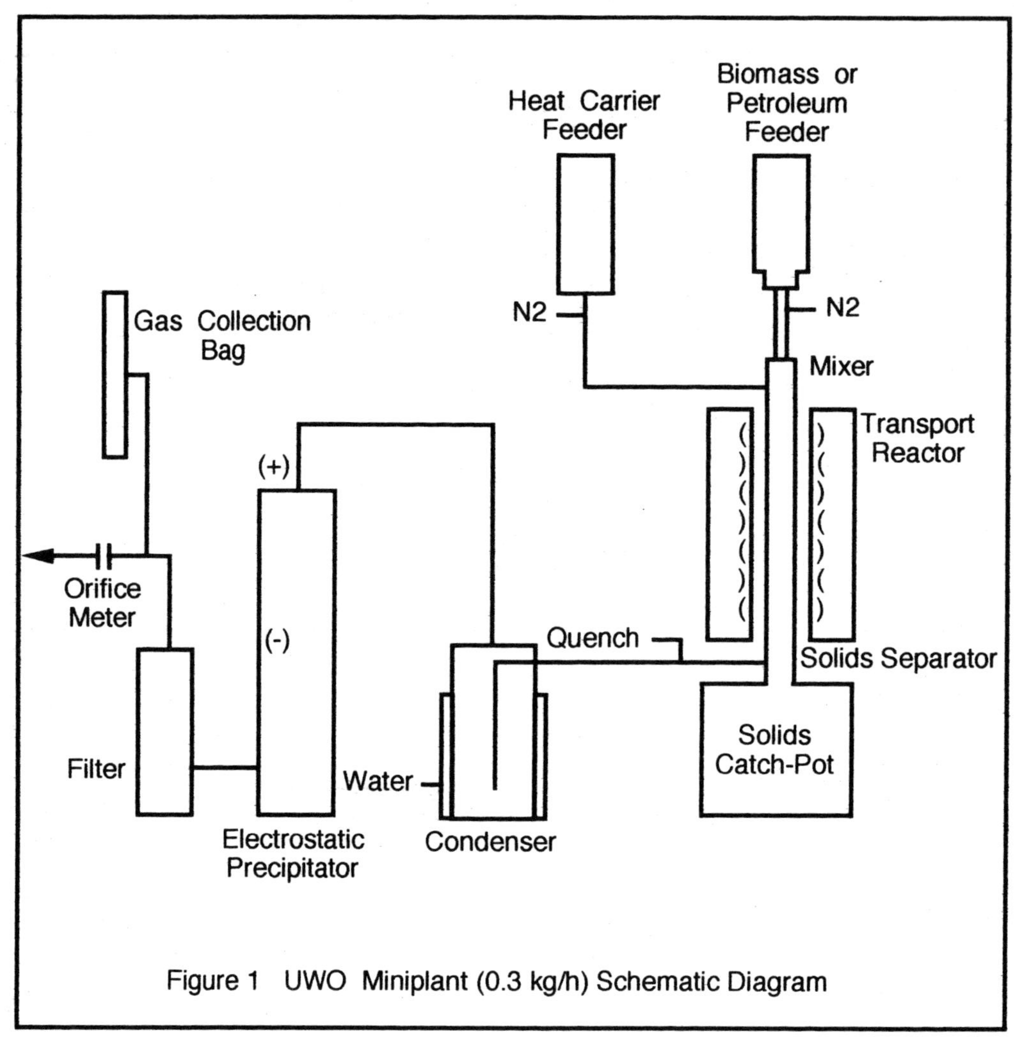

Figure 1 UWO Miniplant (0.3 kg/h) Schematic Diagram

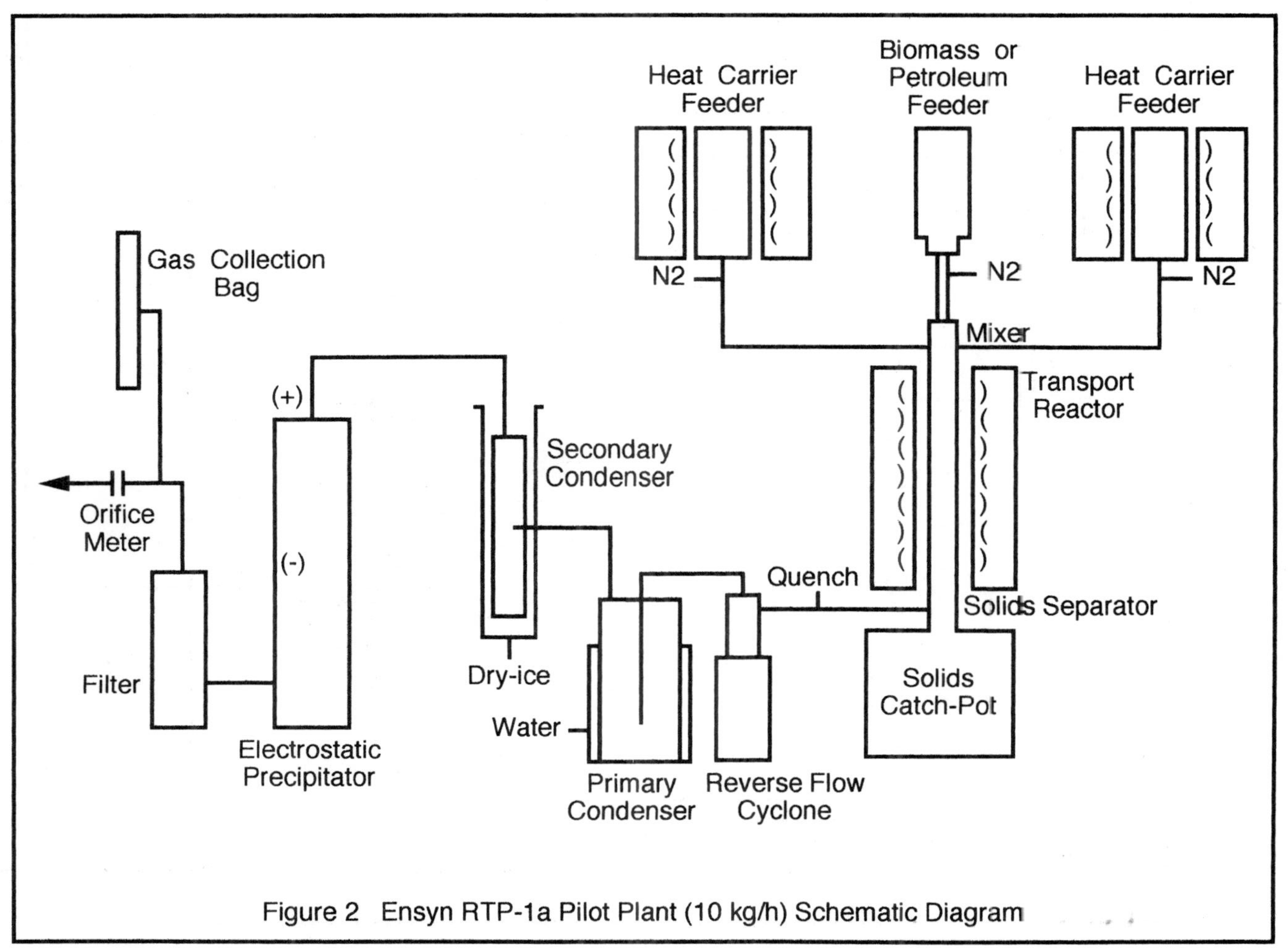

Figure 2 Ensyn RTP-1a Pilot Plant (10 kg/h) Schematic Diagram

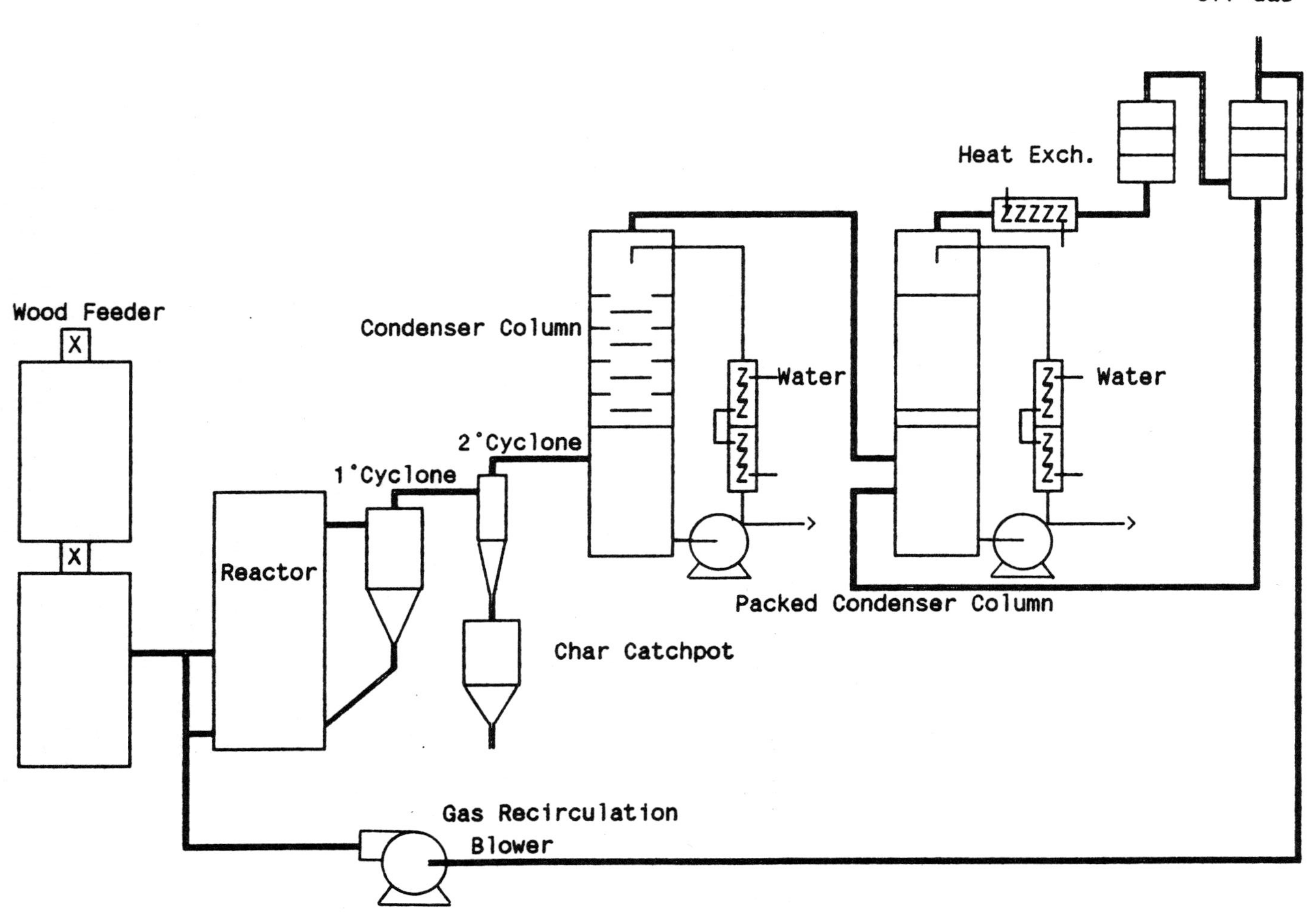

Figure 3 Ensyn RTP-2b Demonstration Plant (2.5 tpd) Scematic Diagram

Figure 4 Ensyn RTP-3 Pilot Plant

New Applications of the Waterloo Fast Pyrolysis Process

J. Piskorz, D.S. Scott, D. Radlein and S. Czernik
Dept. of Chemical Engineering, University of Waterloo

Introduction

The Waterloo Fast Pyrolysis Process (WFPP) employs an atmospheric pressure short residence time fluid bed to achieve high conversions of biomass, or lignocellulosics in general, to liquids. Original studies concentrated on the conversion of forest biomass, and liquid yields of 70% to 85% of the dry biomass were achieved. Later research also investigated the yield obtainable from the fast pyrolysis of several grades of peat. Tests were carried out using either a bench scale unit (30 to 100 gms/hr) or a small continuous pilot plant unit (2 to 5 kg/hr). Complete material balances could be carried out with either unit. Initial tests also showed that the bio-oil produced from forest waste could be readily used as an alternative fuel oil.

In recent years, a number of other possible biomass feedstocks have been tested. In addition to the yields of liquid, gas and char which were produced, the composition of the liquid product was of great interest. Our earlier work had identified the larger yields possible through fast pyrolysis of several low molecular weight carbonyl compounds, especially hydroxyacetaldehyde, acetol, acetic acid and formic acid, as well as levoglucosan and other anhydro sugars or sugars.

We have carried out extensive testing to extend our understanding of the mechanism of fast pyrolysis of cellulose, lignin and hemicellulose. This knowledge has allowed a considerable degree of controlled selectivity to be achieved in maximizing the yields of some of these product compounds. Selectivity can be affected by operating conditions, by catalysts or additives, or by the choice of feedstock.

This report will present the result of fast pyrolysis tests in the WFPP for 18 feedstocks not previously reported, and will discuss the yields obtained. In addition, some new applications of the pyrolysis process and its products will be discussed.

Results

New Feedstocks

Tables 1 to 11 present the pyrolysis yield and the liquid compositions (where measured) for six classes of feedstock. These materials are briefly described below.

A) Agricultural Wastes

1. Wheat Chaff - Grain husks from wheat-sample from Saskatchewan
2. Sunflower Hulls - Seed hulls from sunflower seeds (Manitoba)
3. Flax Shives - Residue of the flax stem after removal of long cellulose fibres

B) Forest Wastes

4. Ulex - a European shrub of the gorse family. Sample from Galicia, Spain
5. Bioshell - Pelleted waste wood from Ontario sawmill waste as produced by the Bioshell process
6. Hog Fuel - Sawmill waste from British Columbia. A mixture of fir and hemlock, approximately 80% bark.

C) Industrial Wastes

7. Newsprint - Newspapers macerated and dried
8. Pulp Mill Waste - Fibrous waste sludge from final clarifier of a pulp mill disposal system
9. Biohol Residue - Residue from the two stage acid hydrolysis Biohol plant project extracted to remove lignin fraction
10. Sewage Sludge - Dried granulated sludge from aerobic sewage treatment system (Hamilton)

D) Agricultural Products

11. Tobacco Leaves - Dried tobacco leaves (Ontario)
12. Corn Starch - Pearl starch (St. Lawrence Starch Co.)
13. Corn Kernels - Dried kernels of feed corn (Ontario)

E) Agricultural Products - Grasses

14. Bagasse - Dried sugar cane bagasse (Jamaica)
15. Wheat Straw - Straw from ripe wheat (Ontario)
16. Corn Stover - Corn stalks from dry ripe corn (Ontario)

F) Miscellaneous

17. Oclansorb - a heat-treated hydrophobic Newfoundland sphagnum peat
18. St. Lawrence Hydrolyzed Wood - Poplar wood hydrolyzed in the St. Lawrence Reactor continuous pilot plant reactor (0.5% HCl at 165°C for 6 minutes) and washed with water.

Discussion of the results shown in Tables A to F will be in terms of group behaviour, where this is characteristic.

Otherwise, yields for individual feedstocks will be examined. All yields are expressed as a percentage by weight of the moisture-free feed (mf), except in cases of very high ash content e.g., wheat chaff, sewage sludge, where a moisture and ash free (maf) basis is used to allow reasonable comparisons to be made.

A) Agricultural Wastes - Liquid yield was highest from the wheat chaff (66.7% vs. 53.8% and 56.1% for sunflower hulls and flax shives, resp.). The differences between wheat chaff and sunflower seed hulls is surprising since both are seed covers. However, the wide differences in ash content and possibly ash composition may be a factor. This is again apparent from the data in Table A, showing very different liquid compositions. Flax shives, on the other hand, is a grass residue, and gives results very similar to wheat straw.

In this group, wheat chaff is particularly interesting because it can be readily harvested, gives good liquid yields high in hydroxyacetaldehyde (HOAc) and acetic acid (HA).

B) Forest Wastes - In this group, the percentage of bark varies Ulex Bioshell Hog Fuel. This variation is apparent in the increasing char yields and decreasing liquid yields, the latter from 71.2% for Ulex to 58.7% for Hog Fuel. A smaller but continuous decrease in CO and increase in CO_2 with bark content increase is also apparent. HOAc content is lower in all three than expected for wood. However, Ulex shows a higher than usual yield of sugar. HA yield appears to decrease steadily as bark content increases, no doubt because of the decrease in holocellulose.

C) Industrial Wastes - These four materials, although all derived primarily from biomass, had very diverse histories and treatment, so that group behaviour is not really relevant. Newsprint gave the results to be expected if the major component is whole ground wood, both in liquid yields and liquid composition. The fibrous pulp mill waste gave 67.1% liquid, but with a higher water content than normal. The very low "water insolubles" which are largely pyrolytic lignin is due to the fact that the pulping process has caused delignification. A good yield of HOAc was obtained from this material. The residue of the two stage acid hydrolysis process, on the other hand, gave a high yield of sugars, as might be expected since the cellulose content was surprisingly high and it had been acid treated. Sewage sludge (on a maf basis) gave good liquid yield (65.1%) and a char yield (ash- free) above normal and a gas yield below normal. Although the ash has a high content of catalytically active metals, these appear to have only a minor effect on product yields.

D) Agricultural Products - Tobacco leaves have a high content of alkaline ash, and this factor probably accounts for the high gas and char yields and low organic liquid yields observed. Only one component occurred in the liquid product in reasonable yield - nicotine. Corn starch and corn kernels were tested to evaluate the behaviour of grain starch in fast pyrolysis. Very high liquid yields and low char yields were obtained (87.2% and 1.9%, similar to cellulose). A good yield of both levoglucosan and HOAc were obtained, with a surprisingly high amount (13%) of insoluble, presumably degraded, product. While whole corn also gave good liquid yields (75.7%), the high oil content makes this liquid product quite dissimilar to other bio-oils. This is reflected in the liquid composition.

E) Grasses - These materials behave similarly on pyrolysis, with organic liquid yield being proportional to ash content, probably to the amount of alkaline cations.

Hence, liquid yield is bagasse wheat straw corn stover, with the gas + char yield the reverse. The high CO_2 yield is typical of this type of feedstock. Lignins in grasses differ chemically from wood or bark lignins, and are primarily polymers of substituted dihydroxybenzenes.

F) Miscellaneous - The Oclansorb gave results typical of a high grade sphagnum peat moss. Liquid yield of 57.3% is among the best obtained from the various peats we have used. The high char content is typical of peats.

The St. Lawrence hydrolyzed wood gave a high liquid yield (81.3%), low in water due probably to the partial removal of hemicelluloses. The yield of HOAc was considerably greater than the sugar yield, despite the acid treatment. We have shown that this result is due to the presence of alkaline cations in the wood derived from the wash water used.

New Catalytic Studies

If the sand in the fluidized bed is replaced by a suitable catalyst, and a hydrogen atmosphere used at atmospheric pressure, and normal WFPP operating conditions, a very different product spectrum can be obtained. In earlier work, the direct single-step methanation of wood was carried out to give a 70% to 75% carbon conversion to methane, and to yield either a gas containing 85% to 90% CH_4 by volume, or a synthesis gas. A new catalytic gasification scheme for wood was proposed on this work.

More recently, we have modified the catalyst, and can now obtain 80% to 85% carbon conversion to methane in a hydrogen atmosphere, with a gas which is 90% to 95% CH_4. In addition, an inert gas has been used with our modified catalyst to give a very high conversion to a synthesis gas with a molar H_2/CO ratio of 2.0. These results for H_2 and N_2 atmospheres are shown in Table 1. This catalytic gasification in inert gas is similar to other catalytic gasification studies of biomass that have been reported, except that in the results shown in Table 1 the temperature was much lower and no steam or other reactive gas was used.

If a different catalyst is used in a hydrogen atmosphere with wood or cellulose then the gas produced can contain little CH_4, but a good yield of C_2+ hydrocarbons. A typical set of results is shown in Table 2. In particular, the high yield of C_4 + hydrocarbons should be noted, and the carbon conversion of about 43% to C_2 + hydrocarbons. A considerable number of catalysts have been screened for their effectiveness in WFPP "hydro-pyrolysis" and this work is continuing.

Uses of Lignin

The pyrolytic lignin which can be precipitated from WFPP pyrolysis liquids has been characterized as a low molecular weight lignin (MW average 900) of high reactivity. It is apparently much less degraded than other lignins obtained from pulping operations, acid hydrolysis processes, etc. We have shown earlier that this lignin can be hydrotreated to give a 60% to 65% yield of hydrocarbon liquids in the gasoline/diesel oil range.

The reactivity of this lignin suggests that it could be used as a constituent of phenol-formaldehyde resins, either as an extender or as a partial replacement for phenol. The market for plywood or other wood product adhesives is very large, and therefore some tests have been carried out (by Dr. R. Kreibisch, an independent consultant) to test the application of our pyrolytic lignin as a component of plywood adhesives. It is apparent that pyrolytic lignin can replace as much as 50% of the phenol in resole formulations with satisfactory results. Further testing is now underway.

Conclusions

The WFPP has shown itself to be very flexible with respect to feedstocks, and to be capable of achieving a good to excellent conversion of biomass organic material to liquids. These liquids can be used either as an alternative fuel oil, or as a source of high-value oxychemicals. New catalysts have been developed which allow conversion of the wood to hydrocarbons or to synthesis gas at WFPP conditions. The pyrolytic lignin has been demonstrated to have the potential to replace much of the phenol in phenol-formaldehyde resin applications.

Acknowledgements

The authors would like to acknowledge the assistance of Piotr Majerski with aspects of both the experimental and the analytical work. The financial support of the Alternative Energy Division of CANMET, of NSERC, of Royal Dutch Shell, of Union Electra Fenosa and of the Canada Centre for Inland Waters are also acknowledged with our thanks.

We would also like to express our thanks to E. Hogan and the late R.D. Hayes of Energy, Mines and Resources for bringing to our attention various opportunities to apply the WFPP to a variety of feedstocks.

Discussion:

N. Bakhshi (University of Saskatchewan): You were talking questions on your catalysts. You use your nickel catalyst, is that just nickel metal or nickel on something? The reason for that is, your operating temperatures are quite high. In those temperatures, searching of catalysts is a major problem.

J. Piskorz (Speaker): We used many different commercial nickels, modified nickels, copper on activated carbon catalysts. We used nickel aluminate $NiAl_2$, as such, and the results were just presented; and we used nickel which was put on many different supports like silica and alumina; many different catalysts.

The limitation here is, that we use the catalyst as a fluid bed medium. It means we have to have catalysts which are resistant to abrasion.

N. Bakhshi: How long were your runs?

Speaker: To do as many as possible tests, we limited ourselves to one in a day, and during the day we can run one, two or three hours for one catalyst.

TABLE 1 (A)

WFPP PYROLYSIS OF AGRICULTURAL WASTES

Feedstock	Wheat Chaff	Sunflower Hulls	Flax Shives
Moisture, wt%	6.9	11.1	16.3
Ash, wt%	22.5	4.0	2.65
Particle Size, mm	-1.0	-1.0	-2.0
Run #	PP 131	PP 122	PP 110
Temp. °C	515	500	500
App. Res. Time, s	0.44	0.5	0.53
Feed Rate, kg/hr	4.64	3.14	2.06
Yields, wt% mf Feed	*		
Gas	15.9	18.7	17.7
Char	17.6	26.3	23.0
Water	15.7	9.4	13.7
Organic Liquid	51.0	44.4	42.4
Total Recovery	100.2	98.8	96.8
CO	5.5	5.3	5.7
CO_2	8.7	12.6	10.3
CH_4	0.75	0.50	0.78
C_2H_4	0.37	0.16	0.26
C_2H_6	0.06	0.12	0.09

* on maf basis

TABLE 2 (A)

LIQUID PRODUCT COMPOSITION

Agricultural Wastes

* *maf basis*

Feedstock	Wheat Chaff	Sunflower Hulls	Flax Shives
Yields, wt%, mf Feed	*		
Cellobiosan	0.40	0.06	0.11
Glucose	0.19	-	-
Fructose (?)	0.70	0.12	0.28
Glyoxal	0.70	0.07	0.21
Levoglucosan	1.95	0.33	0.58
Hydroxyacetaldehyde	6.53	0.78	1.44
Formic Acid	ND	1.01	ND
Formaldehyde	1.30	ND	0.41
Acetic Acid	6.11	2.12	2.86
Ethylene Glycol	0.93	0.27	ND
Acetol	3.20	1.16	1.42
Water Insoluble	15.1	38.4	22.5

TABLE 3 (B)

WFPP PYROLYSIS OF FOREST WASTE

Feedstock	Ulex	Bioshell	Hog Fuel
Moisture, wt%	8.0	8.6	12.0
Ash, wt%	0.44	2.15	2.2
Particle Size, mm	-1.7	-2.0	-1.0
Run #	PP 142	PP 90	PP 126
Temp. °C	506	504	500
App. Res. Time, s	0.49	0.43	0.44
Feed Rate, kg/hr	2.76	2.50	3.50
Yields, wt% mf Feed		*	
Gas	14.0	13.3	13.1
Char	13.0	19.8	31.8
Water	7.7	13.4	12.1
Organic Liquid	63.5	52.4	46.6
Total Recovery	98.2	98.9	103.6
CO	6.7	5.1	4.5
CO_2	6.0	6.6	7.4
CH_4	0.89	0.75	0.50
C_2H_4	0.28	0.23	0.20
C_2H_6	0.06	0.09	0.05

* on maf basis

TABLE 4 (B)

LIQUID PRODUCT COMPOSITION

Forest Wastes

* *maf basis*

Feedstock	Ulex	Bioshell	Hog Fuel
Yields, wt%, mf Feed		*	
Cellobiosan	1.18	0.76	0.34
Glucose	-	0.40	0.26
Fructose (?)	-	0.76	0.39
Glyoxal	1.05	0.59	0.53
Levoglucosan	5.31	2.78	2.16
Hydroxyacetaldehyde	5.15	5.48	3.75
Formic Acid	-	-	1.06
Formaldehyde	-	1.03	0.83
Acetic Acid	4.11	3.27	2.63
Ethylene Glycol	-	0.76	0.44
Acetol	-	1.75	1.56
Water Insoluble	22.0	22.2	27.7

TABLE 5 (C)

WFPP PYROLYSIS OF INDUSTRIAL WASTE BIOMASS

Feedstock	Newsprint	Sulfite Pulp Mill Waste	Biohol Residue	Sewage Sludge
Moisture, wt%	5.9	5.3	4.1	5.9
Ash, wt%	0.6	2.2	0.7	42.2
Particle Size, mm	-1.0	-0.25	-0.5	-1.0
Run #	PP 153	R 45	DD 1	PS 2
Temp. °C	500	500	500	450
App. Res. Time, s	0.5	0.5	0.7	0.60
Feed Rate, kg/hr	2.91	0.027	0.026	2.0
Yields, wt% mf Feed				*
Gas	9.0	9.3	9.0	7.8
Char	15.5	19.5	11.7	22.6
Water	7.4	20.3	10.3	12.9
Organic Liquid	65.8	46.8	63.7	52.2
Total Recovery	97.7	95.9	94.7	95.5
CO	4.0	3.9	3.6	1.4
CO_2	4.1	5.2	3.1	5.2
CH_4	0.41	0.17	0.70	0.33
C_2H_4	0.19	tr	0.11	0.24
C_2H_6	0.05	-	0.05	0.11

* on maf basis

TABLE 6 (C)

LIQUID PRODUCT COMPOSITION

Industrial Waste Biomass

Feedstock	Newsprint	Sulfite Pulp Mill Waste	Biohol Residue
Yields, wt%, mf Feed			
Cellobiosan	1.64	1.12	3.21
Glucose	-	-	0.90
Fructose (?)	-	-	1.10
Glyoxal	-	0.86	1.50
Levoglucosan	2.46	2.58	14.36
Hydroxyacetaldehyde	8.56	10.83	3.54
Formic Acid	4.03	2.32	-
Formaldehyde	-	-	-
Acetic Acid	0.85	-	0.39
Ethylene Glycol	0.34	0.86	0.75
Acetol	2.46	2.49	-
Water Insoluble	12.8	3.0	16.6

TABLE 7 (D)

WFPP PYROLYSIS OF AGRICULTURAL BIOMASS

Feedstock	Tobacco	Corn Starch	Corn Kernels
Moisture, wt%	8.0	7.9	1.9
Ash, wt%	16.3	ND	ND
Particle Size, mm	-0.5	-0.6	-0.6
Run #	T-1	St-L	C-1
Temp. °C	487	515	470
App. Res. Time, s	0.64	0.5	0.5
Feed Rate, kg/hr	0.092	0.089	0.047
Yields, wt% mf Feed	*		
Gas	19.8	5.7	9.2
Char	25.1	1.9	7.8
Water	11.9	6.0	11.5
Organic Liquid	38.8	81.2	64.2
Total Recovery	95.6	94.8	92.7
CO	2.7	3.7	3.7
CO_2	16.5	2.0	5.5
CH_4	0.32	tr	tr
C_2H_4	0.09	-	-
C_2H_6	0.14	-	-

* on maf basis

TABLE 8 (D)

LIQUID PRODUCT COMPOSITION

Agricultural Biomass

* *maf basis*

Feedstock	Tobacco	Corn Starch	Corn Kernels
Yields, wt%, mf Feed	*		
Cellobiosan	0.10	4.99	1.48
Glucose	0.18	-	-
Fructose (?)	0.34	1.34	-
Glyoxal	-	2.22	0.95
Levoglucosan	0.53	14.4	6.5
Hydroxyacetaldehyde	0.49	12.5	4.7
Formic Acid	1.11	-	-
Formaldehyde	0.12	-	-
Acetic Acid	1.55	-	0.5
Ethylene Glycol	0.36	-	-
Acetol	1.70	-	-
Water Insoluble	15.2	13.0	10.5

TABLE 9 (E)

WFPP PYROLYSIS OF AGRICULTURAL BIOMASS

Feedstock	Bagasse	Wheat Straw	Corn Stover
Moisture, wt%	5.5	5.3	9.0
Ash, wt%	ND	3.8	12.2
Particle Size, mm	-0.25	-0.6	-0.25
Run #	B-2	PP 20	C-2
Temp. °C	500	500	550
App. Res. Time, s	0.44	0.75	0.44
Feed Rate, kg/hr	NA	1.92	-
Yields, wt% mf Feed			
Gas	11.1	18.9	15.1
Char	22.0	23.7	37.9
Water	7.0	13.2	11.1
Organic Liquid	59.0	43.0	34.9
Total Recovery	99.1	98.8	99.0
CO	4.3	6.9	3.8
CO_2	6.4	10.7	10.7
CH_4	0.21	0.74	0.23
C_2H_4	0.19	0.20	0.04
C_2H_6	tr	0.13	0.04

TABLE 10 (F)

WFPP PYROLYSIS OF MISCELLANEOUS BIOMASS

Feedstock	St. Lawrence Hydrolyzed Wood	Oclansorb (Peat)
Moisture, wt%	25.4	9.8
Ash, wt%	0.85	4.5
Particle Size, mm	-0.5	-0.5
Run #	S.L.1	St. 69
Temp. °C	488	503
App. Res. Time, s	0.5	0.5
Feed Rate, kg/hr	0.080	0.060
Yields, wt% mf Feed		
Gas	7.6	16.2
Char	10.0	26.3
Water	2.9	8.2
Organic Liquid	78.4	49.1
Total Recovery	98.9	99.8
CO	3.8	4.7
CO_2	3.9	10.7
CH_4	0.35	0.4
C_2H_4	0.10	0.2
C_2H_6	0.04	0.1

TABLE 11 (F)

LIQUID PRODUCT COMPOSITION

Miscellaneous Biomass

Feedstock *Yields, wt%, mf Feed*	St. Lawrence Hydrolyzed Wood	Oclansorb
Cellobiosan	2.38	-
Glucose	0.52	-
Fructose (?)	1.29	-
Glyoxal	1.72	-
Levoglucosan	4.51	5.6
Hydroxyacetaldehyde	9.11	0.7
Formic Acid	-	7.6
Formaldehyde	2.43	-
Acetic Acid	2.84	2.2
Ethylene Glycol	0.59	-
Acetol	1.17	0.5
Water Insoluble	18.8	24.2

TABLE 12

IMPROVED HYDROGASIFICATION PROCESS

POPLAR WOOD, NICKEL CATALYST	H_2		N_2	
YIELDS	Wt. % mf Feed			
Organic Liquid	Trace		7.6	
Water	42.0		3.6	
Gas	59.8		68.1	
Char	5.0		16.1	
	106.8	Vol. %	95.4	Vol. %
H_2	Trace	0.0	3.1	41.9
CO	4.2	4.2	27.1	26.2
CO_2	2.1	1.4	29.9	18.4
CH_4	53.6	94.4	8.0	13.5
C_2+	0.0	0.0	0.0	0.0
		100.0		

C fed converted to CH_4 $\approx$ 83% - 85%

PYROLYTIC LIGNINS IN RESINS

"Pyrolytic lignin" can replace part of the phenol in phenol-formaldehyde resins e.g. adhesives

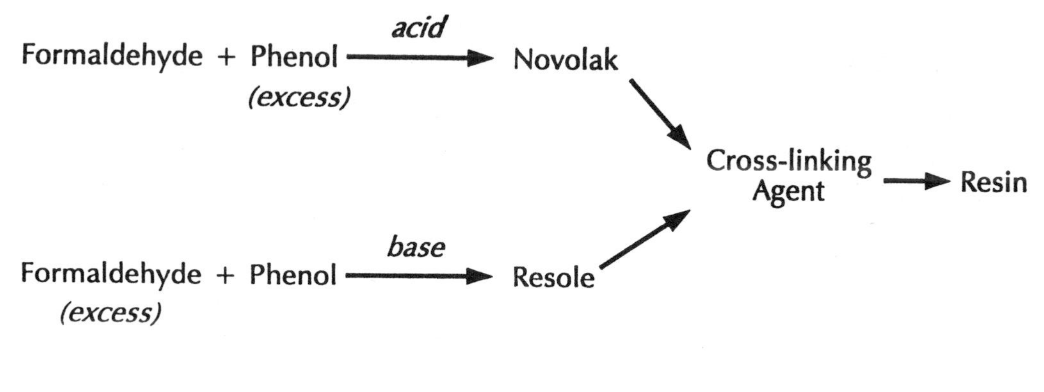

FIGURE 1

PYROLYTIC LIGNINS IN ADHESIVES

Hog Fuel Lignin (softwood bark)

Replaced 50% of phenol in Novolaks

Poplar Wood Lignin (hardwood)

Replaced 50% of phenol in Novolaks

Replaced 50% of phenol in Resole

activity
viscosity
resin characteristics
} comparable to 100% phenol Resole

Plywood panels had good properties

FIGURE 2

Fast Pyrolysis Demonstration Plant

P.B. Fransham
President, Encon Enterprises Inc., Calgary, Alberta

Introduction

Encon Enterprises Inc. was incorporated in 1989 as a research and development company with a mandate to construct and test a 100kg/hr flash pyrolysis demonstration plant. Initially the company was involved solely in pyrolysis technology developed by the University of Waterloo, but has recently branched out to include research on biomass preparation and is actively looking for combustion technology suitable for burning pyrolysis derived fuels. Encon also funded a study at the University of Waterloo to evaluate the potential for upgrading of pyrolysis char into activated charcoal.

Figure 1 shows the overall research and development activities of Encon. Encon's activities can be divided into four distinct and separate areas. The first area deals with feedstock preparation using a Harris Dryer. The second is flash pyrolysis and covers the mechanics of converting biomass to other forms. The third area is activated charcoal and the forth is combustion technology. Significant advancements have been made by Encon in the first three areas. Combustion technology is beyond Encon's capabilities and other than identifying research being carried out by AVCO Research as a possible companion technology, no additional activities are planned.

Feedstock Preparation

Rapid biomass pyrolysis would be largely an academic pursuit if it were not for adequate means of preparing the feedstock. It is generally accepted that rapid pyrolysis does not produce acceptable results unless the particle size is in the order of 1 to 2 millimetres in diameter and the moisture content is below 10%. Most suitable for flash pyrolysis does not meet this criteria and size reduction and drying is required.

Over the past ten months, Encon has been testing a unique drying system developed by Harris Dryer Corp. The Harris Dryer reduces grinding costs and energy costs required to remove excessive moisture by a patented dual chain flail drum. Biomass is conveyed into the drums and is immediately picked up by the chains and suspended in a highly turbulent flow of hot gas. Particle collisions and impact with the chains reduces the grain size. Particles too large or heavy to suspend in the air stream are removed from the bottom of the dryer via an auger, while the fine particles are separated from the gas stream in a cyclone.

Two small laboratory scale dryers have been extensively used to test the drying characteristics of biomass. Scale up to a 2.7 m and a 3.6 m diameter drums was accomplished without performance loss. Harris Dryers have been tested on a variety of materials including: peat, wood, agricultural wastes, bentonite, coal, and sewage sludge.

Single and triple pass drum dryers are perhaps the most common method for removing moisture from biomass. A well maintained rotary dryer requires 4600 kJ to evaporate 1 kg of water. In a similar application the Harris Dryer requires only 3200 kJ to evaporate the same 1 kg of water. In most applications the Harris Dryer will use approximately 30% less energy than other drying technologies. Since feedstock preparation can account for a significant energy loss, savings of 30% in drying improves pyrolysis economics. Additional savings in grinding will also be afforded the users of this technology. Since the technology is relatively simple, capital costs are approximately the same as other dryers.

Flash Pyrolysis

The general plant layout of Encon's 100 kg/hr mobile plant is shown in Figure 2. The technology centres around a fluidized sand bed and recycle gas to provide rapid pyrolysis under an oxygen free atmosphere. Hot gasses emanating from the single fluidized bed reactor are rapidly quenched by recirculating the aqueous phase. The gas stream is subsequently scrubbed using organic liquids sprayed into a venturi. Final cooling and condensation of the water phase is accomplished with a single parallel plate heat exchanger. Since the demonstration plant is a research tool, direct quenching and wet scrubbing was chosen to provide maximum flexibility in the design. Small changes in the volume of quench water have been shown to have a rapid lowering or raising of the temperature. Exchanger plates can also be added or removed to provide more or less heat removal.

A second generation reactor based on the Harris Dryer technology is presently under construction. This reactor should allow for larger feed particles and should have a higher throughput than a corresponding fluidized bed. The reactor is compatible with the mobile plant and will be tested without significant modification to the 100 kg/hr system.

The terms of reference for pyrolysis research and development activities called for moving the plant to Bishop's Falls, Newfoundland for testing on locally available sphagnum peat moss. Extensive testing of the plant in Newfoundland identified two major design flaws. The first was the rapid fouling of the parallel plate heat exchanger with peat wax. As a direct result of the fouling, the plant could not be operated for more than seven hours without a major shutdown. A wax removal system has been designed, ordered and will be installed before the end of 1990. The second design flaw was the char handling system. A gas leak at the char screw eliminated the possibility of mass balance calculations. A new char system has been designed and the components are presently being assembled.

Activated Charcoal

Pyrolysis of peat results in about 40% of the feed stock being recovered as a fine grain char. The char is flammable and is an excellent energy source. However, fuel is the lowest value product derived from pyrolysis char. For example, char used as a fuel commands a $100/tonne value, while activated charcoal commands $1600/tonne. Adding value to pyrolysis products has a marked effect on the economics and helps to provide an adequate rate of return on investment.

Encon Enterprises sponsored research conducted by Matt Siska at the University of Waterloo **(2)**. The results of Siska's work indicated that peat and hog fuel char can be upgraded to a middle grade activated charcoal. The powdered activated charcoal derived from pyrolysis char can be further upgraded through agglomeration to granular activated charcoal. Research is presently underway to determine the optimum agglomeration method and binder.

Combustion

A unique and efficient solid/liquid fuel burner has been developed by Avco Research TEXTRON **(1)** that appears to be ideal for biomass combustion. The advantage of the Avco system is its potential to be retrofitted into existing heat generating systems without significant modification. Once the mobile plant has been fully debugged, char and oil will be shipped to Avco for combustion testing. Avco combustion technology could be used on the Harris Dryers and on pyrolysis plant furnaces.

Project Sponsors

The following is a list of the project costs as of October 1990 and the contribution by Government agencies and private enterprise:

Project Costs	
Capital Costs	$400,000
Materials and Supplies	$300,000
Engineering and Labour	$300,000
TOTAL	$1,000,000

Sponsors	
Energy Mines and Resources Canada	$400,000
National Research Council	$230,000
Department of Energy	$90,000
Culmen Enterprises Inc.	$280,000

Bridge and receivable financing was provided by the Royal Bank of Canada.

References

1. Stankovics, J.O.A., Peat Combustion Tests in Toroidal Flow Coal Combustor for Boiler Retrofit", *Symposium '89 - Peat and Peatlands*, Quebec City, Quebec, (1989) August 6-10.

2. Siska, M., *Activation of Chars Produced by Fast Pyrolysis*, Master of Applied Sciences, University of Waterloo, (1990).

Discussion:

D. Beckman (Zeton Incorporated): I am wondering in your system, where the char is coming off your cyclone, are you cooling the char in that char screw? I couldn't tell from the slide.

P. Fransham (Speaker): Yes, we are. We drop it from its 500 degree temperature down to about 30 degrees. Whatever the water is in the cooling system, which is generally running around 30 degrees.

D. Beckman: So, it is an indirect cooler jacket around the char screw?

Speaker: That's right. We totally underestimated the chars volatility. We ended up with fires because it is an extremely potent material. The liquids that we produced, we dried down to zero moisture content, and they spontaneously combusted also. So, we learned to have a fair respect for the fuels that were generating from this pilot plant.

Table 1. Cashflow summary

ENCON ENTERPRISES INC - PYROLYSIS PLANT

Year	1991	1992	1993	1994	1995	1996	1997	1998	1999	2000	TOTAL
	1	2	3	4	5	6	7	8	9	10	
Production profile											
Char tonnes/year	0	0	1984	3066	1318	1318	1318	1318	1318	1318	12958
Oil BBL/year	0	0	19277	33057	33057	33056	33056	33056	33056	33056	250671
Gas MCF/year	0	0	36372	62372	62372	62370	62370	62370	62370	62370	472966
Activated charcoal kg/yr	0	0	190000	684000	1368000	1368000	1368000	1368000	1368000	1368000	9082000
Capital costs: ($CDN)											
Plant design	10000	0	0	0	0	0	0	0	0	0	10000
Bog selection	10000	0	0	0	0	0	0	0	0	0	10000
Development plant	10000	0	0	0	0	0	0	0	0	0	10000
Equipment purchase	80000	0	0	0	0	0	0	0	0	0	80000
Plant equipment	0	1500000	0	0	0	0	0	0	0	0	1500000
Building construction	0	100000	0	0	0	0	0	0	0	0	100000
Harvesting equipment	0	500000	0	0	0	0	0	0	0	0	500000
Field ditching	0	41000	0	0	0	0	0	0	0	0	41000
Perimeter ditching	0	60000	0	0	0	0	0	0	0	0	60000
Site location/prep/erect	0	50000	0	0	0	0	0	0	0	0	50000
Progress assembly	0	800000	0	0	0	0	0	0	0	0	800000
Final surface preparation	0	91400	0	0	0	0	0	0	0	0	91400
Start up	0	50000	50000	0	0	0	0	0	0	0	100000
Total capital ($CDN)	110000	3192400	50000	0	0	0	0	0	0	0	3352400
Operating costs: ($CDN)											
Labour	0	0	113481	191637	234468	234468	234468	234468	234468	234468	1711926
Peat costs	0	0	145488	249488	249480	249480	249480	249480	249480	249480	1891856
Factory overhead	0	0	238153	644793	1096233	1096233	1096233	1096233	1096233	1096233	7460344
Selling and admin	193256	278849	328176	328176	367176	367176	367176	367176	367176	367176	3331513
Total op costs ($CDN)	193256	278849	825298	1414094	1947357	1947357	1947357	1947357	1947357	1947357	14395639
Prices:											
Char $CDN/tonne	100	100	100	100	100	100	100	100	100	100	
Oil $CDN/BBL	12	12	12	12	12	12	12	12	12	12	
Gas $CDN/MCF	1	1	1	1	1	1	1	1	1	1	
Activated charcoal $CDN/kç	2	2	2	2	2	2	2	2	2	2	
Total plant revenue											
Revenue:	0	0	751096	1791656	2642856	2642842	2642842	2642842	2642842	2642842	18399818
Financial assumptions											
Inflation factor (%)	0	0	0	0	0	0	0	0	0	0	
Discount rate nominal (%)	11	11	11	11	11	11	11	11	11	11	
Tax calculation:											
Revenue ($CDN)	0	0	751096	1791656	2642856	2642842	2642842	2642842	2642842	2642842	18399818
Op costs ($CDN)	193256	278849	825298	1414094	1947357	1947357	1947357	1947357	1947357	1947357	14395639
Depreciation	11000	330240	335240	335240	335240	335240	335240	335240	335240	335240	3023160
Profit ($CDN)	-204256	-609089	-409442	42322	360259	360245	360245	360245	360245	360245	981019
Loss carried forward	-204256	-813345	-1222787	-1180465	-820206	-459961	-99716	0	0	0	-4800736
Taxable income	0	0	0	0	0	0	0	260529	360245	360245	981019
Total taxes	0	0	0	0	0	0	0	78159	108074	108074	294307
Unburden cashflow	-292256	-3141009	211038	712802	1030739	1030725	1030725	1030725	1030725	1030725	3674939
Total taxes	0	0	0	0	0	0	0	78159	108074	108074	294307
Net cashflow	-292256	-3141009	211038	712802	1030739	1030725	1030725	952566	922652	922652	3380634
Cum cashflow	-292256	-3433265	-3222227	-2509425	-1478686	-447961	582764	1535330	2457982	3380633	
NPV (unburdened)	683666										
IRR (unburdened) (%)	16.4										
NPV (after tax)	569440										
IRR (after tax) (%)	15.63										

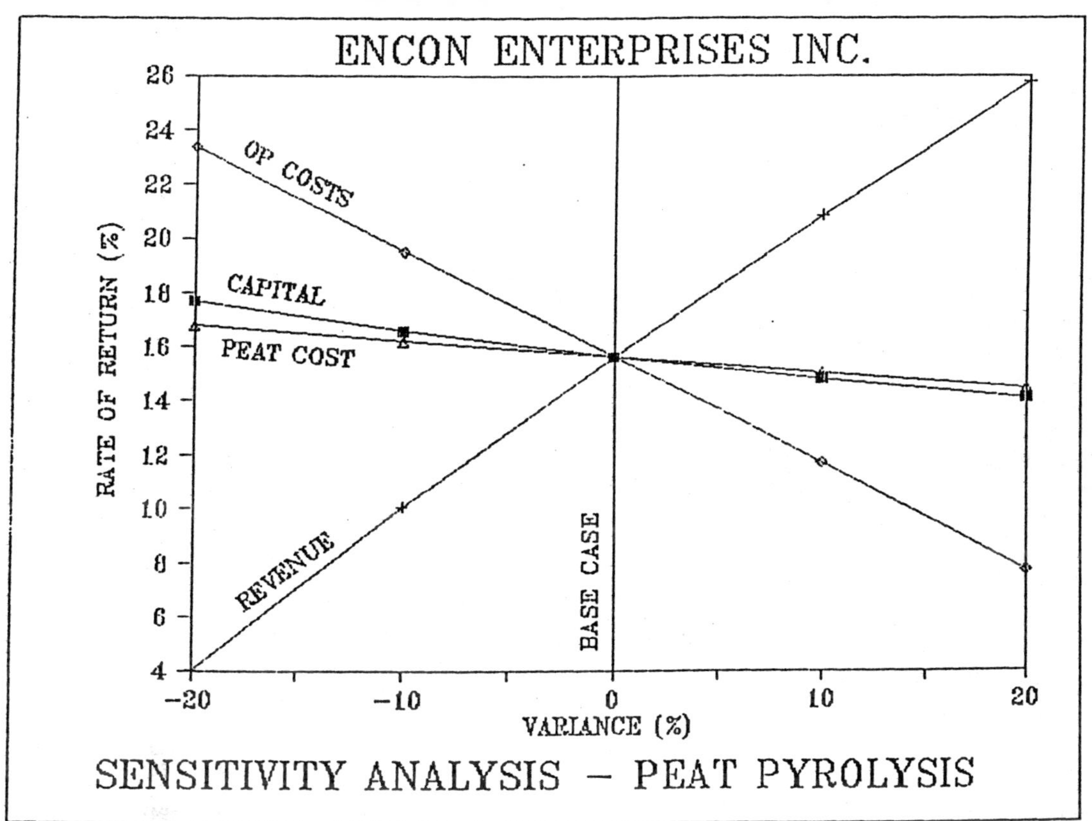

Figure 1 Sensitivity Analysis

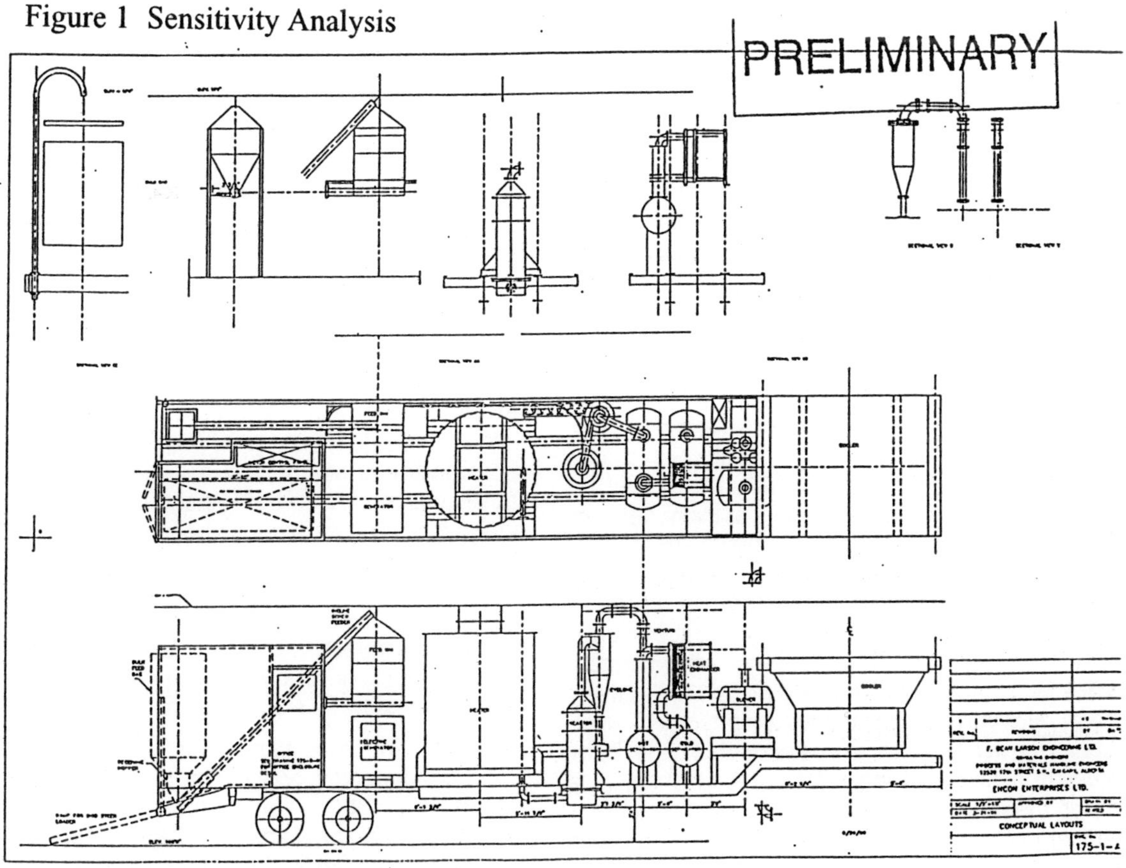

PRELIMINARY

ENCON ENTERPRISES LTD.

CONCEPTUAL LAYOUTS

175-1-A

Figure 2 PDU Conceptual Layout

Converting Sludge to Fuel - A Status Report[1]

H.W. Campbell
Environment Canada, Wastewater Technology Centre, Burlington, Ontario, Canada

Environment Canada has been assessing the viability of thermally converting sewage sludge to liquid and solid fuels, over the past six years. The experimental evaluation has been completed on a 1 kg/h bench-scale reactor and a 40 kg/h pilot plant. Oil yields have ranged from a low of 13% for an anaerobically digested sludge to a high of 46% for a mixed raw sludge. Char yields have ranged from 40 to 73% at the optimum operating temperatures. Fundamental mechanisms involved in the thermal liquefaction of sewage sludge are discussed and the economics of conversion are compared with those of incineration.

Au cours des six dernières années, Environnement Canada a évalué dans quelle mesure il était rentable de convertir par traitement thermique les boues d'épuration en combustibles liquides et solides. L'évaluation expérimentale a été réalisée dans un réacteur de laboratoire d'une capacité de 1 kg/h et dans une installation pilote d'une capacité de 40 kg/h. Les rendements en huile variaient de 13 % pour une boue obtenue par digestion anaérobie à 46 % pour une boue brute mixte. Les rendements en produits de carbonisation variaient de 40 % à 70 % aux températures optimales de traitement. On examine les mécanismes fondamentaux de la liquéfaction des boues d'épuration, et on compare les aspects économiques de la conversion et de l'incinération.

[1]Presented at the 19th National Conference on Muncipal Sewage Treatment - Plant Sludge Management, New Orleans, LA, May 31-June 2, 1989.

Introduction

Sewage sludge disposal represents an ever increasing problem for modern municipalities. The volume of sludge continues to grow while the options for its cost effective disposal remain limited. These limitations may be due to the size of the municipality (i.e., some technologies are neither technically nor economically feasible above a certain capacity), the imposition of legal constraints or public opposition to specific treatment options. Large municipalities are being pressured by these circumstances towards considering some form of thermal treatment which results in the minimum amount of material (i.e., ash), for ultimate disposal. Currently, incineration represents the most widely accepted state-of-the-art technology which satisfies this criterion. In countries, such as the United Kingdom, where incineration has played a very minor role in sludge disposal over the past 15 years, there is an increasing interest in determining how incineration could fit into future overall sludge disposal practices **(1)**. At the same time, because of the historically high cost of incineration, municipalities are interested in looking at ways to improve the cost effectiveness of incineration or at new, equivalent technologies. The approach the City of Los Angeles has taken to reduce incineration costs, involves recovering the energy in the form of electricity and selling this to the local utility **(2)**. The cost effectiveness of this program has not as yet been demonstrated. Cities like Melbourne, New York and Boston are actively looking at alternative technologies to incineration which are less expensive and offer a greater degree of flexibility in the use of recovered energy. One of the technologies which has the potential to satisfy these requirements is low temperature conversion of sludge to liquid and solid fuel products.

This paper summarizes the progress of Environment Canada's program over the past six years to assess the viability of the oil from sludge technology. Although the discussion focuses on the results of the bench and pilot scale studies, other parallel activities aimed at commercialization of the technology are also identified. The results presented in this paper are general in nature but more detailed results and discussions are available in the papers listed in the reference section.

Background

The basic concept of low temperature conversion of sewage sludge to produce fuel products has been known for many years **(3)**. Recently, German researchers have made significant advances in understanding the mechanisms by which sludge is converted to oil **(4)**. They heated dried sludge to 300 - 350°C in an oxygen free environment for about 30 minutes. The researchers postulated that catalysed vapour phase reactions converted the organics to straight chain hydrocarbons, much like those present in crude oil. Analysis of the product confirmed that aliphatic hydrocarbons are produced, in contrast to all other pyrolysis processes, which produce aromatic and cyclic compounds, regardless of the substrate (e.g., sludge, cellulose or refuse). The German researchers demonstrated oil yields ranging from 18-27% and char yields from 50-60%. The oil had a heating value of approximately 39 MJ/kg and the char about 15 MJ/kg.

Based on the potential of this process, Environment Canada conducted preliminary batch, bench-scale experiments in 1982 to confirm the results. After confirmation, Environment Canada designed, fabricated and patented a continuous-flow bench-scale reactor and embarked on a major technology development and demonstration program.

Bench-scale Studies

The bench-scale reactor system is shown schematically in Figure 1. The system has a capacity of 1 kg of dried sludge per hour and has been used to evaluate process performance and to generate design information.

The reactor is subdivided by a helical gas seal into a volatilization zone and a char/gas contact zone. Solids retention time (SRT) in the reactor is controlled by varying both the sludge feed rate and the reactor inventory. Sludge is fed to the reactor by a calibrated screw conveyor and travels through the reactor by means of the reactor conveyor. Volatilized material is withdrawn in the first zone and is contacted with the char in either a co-current or a counter-current mode in the second stage. Product vapours are condensed externally in a water cooled condenser and separated into oil and reaction water. The non-condensable gas (NCG) is vented to a stack. Inert gas is used to purge the system of oxygen and the operating pressure is generally less than 2000 pascals. The system has been operated using twenty-two different sludges from Canada, USA and the United Kingdom. A summary of representative process performance data is shown in Table 1. Additional details on the results achieved during this phase of the development are available in the literature **(5,6,7)**.

Table 1. Bench-Scale Operating Conditions and Results

	Raw	Digested
Feedrate (g/h)	750	750
Temperature (^{o}C)	450	450
Solids Retention Time (min)	20	20
OIL		
Yield (%)	22-46	13-29
Viscosity (cstks)	23-168	9-87
Calorific Value (MJ/kg)	33-38	32-42
CHAR		
Yield (%)	40-66	41-73
Calorific Value (MJ/kg)	7-23	6-17
NCG		
Yield (%)	3-12	4-12
Calorific Value (MJ/kg)	2-9	4-8
REACTION WATER		
Yield (%)	3-15	7-16

The information in Table 1 has been separated into mixed raw and anaerobically digested wastewater sludges. The results show that the oil yield from raw sludge is generally greater than that from digested sludge. This is as expected because the volatile material which is destroyed through the digestion process represents precursors for the oil which would have been generated in the conversion process. Oil yields range from a low of 13% for an anaerobically digested sludge to a high of 46% for a mixed raw sludge, and are primarily a function of the operating temperature and the sludge source. Char yields vary from 40% to 73% at the optimum operating temperature. NCG yields vary from 3% to 12% and are primarily a function of operating temperature. Production of reaction water varies from 3% to 16% and does not appear to be directly related to any of the operating variables. Thermal efficiencies of greater than 95% are routinely obtained with the bench-scale reactor.

From an operational point of view, the reactor has proven to operate equally well with either raw or digested sludge. The main exceptions are the amount of oil produced, (higher quantities from raw sludge) and the oil from a digested sludge has a lower viscosity than the oil from raw sludge from the same source. In general, oils from digested sludges tend to exhibit lower viscosities than those from raw sludges.

The elemental characteristics of the oil are quite stable over a wide range of operating conditions. Typical values for C, H, O, N and S are 76%, 11%, 6.5%, 4% and 0.5%, respectively. Generally the oil contains less than 7% oxygen, but levels as low as 2% have been achieved.

The relationship between temperature and oil yield, is illustrated for a number of sludges in Figure 2. Oil yield is calculated based on volatile solids rather than total solids which normalizes the results to account for the fact that raw sludge generally has higher volatile solids than digested sludge. Even with this normalization, the yield from raw sludge is still generally higher than that from digested sludge. This indicates that the volatile solids in the raw sludge contain more of the components which are ultimately converted to oil than do those in digested sludge. The volatile solids which are destroyed during digestion constitute a very high percentage of the oil precursors.

Figure 2 shows that although the oil yield varies significantly from sludge to sludge, the effect of temperature is identical for all sludges. The oil yield increases with increasing temperature until it reaches a maximum in the 400 to 450°C range. Above the optimum temperature the yield begins to decrease as conditions favour the formation of increasing quantities of non-condensable gas. Sufficient data is not available to pinpoint the exact shape of the temperature/yield curve, but it is obvious that the optimum, for all practical purposes, is a range as opposed to one specific point. In engineering terms, this is extremely important because it indicates a very stable operating system (i.e., if the temperature fluctuates to some degree due to other perturbations within the system, the process will not fail). No clear relationship between temperature and oil calorific value has yet been identified but the viscosity of the oil has been found to decrease with increasing temperature, indicating that some thermal cracking of the oil is occurring.

Pilot-scale Studies

In 1985 Environment Canada contracted with Petro Sun International Incorporated (PSI) to design, construct and operate a one tonne per day pilot-scale conversion reactor system to confirm the laboratory-scale results and the projected energy savings from the oil from sludge technology. A schematic of the pilot plant system is shown in Figure 3. It consists of a 1 m^3 sludge storage bin, a sludge feed system, a 25 cm diameter x 3.2 m long reactor, a char discharge system, a packed-tower direct-contact condenser, an oil-water separator and a self-contained propane/NCG fired boiler to provide hot flue gas for reactor heating. The configuration and operation of the reactor is exactly the same as described for the bench-scale reactor with the exception of the method of heating. The entire pilot plant is mounted on a 1.7 m x 7.4 m skid and is readily transportable.

The system was delivered to the Regional Municipality of Hamilton-Wentworth, Woodward Avenue Sewage Treatment Plant in September 1986. Since that time, a number of modifications have been made to improve both performance and ease of operation. Although fine tuning of the system continues, the current configuration has proven to satisfy the original design criteria. The pilot plant has been operated on three different sludges. The results have generally confirmed that the system is capable of transporting both the sludge and the char, sufficient heat capacity is available to reach conversion temperatures at relatively low solids retention times, the throughput capacity of 40 kg/h can be reached, the products can be separated and performance is similar to that achieved with the bench-scale system. The only aspect of the pilot plant which was not satisfactory was the separation of the oil and reaction water. When the density of the oil was very close to that of water, problems were encountered trying to discharge the oil and water separately. As a result, the gravity decanter was replaced by a disc centrifuge and the problem was alleviated.

Table 2 shows results from the pilot plant for one raw and two digested sludges. For a specific sludge, the data from the bench-scale and the pilot-scale systems compare well in terms of product yield and quality. As predicted by the bench tests, the oil yield from the raw sludge is higher than from the digested sludge while the char yield is lower. The oil from digested sludge also has a significantly lower viscosity. Elemental analysis shows that oils from the same sludge but generated on different systems (i.e., bench and pilot scale), are very similar.

Table 2. Pilot-Scale Operating Conditions and Results

	Atlanta (Raw)	Rockford (Dig)	Hamilton (Dig)
Feedrate (kg/h)	24	24	24
Temperature (^oC)	450	450	450
Solids Retention Time (min)	25	25	25
OIL			
Yield (%)	27	11	12
Viscosity (cstks)	62	20	24
Calorific Value (MJ/kg)	31	34	40
CHAR			
Yield (%)	43	71	69
Calorific Value (MJ/kg)	16	8	6
NCG			
Yield (%)	9	6	7
Calorific Value (MJ/kg)	5	7	7
REACTION WATER			
Yield (%)	15	12	12

Additional pilot-scale studies have been carried out by Campbell Environmental Ltd. in Perth, Australia using the Environment Canada conversion technology. The pilot plant in Perth has a capacity of 40 kg/h and is identical to that of Environment Canada, with one exception. The reactor is heated by burners mounted directly on the reactor shell rather than by hot flue gas from a propane fired boiler. Testing has been completed on a digested sludge and a raw sludge is currently being dried for future work. The pilot plant will be used to demonstrate the technology to the Australian market.

During the bench and pilot-scale studies, a number of other activities were carried out concurrently. These studies fell into two major categories; process design and engineering/cost studies.

Process Design Studies

A study by the University of Waterloo was conducted to evaluate the potential of using fluid-bed technology for the sludge conversion process **(8)**. The results indicated that comparable yields could be obtained for the various products using either the WTC reactor or a fluidized bed. The study did not demonstrate any obvious advantages to using the more complex fluid-bed equipment.

In 1984 a contract was awarded to Chemical Engineering Research Consultants Ltd. (CERCL) to evaluate the fundamental mechanisms involved in the thermal liquefaction of sewage sludge. The study concluded that thermally produced sewage sludge oils are mainly lipid in origin, although chemical reactions do lead to incorporation of nitrogen from protein and nitrogenous material **(9)**. Raw sludges produce higher yields of oil, but these usually have a higher viscosity than oils from digested sludges. The raw sludge oils, unlike the oils from digested sludges, contain considerable carboxylic acids. Surfactant type molecules complicate the removal of water from the oil. A simple thermal process operating at reactor temperature was developed which enhances the oil/water separation. The process results in a low viscosity, homogeneous oil which contains less than 1% water. The nitrogen content can be reduced to 2% or less, whereas the sulphur content remains at 0.5%. Approximately 10% of the oil may be lost in this process. The product is miscible with diesel fuel in all proportions. A very slight precipitate, which results on blending the oil and diesel fuel, can easily be removed by filtration.

A study is also underway with the National Research Council of Canada to evaluate the diesel qualities of the oil. Typical diesel tests such as cetane number, flash point, pour point, etc., are being conducted on oils as received, after upgrading with the CERCL process and in various blends with commercial diesel fuels. A preliminary comparison between the quantity of diesel fuel used by a municipality and the quantity of sludge oil which could potentially be produced, will be conducted to determine the sludge oil to diesel fuel ratio that would be necessary for the municipality to utilize all of the sludge oil in-house. Results from this study are not yet available.

Engineering And Cost Studies

Based on the positive results achieved with the 1 kg/h bench-scale reactor system, Environment Canada contracted with Zenon Environmental Inc. in 1984 to assess the commercial viability of the technology. This study generated a detailed process design for a 25 tonne per day conversion plant and assessed the economics in comparison to existing sludge management options. A detailed discussion of the alternative sludge disposal options is available in the literature **(7)**. The study found that the technology based on the lab results appeared to be economically viable and warranted further development.

One of the concerns identified in the Zenon study, was the degree of accuracy related to existing sludge incineration costs. In many cases sludge facilities have been constructed in a number of phases over several years and plant personnel do not have a good estimate of the overall costs. In order to address this problem, Environment Canada contracted with Proctor and Redfern Ltd. to develop total sludge treatment costs at four integrated sludge management plants in Ontario **(10)**. The sludge treatment train was considered to include digestion, thickening, dewatering, conditioning, incineration and ash disposal, although not all unit processes were present in each plant. Capital costs were taken from actual construction contracts and updated to July 1986 by applying the Engineering News Record Construction Cost Index - Toronto. Operating and maintenance costs were determined from plant records and discussions with plant personnel. The costs were presented in a variety of ways but the most useful method was determined to be total cost (O&M plus amortized capital) per unit tonne of raw sludge generated. The capital component was calculated by amortizing the total capital cost over 20 years at 10% interest. At existing flows, which range from 36 to 67 tonnes per day (on a dry solids basis), the total sludge costs range from $350 to $1042 (1986 Cdn$) per tonne. This represents 35%, 50%, 51% and 52% of the total wastewater treatment costs, for the four plants surveyed. The sludge operating costs represent 48 to 60% of the total plant operating costs while the capital costs related to sludge, account for 32 to 51% of the total plant capital costs.

Projected costs for the oil from sludge technology have been developed in a variety of ways. Preliminary costs, based on data from the lab-scale unit, were prepared for a 25 tonne per day plant in 1985 as part of the Zenon study. These costs were updated by Petro Sun in 1986 based on additional data and scaled up to 65 tonnes per day. The most accurate costs currently available are from a proposal by Campbell Environmental Limited to the City of Melbourne, Australia for a 45 tonne per day, oil from sludge facility. The capital cost, which included dewatering, conditioning, drying, conversion, condensing, oil/water separation, char combustion and ash disposal was estimated at $12.5 million. Based on an amortization rate of 10% per year and an annual operating cost of $2.36 million per year, this results in a total unit cost of $240 per tonne of dry solids. This cost, which does not include any value for the oil, compares favourably with the minimum cost of $350 per tonne reported by Proctor and Redfern for incineration. If a value of $0.30 per litre is attributed to the oil, then the net total cost for sludge treatment by conversion is reduced to $138 per tonne.

It should be noted that the costs generated by Proctor and Redfern represent the costs required to build the plant as it existed in 1986. It does not account for the fact that if that plant were being designed today, many improvements and operating strategies, which represent current state-of-the-art, would be incorporated into the design and result in a significantly lower cost. However, the projected cost savings with conversion are of a sufficient magnitude to warrant demonstrating the technology at full-scale and confirming whether these estimates can be achieved in actual practice.

Current Activities

The pilot plant is being operated for an extended period of time to generate large volumes of oil for market evaluation. The oil will be used for an extensive engine testing program both as received and blended with commercial diesel fuels.

Canadian Patents and Development Limited have licensed the technology to a Canadian company, Enersludge, which is owned 50% by SNC of Montreal and 50% by Campbell Environmental Limited of Perth, Australia.

Patents have been issued in Canada and the United States and are pending in several other countries.

Efforts are underway to select a site for a full-scale demonstration plant. A proposal for a 45 tonne per day, oil from sludge plant, is currently being prepared for Metro Toronto. If accepted, it is expected that the plant would be operational by 1991.

Acknowledgements

The author wishes to acknowledge the technical support of the contractors on various components of the program and the financial support provided by the Department of Supply and Services and the Federal Panel on Energy Research and Development.

References

1. Frost, R.C., Developments in Sewage Sludge Incineration, presented at *IWEM Metropolitan Branch*, England, 1988.
2. Haug, R.T. and H.M. Sizemore, Energy Recovery and Optimization: The Hyperion Energy Recovery System, presented at the *International Conference on Thermal Conversion of Municipal Sludge*, Hartford, Conn., 1983.
3. Shibata, S., Procede de Fabrication d'une Huille Combustible a Partir de Boue Digeree. *French Patent 838,063*, 1939.
4. Bayer, E. and M. Kutubbudin, Low Temperature Conversion of Sludge and Waste to Oil. *Proceedings of the International Recycling Congress*, Berlin, West Germany 1982.
5. Bridle, T.R. and H.W. Campbell, Liquid Fuel Production from Sewage Sludge, presented at the *ENFOR Third Canadian Biomass Liquefaction Experts Meeting*, Sherbrooke, Quebec 1983.
6. Bridle, T.R. and H.W. Campbell, Conversion of Sewage Sludge to Liquid Fuel, presented at the 7th *Annual AQTE conference*, Montreal, Quebec, 1984.
7. Campbell, H.W. and T.R. Bridle, Sludge Management by Thermal Conversion to Fuels. *Proceedings of Conference "New Directions and Research in Waste Treatment and Residuals Management"*, Vancouver, B.C. 1985.
8. Piskorz, J., D.S. Scott and I.B. Westerberg, The Flash Pyrolysis of Sewage Sludge. *Ind. & Eng. Chem., Process Design & Development* Vol. 25, pp 265-270, 1986.
9. Boocock, D.G.B., F.A. Agblevor, F. Chirigoni, T. Crimi, A. Khelawan and H.W. Campbell, The Mechanisms of Sewage Sludge Liquefaction During Thermolysis, presented at *Research in Thermochemical Biomass Conversion: An International Conference*, Phoenix, Arizona 1988.
10. Proctor and Redfern, *Development of a Methodology to Investigate the Cost-Effectiveness of Various Sludge Management Systems.* Final Report (DSS-UP-205) to Environment Canada 1987.

Discussion:

G. Grassi (Commission of European Communities): One question, please? Do you find the concentrate in the oil?

H. Campbell (Speaker): No. All of the conservative metals are in the char, as you would expect. We are only looking at temperatures of 450 degrees C; there are very, very small quantities of metal in the oil, essentially non-detectable.

G. Grassi: So, does it remain the charcoal, in the solid part, the content of a heavy metal?

Speaker: Essentially, the bulk of the heavy metals are in the char, there are some in the reaction water.

J. Piskorz (University of Waterloo): You mentioned that this concept is based on German technology.

There is a major effort now in Germany to go into the fast pyrolysis of sewage sludge using a fluid bed reactor. Dr. Kominsky grew up in Hamburg and to my knowledge the papers are being published.

Are you going to do some research following the German lead towards conversion of sewage sludge in fast pyrolysis? What are you going to do with your solids residue, and what is the nitrogen history in your process?

Speaker: Okay. No, we have no plans to do any work on the fast pyrolysis. Obviously we could, but we don't. We are going to continue to support the work that we have done. What was the second question?

J. Piskorz: Do you decompose your proteins, those that contain nitrogen?

Speaker: Yes.

J. Piskorz: What can you tell us about the nitrogen? Where it goes, to oil, to char, to gas?

Speaker: The nitrogen - there is a significant portion anywhere from 3 to 7 percent nitrogen in the oil. There are high levels of nitrogen in the pyrolytic water and that basically accounts for it; there is very little left in the char.

J. Piskorz: With the high metal content of your char, do you have some utilization of the char in mind?

Speaker: In the current work that we are doing on the Houston sludge, we are going to look at an extensive leeching program on both the char and the ash.

We are not particularly concerned with the heavy metal in the char because we have to burn the char in the process to generate the heater for drying. When we burn the char we are going to end up with exactly the same ash as we would if we went with incineration in the first place. Basically, incinerator ashes from sludges are non-hazardous and there is essentially no disposal problem.

P. Fransham (Encon Enterprises): Your water that is produced with the 3 to 7 percent nitrogen, is this sufficiently benign to be used as a fertilizer for, say, spraying on agricultural fields?

Speaker: I am not sure that I can comment on that, but it's not a very plausible application; the odour is rather severe. I don't think you would make many friends.

G. Grassi: I would like to inform you that in our program we have an activity with a German university and a company, Kruger, from Denmark, and we are only on a laboratory scale, not as yet a pilot plant. We have been able to neutralize completely the sludge from heavy metal, also to control the potassium and the phosphor in such a way we have been able, only on a laboratory scale, to produce ammonia. With our idea of intergrated projects we could treat the sludge, in time, to produce nitrogen fertilizer, after, I say, the complete elimination of heavy metal control is erased. And so, it could be recycled and be a producing area of biomass. Our intention is to implement the first pilot plant in two years time.

P. Laborde (CQVB): Your reactor at the pilot plant, it seems to be six meters long, something like that.

Speaker: Five meters.

P. Laborde: If you go to a full-scale plant, do you have the same design, or do you need to have all of that kind of design? First, on the type of reactor and inside, the type of screw and things like that?

Speaker: Basically, the design that we are going with for the first full-scale is the same functional design except that it is split into two reactors and basically I think they are about two meters in diameter. I am not sure what the lengths are, but basically the internals are essentially the same as you saw in the diagrams.

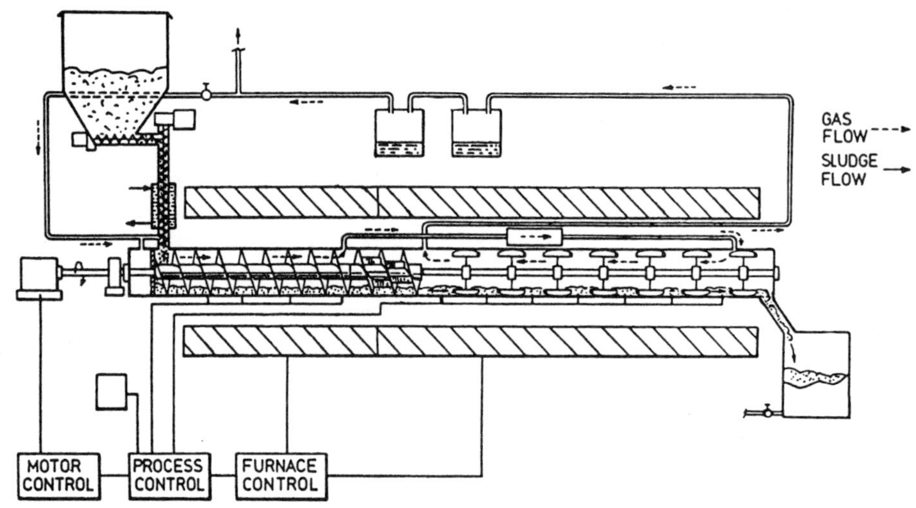

Figure 1. Schematic of bench-scale reactor

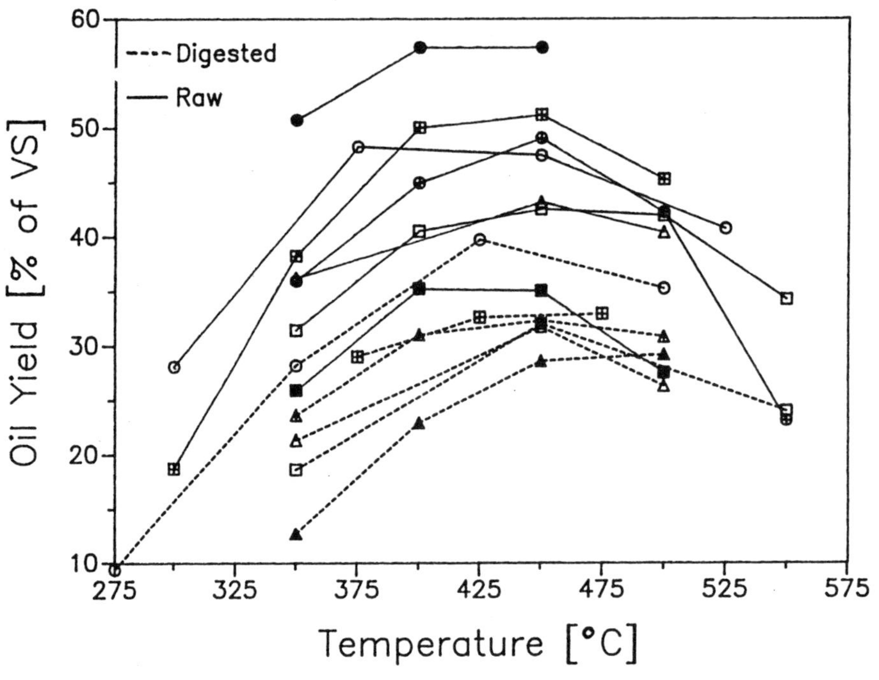

Figure 2. Relationship between temperature and oil yield

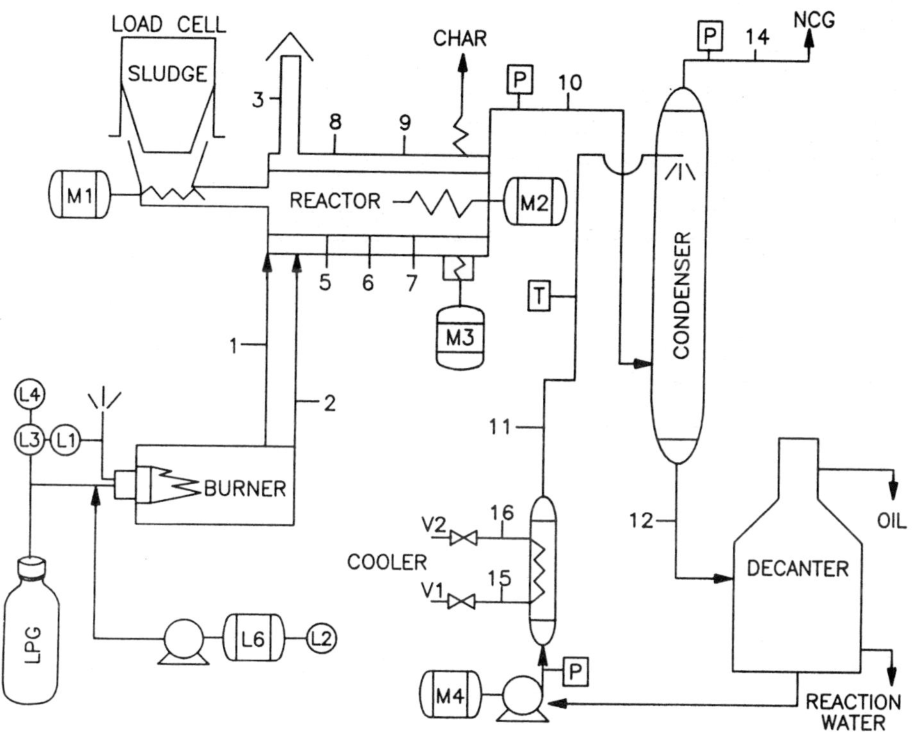

Figure 3. Schematic of pilot scale conversion system

Hydrolysis, Liquefaction, Solvolysis and Fractionation: an Overview

E. Chornet and R.P. Overend
Département de génie chimique, Université de Sherbrooke, Sherbrooke, Qué., Canada, J1K 2R1

This presentation aims at providing a critical review of the liquefaction approaches and their evolution towards the unifying concepts of pretreatment and fractionation. The latter two have progressively evolved as the central pieces of a generic strategy which is common both to biochemical and thermochemical upgrading of biomass into fuels, chemicals and fiber products.

La présentation a pour but donner un aperçu critique des procédés de liquéfaction et de son évolution vers des concepts d'unification des techniques de prétraitement et de fractionnement. Ces techniques ont évolué en tant que fondement à une stratégie commune aux approches biochimiques et thermochimiques quant à la transformation de la biomasse en carburants, en produits chimiques et en fibres.

Introduction

Liquefaction processes have traditionally followed two distinct routes:

- saccharification of the carbohydrates, via either acid or enzymatic hydrolysis has been the necessary step to convert residual lignocellulosics into "low cost sugars" used as fermentation feedstocks whose most universal end product is ethanol. The key catalytic act during saccharification is the hydrolysis of the glucosidic bonds. It is now well known that enhanced accessibility of the catalysts to the glucosidic sites is essential to conduct a reproducible and efficient hydrolysis of the polysaccharides.

- high severity thermal liquefaction of lignocellulosics has long excited the thermochemical community as a possible route to convert residual lignocellulosics into oils to be used as feedstock for fuels and chemicals. Three parallel approaches have emerged:

(i) *aqueous liquefaction* using base catalysts of which NaOH has been the preferred choice. This approach is the high severity extension of known "soda pulping" processes. Extensive hydrolysis of the C-O-C bonds in carbohydrates and lignin, as well as "peeling" reactions lead to a complex mix of acids and hydroxyacids;

(ii) *solvent-based liquefaction* was essentially pursued as an extension of low rank coal conversion. A generic liquefaction process consists of: grinding the lignocellulosics to a suitable size (< 0.5 mm); slurrying the wood meal in a recycle oil; treating the slurry at high temperatures (> 350°C) for a prescribed length of time (of the order of min); separating the products by combinations of solid/liquid and vapour/liquid unit operations. A fraction of the separated liquid is used as recycle oil. Such an approach was tested during the severities at the pilot plant level in Albany, Oregon. It failed both technologically and also by way of its products. A key lesson was learned: "biomass is not coal after all". Two important chemical features of solvent-based liquefaction are: (a) the interaction between the solvent and the matrix which requires accessibility of the solvent to the inner structures as well as chemical affinity between constitutive polymeric families and the solvent; and (b) pyrolytic rupture or weakening of C-O-C and C-C bonds. In the decade of the eighties solvent-based liquefaction of lignocellulosics went through a reassessment of both its structural and physicochemical basis as well as its process options. The link with organosolv pulping processes was established.

(iii) *rapid pyrolytic processes*, aimed at high yields of liquids have gained widespread acceptance to convert a large fraction of the lignocellulosics into a highly oxygenated oil. A variety of process configurations have been proposed and are under development at the levels of either pilot plant or demonstration units. The complexity of the oil obtained is clearly due to the non-selective rupture of the C-O-C and C-C bonds induced by the application of heat to the lignocellulosic matrix.

The initial goal of *high severity liquefaction* processes was to obtain a liquid oil to be used as feedstock for further catalytic upgrading (i.e., 0 removal and molecular rearrangements) to yield substitute hydrocarbons. Such a goal has proved difficult even if it has been demonstrated that gasoline-like products can be derived using shape selective catalysts. However, the highly functional and very diversified oxygenates found in liquefaction oils has rendered the hydrocarbon option virtually unreachable. An alternate line of thought thus emerged: (1) a better selection and preparation of initial feedstock used; (2) aim at oxygenates having a potential value as chemicals; and (3) use the residual oxygenates (i.e., those with no apparent chemical use) as fuel oil. Prerequisites (1) and (2) are also encountered in *saccharification* processes. The key point is in having significant product improvement, i.e. higher yields of desired intermediates and a "cleaner" liquour or oil.

From a market perspective, the initial thrust on saccharification and high severity liquefaction processes towards energy-products-only has shifted to a co-product strategy where energy products are to be complemented with chemicals and perhaps fiber. The concept of biomass refinery, long ago proposed, is revived for residual lignocellulosics.

Finally, the environmental constraints imposed nowadays on any new chemical process force designers to aim at zero discharge technologies. This is a very stringent specification for any process and forces a better and more selective control of all the stages of a given conversion. Both biochemical and thermochemical biomass conversion approaches must consider the enivonmental consequences of the different discharge streams. This calls for improved technological strategies.

Structural and chemical considerations

Since lignocellulosics are complex structural and chemical systems, their direct conversion to liquids (either via saccharification or high severity liquefaction) leads, inevitably, to complex and, often, dilute mixtures. Improvements of this situation can be made by giving proper consideration to the structural organization of the lignocellulosic matrix as well as the constitutive chemical molecules (i.e., macromolecules).

An ideal biomass conversion process will have to consider:

- destructuring the matrix
- disaggregation/defibration to free cells
- defibration/defibrillation to fibrilar substructures
- controlled depolymerization of the constitutive macromolecules via either hydrolysis, solvolysis, pyrolysis or their combinations.

The first three steps are commonly known as *pretreatment*. In fact, the first two steps are also the heart of thermo-mechano-chemical pulping processes (TMP and CTMP processes).

The controlled depolymerization of the constitutive macromolecules has the limitation that hemicelluloses, lignin and cellulose, the three main components of wood, have distinct structures and chemical composition as well as overall reactivity. It is thus unlikely that a single depolymerization strategy can be effectively applied to the three major components. The simple idea is then, why not separate them and convert or use them individually? Such a strategy is commonly know as *fractionation*.

The unifying concept for both saccharification and high severity liquefaction process is then the *pretreatment*. It is furthermore very significant that this pretreatment strategy joins the relatively recent thermo-mechano pulping approaches being progressively adopted worldwide by the pulp and paper industry.

The bottom line is that direct or integral treatment of residual lignocellulosics aiming at liquid fuels and co-product chemicals is seen as being logically displaced by a relatively severe pretreatment of the raw material leading to fractionation and further use or conversion of the constitutive chemical entities present in the lignocellulosics.

Pretreatment and fractionation

Current approaches to pretreatment are well established. They are essentially aqueous/steam technologies whose severity dictates the extent of destructuring, disaggregation, defibration and initial depolymerization. We have shown how a simple phenomenological kinetic model (the reaction ordinate approach) can determine optimum operating zones for pretreatment leading to fractionation.

Finely divided lignocellulosics, irregular shavings, and regularly shaped chips can all be subjected to aqueous/steam pretreatments using essentially variations of two basic approaches: (a) medium consistency aqueous suspensions (10-15% solids) and (b) batch or continuous steam exposure of partially wetted biomass. Both approaches can be carried out within a wide range of temperature, time and impregnating agents. At the end of the pretreatment, three streams are collected: the steam condensate (extractive-rich); a wetted fiber mass (to be washed) and an aqueous liquour (which is reused to wash the wetted fiber. This requires additional water). The washing of the wetted fiber provides a liquour rich in hemicelluloses. Typically, 60-75% of the hemicellulosic sugars can be recovered in the liquour predominantly as oligomers. Impurities related to extractives, soluble lignin fragments and acidic and uronic acids are present in the liquour. The significant point is, however, the rather limited range of compounds present in this fraction.

The washed wetted fiber can be delignified by a variety of methods, all borrowed from established or emerging methods. The traditional approach is to use an alkaline cooking at conditions similar to those used in Kraft pulping. Emerging technologies are organosolv in nature. Ethanol, methanol, acetone ethylene glycol, amines, etc. are all candidates for the delignification step which, in all cases, yields very pure lignin and a cellulosic residue.

Thus, pretreatment/fractionation results in fractions which are chemically distinct and relatively pure for further upgrading. This strategy could be advantageously used in both biochemical and thermochemical upgrading routes.

Upgrading the Fractions: the Liquefaction Options

The hemicellulose-rich fraction is obtained as a liquid. Simple acid or enzymatic post-treatment transforms the oligomers into sugars. The upgrading strategies are then pursued depending on market opportunities. Using this fraction as feedstock for fermentation to ethanol makes a clear link with an energy objective. Extensive research efforts are underway to limit the extent of dilution during the washing steps needed to isolate this fraction, to "optimize" the micro-organisms used for fermentation and to improve the energy requirements of the distillation step.

The lignin fraction, being of polyphenolic nature, is a prime candidate for subsequent upgrading. Here, thermochemical approaches have a clear lead over biochemical strategies. The depolymerization of the lignin down to monomers is an essential step if fuel-like or phenolic-like products are to be produced. Intensive research is underway and liquefaction/pyrolysis approaches appear as prime technological approaches for depolymerization. The latter is essentially a selective rupture of specific C-O-C and C-C bonds. Catalysis will doubtless help in improving pure thermal or solvolytic processes.

The cellulosic residue has multiple possibilities directly as fiber, as a polymer matrix which can be easily derivatized or as a substrate to undergo controlled depolymerization via thermochemical or biochemical catalysis. The latter option leads to fermentable sugars. Despite significant improvements in the hydrolytic and fermentation processes, the ethanol production from the cellulosic residue from fractionation has not been, so far and under current economic laws, attractive to investors. An alternate liquefaction strategy for the cellulosic residue is the high severity treatment (liquefaction or pyrolysis) to yield oxygenates of which anhydrosugars and/or hydroxyacids are the predominant components. Catalytic transformation of these products into fuel-compatible chemicals will require extensive de-oxygenation and probably original and shape selective catalytic preparations.

Concluding Remarks

We have tried to demonstrate that the central strategy on biomass conversion to fuel, chemicals and fiber is the pretreatment step leading to fractionation. The pretreatment becomes thus a focal strategy for both biochemical and thermochemical conversion routes.

Pretreatment leads to fractionation of the constitutive chemical entities of lignocellulosics. The fractions, being relatively pure, can be used as feedstocks for more specific end products.

In an overall strategy aiming at an eventual biomass refinery, whose objective will always be to maximize revenue, the hemicelluloses as well as the cellulosic fines should be considered as prime feedstock for the production of sugars. The link with fuel alcohol is then made. The lignin fraction could also be targetted to eventual fuel additives and petrochemical intermediaries. A clever catalytic approach is however needed for the selective depolymerization and correct functionalization of the monomers. The cellulosic residue will likely have a high value as fiber. Only the screened and separated fines will be of immediate interest for its eventual conversion into sugar and ethanol.

The goal of a simple and direct conversion process for the conversion of lignocellulosic biomass into energy has proven very difficult to reach via saccharification or liquefaction/pyrolysis. A co-product strategy appears to be a better option. If the energy-only option is desired, direct combustion, in combination perhaps with gasification/cogeneration, still appear as the simplest, most reliable and energy efficient technology.

Acknowledgments

The authors are indebted to the CQVB and EMR for support of the fractionation/solvolysis/liquefaction program.

References

1. Beckman, D., Elliot, D.C., Gevert, B., Hörnell, C., Kjellström, B., Ostman, A., Solantausta, Y., Tulenheimo, V., Technoeconomic assessment of selected biomass liquefaction processes. ISBN 951-38-3719-S. Research report no. 697, Technical Research Center of Finland (VTT), Vuorimiehentie 5, SF-02150 Espoo, Finland (1990).

2. Overend, R.P. and Chornet, E., Fractionation of lignocellulosics by steam-aqueous pretreatments. Phil. Trans. R. Soc. Lond. (1987) A321, 523-536.

3. Aravamuthan, R., Chen, W., Zargarian, K., April, G., Chemicals from wood: prehydrolysis/-organosolv methods. Biomass, (1989) 20, 263-276.

4. Bouvier, J.M., Gelus, M., Maugendre, S., Wood liquefaction - an overview. Applied Energy (1988) 30, 85-98.

5. Piskorz, J., Radlein, D. St.A.G., Scott, D.S., Czernik, S., Pretreatment of wood and cellulose for

production of sugars by fast pyrolysis. J. Anal. App. Pyrolysis (1989), 16, 127-142.

6. Niemela, K., Low molecular-weight organic compounds in birch kraft black liquour. Dissertation. Doctor of Technology. ISBN 951-41-0633-4. Helsinki University of Technology. Laboratory of Wood Chemistry. SF-02150 Espoo. Finland (1990).

7. Soltes, J., ed., Wood and Agricultural residues: research on use for feed, fuels and chemicals, Academic Press, N.Y. (1983).

Developments in the Steam/Water Liquefaction of Wood

S.G. Allen, D.G.B. Boocock and A.Z. Chowdhury
Dept. of Chemical Engineering and Applied Chemistry, University of Toronto

The 'cascade' reactor at the University of Toronto uses steam and water at temperatures of 330-350°C to convert wood (e.g., poplar up to 7.5 cm diameter - limited by diameter of reactor) into acetone-soluble oil, gas and water soluble organic material in about 1-3 minutes. The oil yield is about 45% and the oil contains 20-25% oxygen. The oil, which contains 60% of the wood carbon, is thermally stable up to 200°C.
When the steam, together with entrained water, initially enters the preheated reactor, it expands and cools somewhat. It also condenses on cool surfaces including those inside and outside the wood. Latent heat is given up and there is an initial fast conversion (within about 15 s) of the wood. Such factors as wetness of wood, water-impervious bark, and the reduced porosity of softwoods retard the conversion. A second slower conversion rate follows in which heat is transferred from the heated walls of the reactor. Gas production is seen as a pressure rise at a time when the reactor temperature is constant. Large dowels with diameters close to that of the cylindrical reactor undergo anomalously high conversion rates. This is due to the good heat transfer across the relatively small annular space which exists between the reactor wall and the dowel. All yields are over 50% in this case. Organosolve lignin is easily converted to oil with yields of over 80%.

On utilise, dans le réacteur en "cascade" de l'Université de Toronto, de la vapeur et de l'eau à des températures de 330 - 350 °C pour convertir, en environ 1 - 3 minutes, du bois (peuplier jusqu'à 7,5 cm de diamètre, limite imposée par le diamètre du réacteur) en une huile soluble dans de l'acétone, un gaz et une substance organique soluble dans l'eau. Le rendement en huile est d'environ 45 %, et l'huile obtenue contient 20-25 % d'oxygène. L'huile, qui renferme 60 % du carbone présent dans le bois, est stable jusqu'à une température de 200 °C.
Lorsqu'elle pénètre dans le réacteur préchauffé, la vapeur, avec l'eau qu'elle entraîne avec elle, se dilate et se refroidit quelque peu. Elle se condense aussi sur les surfaces froides, y compris celles à l'intérieur et à l'extérieur du bois. La chaleur latente est absorbée par le bois qui subit alors une conversion initiale rapide (en environ 15 secondes). Des facteurs, comme la présence d'eau dans le bois, la présence d'écorce imperméable à l'eau et une diminution de la porosité des bois de conifères, retardent la conversion. Cette première conversion est suivie d'un second processus plus lent au cours duquel il y a transfert de chaleur à partir des parois chauffées du réacteur. La production de gaz se traduit par une augmentation de pression pendant une phase où la température dans le réacteur est constante. Les gros morceaux dont le diamètre s'approche du diamètre du réacteur cylindrique sont transformés à une vitesse anormalement élevée et ce, en raison du transfert efficace de chaleur à travers le volume relativement faible de forme annulaire séparant la parois du réacteur de la surface du bois. Dans un tel cas, tous les rendements sont supérieurs à 50 %. La lignine organosoluble est rapidement convertie en huile avec un rendement de plus de 80 %.

Introduction

In the course of our research into the catalytic liquefaction (nickel/hydrogen) of powdered poplar wood in water, it became evident that the catalyst and reducing gas served only to stabilize the product from, what was otherwise, direct non-catalytic liquefaction (Boocock *et al.*, 1980). Subsequently, it first was shown, using tubing bomb reactors that if powdered wood was heated to 350°C in 1-3 minutes and immediately quenched, then liquefaction resulted (Beckman and Boocock, 1983). Further research demonstrated that square cross-sectioned poplar sticks (6.5 mm) could be similarly liquefied (Boocock and Porretta, 1986). Penetration of water and steam into the sticks swelled them prior to their breakdown and liquefaction. Scanning electron microscopy also identified physical changes which occurred outside and inside dry wood (5% moisture) during the liquefaction (Boocock and Kosiak, 1988).

As a result of the research a unit was built for the study of larger wood pieces such as chips and dowels. The major difference between this unit and the equipment used previously was the provision for direct steam and water injection into the reactor. Previously, heat was provided only to the reactors from outside.

Design and Operation of 'Cascade' Unit

The cascade unit was so named because the substrate/product flow was from top to bottom, and where pressure above was not sufficient, gravity provided for a downward flow. The originally conceived design incorporated a vertically oriented cylindrical reactor having inlet (at the top) and outlet (at the bottom) ball-valves with the same diameter as the reactor. The feed in a cylindrical mesh basket was to be introduced to the preheated reactor, and after the closure of the inlet valve, steam (or steam and water) was to be injected. After reaction the mesh basket and products were to exit the reactor at the base and enter a cooling vessel. The products would then have entered a product separation vessel through a third ball valve. Unfortunately, cost considerations precluded the purchase of the three necessary ball-valves. Therefore a design compromise was made in which only the inlet valve had the necessary diameter and thus, only liquid and gaseous products would exit the reactor and any unreacted wood and some residual oil would remain in the basket. After the reactor was cooled sufficiently, the basket would be retrieved through the inlet valve using a hook device. A second cost compromise was made in which a 2 - L autoclave from a previous liquefaction unit was used as the steam vessel. A steam/water separator was not placed between the generator and the reactor.

The reactor (Figure 1) was equipped originally with 3 thermocouples although only the one near the top and in the wall, and the one at the base in the centre have been used to collect data. The reactor is also connected to a pressure transducer. A computerized data acquisition system allowed the temperatures and pressure readings to be stored for further use (1 or 5 s sampling times were used). Three one-eighth of an inch steam lines and inlets were equally spaced along the reactor body. External heating was provided by heating elements, clamped and helically wound around the reactor. These heaters are capable of drawing 4 KW and were used primarily for preheating the reactor.

Operation of Unit

In a typical operation the feed basket containing the wood was introduced to the preheated, nitrogen-purged reactor. The upper reactor wall temperatures were 400°C, whereas close to the base the temperature was about 330°C. After the closure of the inlet valve the steam valve was opened for 7 s. At the end of the reaction time (usually 2-3 min.) the contents of the reactor were discharged into the product receiver via the cooling lock. The "oil" solidified on cooling and was separated mechanically from the aqueous phase. After the reactor was cooled, the feed basket was retrieved through the inlet valve, and more oil could be obtained by acetone washing the reactor. Gas production was measured by displacement over brine. Measurements were eventually discontinued when it was found that routinely 8 L of gas, containing 90% CO_2 and 10% CO, were collected in the conversion of 100 g of poplar (7% moisture).

Some Basic Considerations

Some calculations of the relative volumes of water and vapour both in the steam and reactor vessels under testing and operating conditions have been made. For example under typical operating conditions (steam vessel 350°C and containing 1000 g of water, reactor temperature 420°C on upper wall, 345°C at the base, 100 g wood) at steady state the upper and lower reactor temperatures were both close to 330°C. At this temperature about 65 g of water vapour would have saturated the empty reactor (steam density 0.08 g mL^{-1} at 330°C). A simple steam test under the same conditions (no wood) resulted in about 200 g of water passing to the reactor. Of this amount, liquid water (0.6 g mL^{-1}) would have occupied 260 mL of the reactor and the other 540 mL would have been occupied by vapour (0.08 g mL^{-1}). Therefore in the steam test, considerable entrained water apparently passed to the reactor, even allowing for cool spots at the base of the reactor and at the ball of the inlet valve.

In the steam vessel the 1000 g of water at 350°C virtually occupied the total volume (vapour density 0.1 g mL^{-1}, water density 0.5 g mL^{-1}). Thus when the pressure was released entrained water would again be expected to pass to the reactor. It had been shown previously that if the amount of water in the steam vessel was lowered much below 1000 g then incomplete wood conversion resulted, particularly in the case of dowels. This probably showed that liquid water was required as a heat transfer medium in order to complete the liquefaction. Porous firebrick cubes which had a specific heat of about 1.05 J/g were used to represent a non-reactive medium. The result was that 225 g of water, instead of 200 g, passed to the reactor. The extra 25 g was close to the amount of steam (30 g) required, by condensation at 330°C, to raise the temperature of the brick from 25°C to 330°C. The same mass of poplar chips (specific heat 1.88 J/g) resulted in 250 g of water passing to the reactor - and this is consistent with the firebrick result, taking into account the higher specific heat. It should also be noted that when 100 g of chips were pressurized in the presence of water they could be saturated to hold a maximum of 175 g of water.

The steam test (Figure 2) showed that there was a rapid cooling of both thermocouples but the lower thermocouple cooled below the final steady state temperature of the reactor. This cooling observation was missed if data acquisition was taken at 5 s intervals. The cooling was probably due to the steam expansion into the reactor, but heat stored within the walls rapidly reversed the trend. The presence of porous firebrick or dry chips in the reactor (Figures 3 and 4) delayed the cooling of the thermocouples by up to 4 s. However, when wet chips (41% moisture) were used the maximum cooling again occurred at about 1 s (Figure 5). Porous material therefore intercepts the steam, delaying its arrival at the thermocouples.

Comparison of Figures 2 to 5 also show that gas production occurred when the poplar was liquefied. This was particularly evident from Figure 3 where the pressure is rising when the water temperature is constant. Gas production appears to be connected to those reactions which occur after the wood has physically collapsed.

Results and Discussion

Previously it had been shown that the liquefaction of poplar chips in the "cascade" reactor typically gave oil yields of 40-45%. More significantly, approximately 60% of the feed carbon appeared in the oil, which also contained 20-25% oxygen compared to the 45% in the wood. This oil, although solid at room temperature, melted and flowed at about 75°C. It was also thermally stable up to 200°C.

Runs have been made in the unit which allow comparison of both the conversion and the conversion rates of poplar chips sticks and dowels. The sticks were irregular cross-sectioned offcuts from the rough trimming of dowels and had widths between 0.5 and 1.5 cm. Figure 6 shows mass conversions of chips, sticks and two dowel sizes as a function of time (reaction conditions as stated in 'typical operation' described earlier). The feed mass for the chips and sticks was about 93 g on a dry basis whereas the 2.5 cm dowels had a dry mass of almost 110 g. The 3.0 cm diameter dowels had a mass of just over 160 g. It can be seen that when the steam/water was first injected there was a fast mass conversion. Few data points were obtained inside the 30 s reaction time and when they were, the results tended to be erratic. Only in the case of 2.5 cm and 3.0 cm diameter were data points obtained in the fast conversion region and then only because the conversion rate was slower than for the chips and sticks. The 3.0 cm diameter dowels are a special case and will be discussed later.

It can be seen that the conversion of the chips was essentially over within 15 s. The sticks also showed an initial fast conversion rate - almost the same as for the chips - but a slower rate showed itself from 15 s onwards. The 2.5 cm dowels also had an initial fast conversion rate (but slower than the chips or sticks). Of particular significance was that the subsequent slower conversion rate for the 2.5 cm dowels was very similar to that for the sticks as well as to that for the 3.0 cm dowels between 60 and 120 s. The results suggested that the initial faster rate could be due to the initial high heat transfer particularly from the steam component of the injected mixture. Presumably this heat transfer would be limited by the available surface area of the feed. The second slower reaction which appeared to be independent of feed size could then be explained by heat transfer control from the reactor walls via the water to the broken down substrate.

In order to test the first part of this theory three different dowel sizes were used in the reactor. These dowels had diameters of 1.3, 1.75 and 2.5 cm. One other change was made in that whereas the dowel loading for the two smaller dowels was 50 g that for the largest size dowels was 92 g. The purpose of this was to determine if the weight percent conversion of the dowels was proportioned to the surface area/volume ratio as would be the case if mass conversion was proportioned to the initial surface area. We were uncertain if this relationship would hold as, particularly for the smaller dowels, the conversion after 30 s would be relatively high and the dowels would have degraded and changed shape and/or size. Figure 7 shows that there was a linear relationship between percentage conversion and the surface area/volume ratios. Two conversions are plotted. One took into account only acetone-insoluble material remaining in the feed basket. However, when the feedstock was smaller, the initial attack of steam and water produced a considerable amount of fine material which escaped the basket. This material which was converted to oil by further reaction, could be filtered from the aqueous phase. In the case of chips up to 40 wt. percent of these solids have been isolated for reaction times of 15 s. Approximately half this material was acetone insoluble and when the conversion was corrected for this material the three lower conversion points were obtained. Both sets of data give straight lines which converge around a surface area/volume ratio of approximately 1.7 cm^{-1}. This reflects the relatively negligible amount of the fine material formed after 30 s from the 2.5 cm diameter dowels. Theoretically, for an infinitely large diameter the surface area/volume ratio goes to zero and the percent conversion should also be zero. Significantly, it is the line drawn through the data points for the total acetone-soluble material which passes through the origin. The other line is thus a deviation due to fast formation of the powdered material from the smaller sized feedstocks.

We can now return to the 3.0 cm dowels and the mass conversions shown in Figure 6. One reason these dowels convert faster than the 2.5 cm diameter dowels is that 160 g of them have a larger surface area than the 100 g of smaller dowels. Calculations predict that they should initially convert 1.3 times faster than the 2.5 cm dowels, whereas they actually convert 2.2 times faster. The reason for this is fairly clear. These large dowels are only just smaller than the diameter of the reactor. The flashing of steam and water must result in a rapid filling of the small annular space between the dowels and the reactor wall with water. In addition, the outer layers of the dowel certainly swell due to the absorption of water and this forces substrate against the hot reactor walls. In fact this swelling phenomenum holds for most of the substrates. The initial heat transfer to the large dowels is therefore higher than anticipated. Once the dowels collapse the conversion rate becomes the same as the slower rate for the other substrates.

Organosolv lignin, produced from an ethanol pulping process, has been examined as a feedstock in a preliminary way. The yields are very high (over 80 wt %) and the molecular weight of the product is lower than the produced from whole wood. It is not clear to what extent demethylation or demethoxylation of the aromatic rings may have occurred although neither process is complete.

References

1. Boocock, D.G.B., D. Mackay, H. Franco and P. Lee, "The Production of Synthetic Organic Liquids from Wood Using a Modified Nickel Catalyst", Can. J. Chem. Eng. 58, 466-469 (1980).
2. Beckman, D. and D.G.B. Boocock, "Liquefaction of Wood by Rapid Hydropyrolysis", Can. J. Chem. Eng. 61, 80-86 (1983).
3. Boocock, D.G.B. and F. Porretta, "Physical Aspects of the Liquefaction of Poplar Chips by Rapid Aqueous Thermolysis", J. Wood Chem. and Technol. 6, 127-144 (1986).
4. Boocock, D.G.B. and L. Kosiak, Can. J. Chem. Eng. 66, 121-126 (1988).
5. Boocock, D.G.B. and A.Z. Chowdhury, "The Steam/Water Liquefaction of Non-Powdered Wood etc.", Reprint of DSS Contract File No. 48 52 23283-8-6117, Bioenergy Development Program, Renewable Energy, Energy Mines and Resources Ottawa, Ontario (1988).

Discussion:

J. Piskorz (University of Waterloo): Dave, can you tell us something about the chemical nature of your oil?

G.B. Boocock (Speaker): Yes, we have done quite a lot of work on the aromatic content, about 40 or 50 percent. We don't see the exotic chemicals that you get with your fast pyrolysis process. We don't see things like cellulose and many other things which are water soluble, because they simply go into the aqueous phase.

We haven't done a whole lot of work on the aqueous phase. I must admit that we have not focused on that; it has been looked at by B.C. Research, but that is where the exotic chemicals go that you might be interested in. There is obviously a lot of phenolic content in this particular oil.

N. Bakhshi (University of Saskatchewan): Following the same question, has anybody treated, catalytically, the oils which you have made?

Speaker: Yes, we have tried it and it turns out that even with a thermal stability up to 220, you are still fighting repolymerization. You know where these catalysts start to operate and we are sitting right on the borderline. We would love to have about another 20 degrees of stability in the product to be able to deoxygenate and upgrade this material. So, any suggestions from you Narendra, would be most welcome.

N. Bakhshi: I think we have it for you.

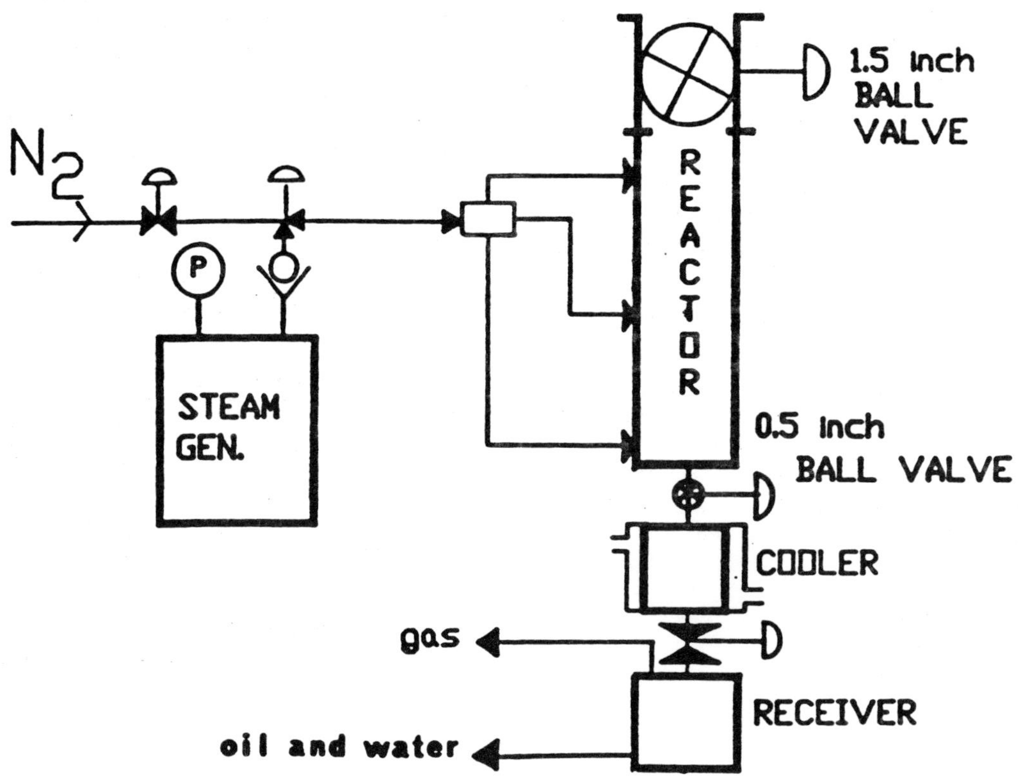

FIGURE 1. GRAVITY FED/DISCHARGED LIQUEFACTION UNIT

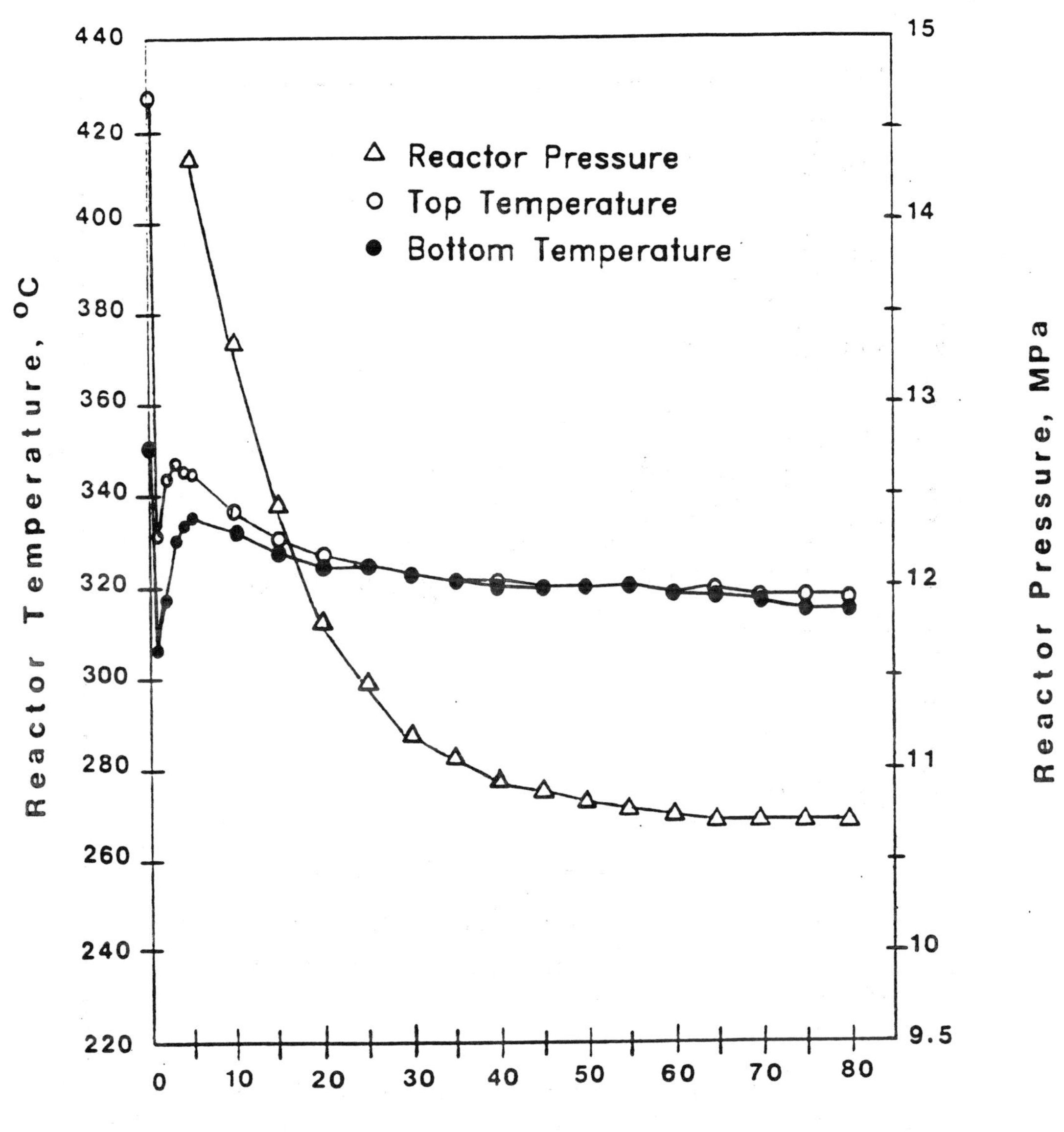

FIGURE 2. REACTOR TEMPERATURE & PRESSURE PROFILES FOR STEAM TEST

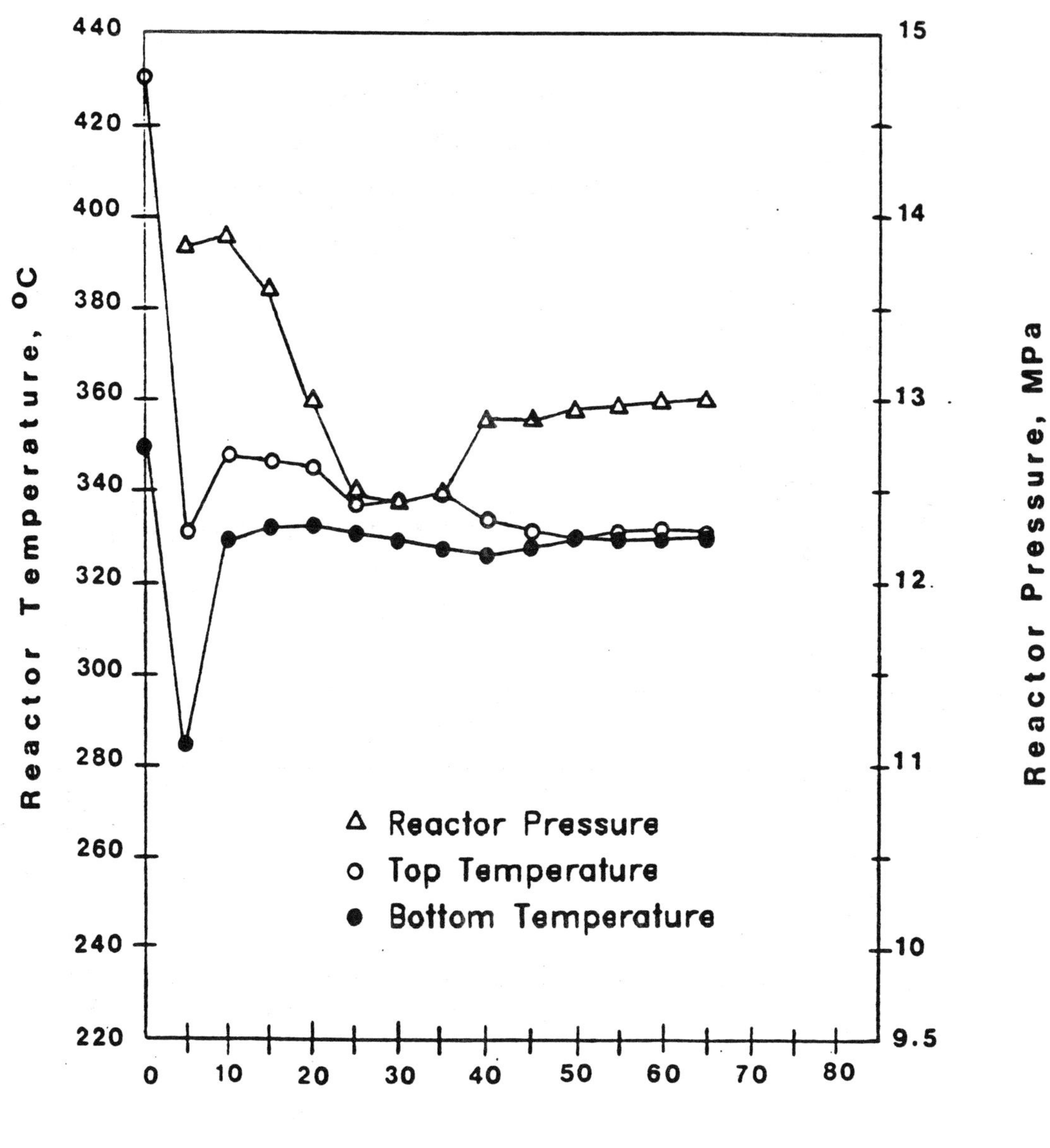

FIGURE 3. REACTOR TEMPERATURE & PRESSURE PROFILES FOR LIQUEFACTION OF POPLAR WOOD CHIPS

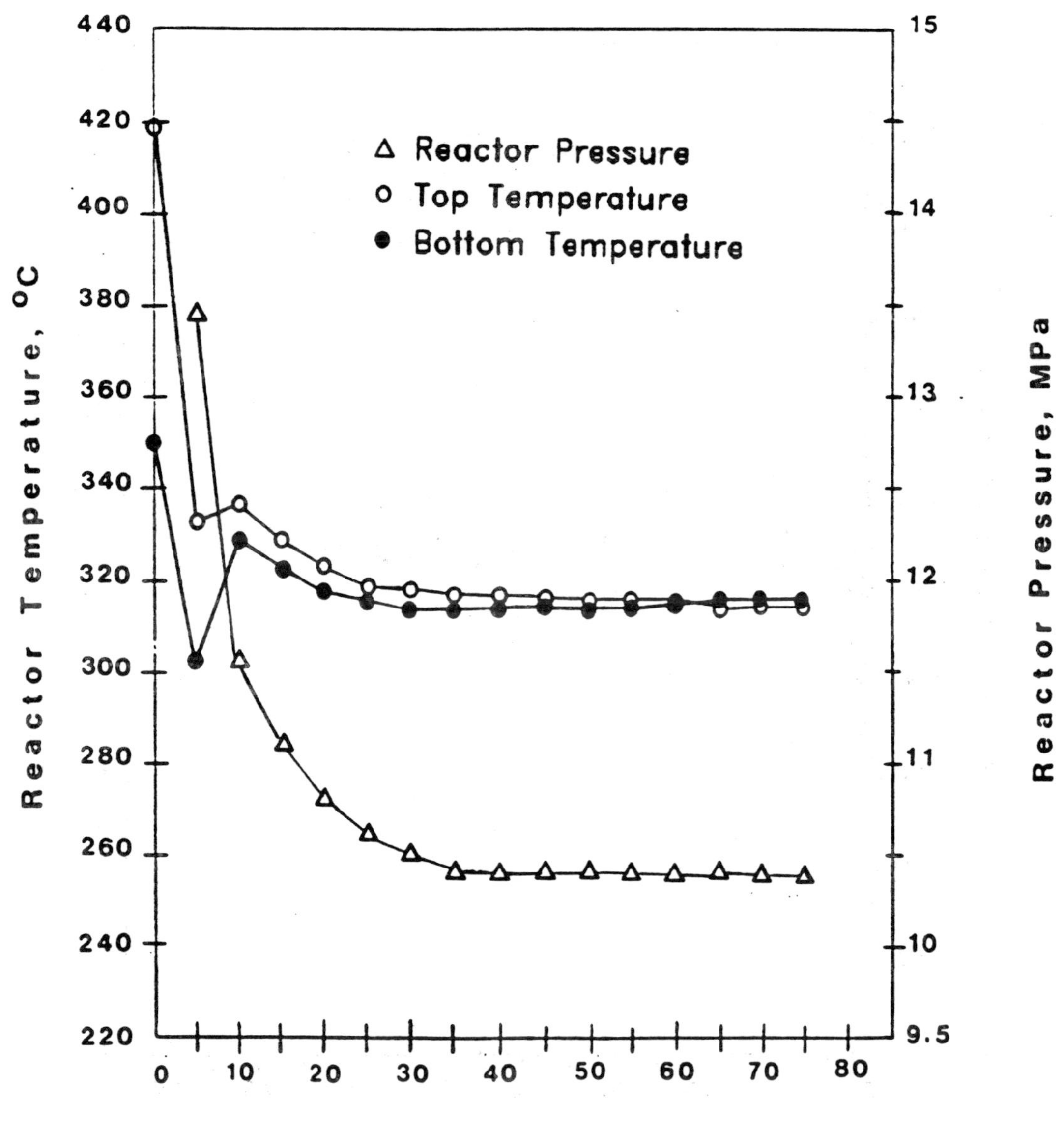

FIGURE 4. REACTOR TEMPERATURE & PRESSURE PROFILES (FIRE BRICK CUBES AS FEED)

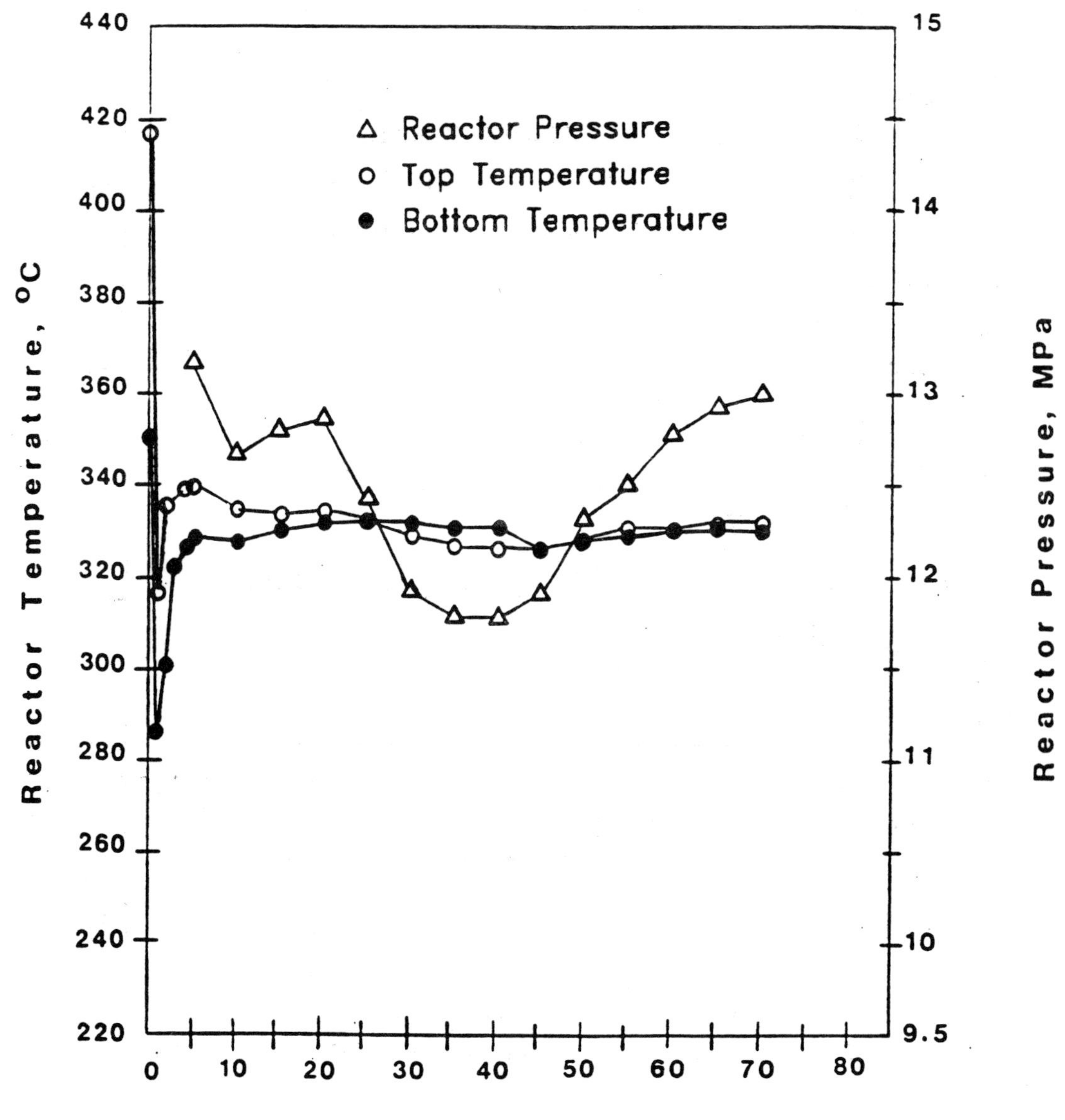

FIGURE 5. **REACTOR TEMPERATURE & PRESSURE PROFILES FOR THE LIQUEFACTION OF WET POPLAR CHIPS (41% MOISTURE)**

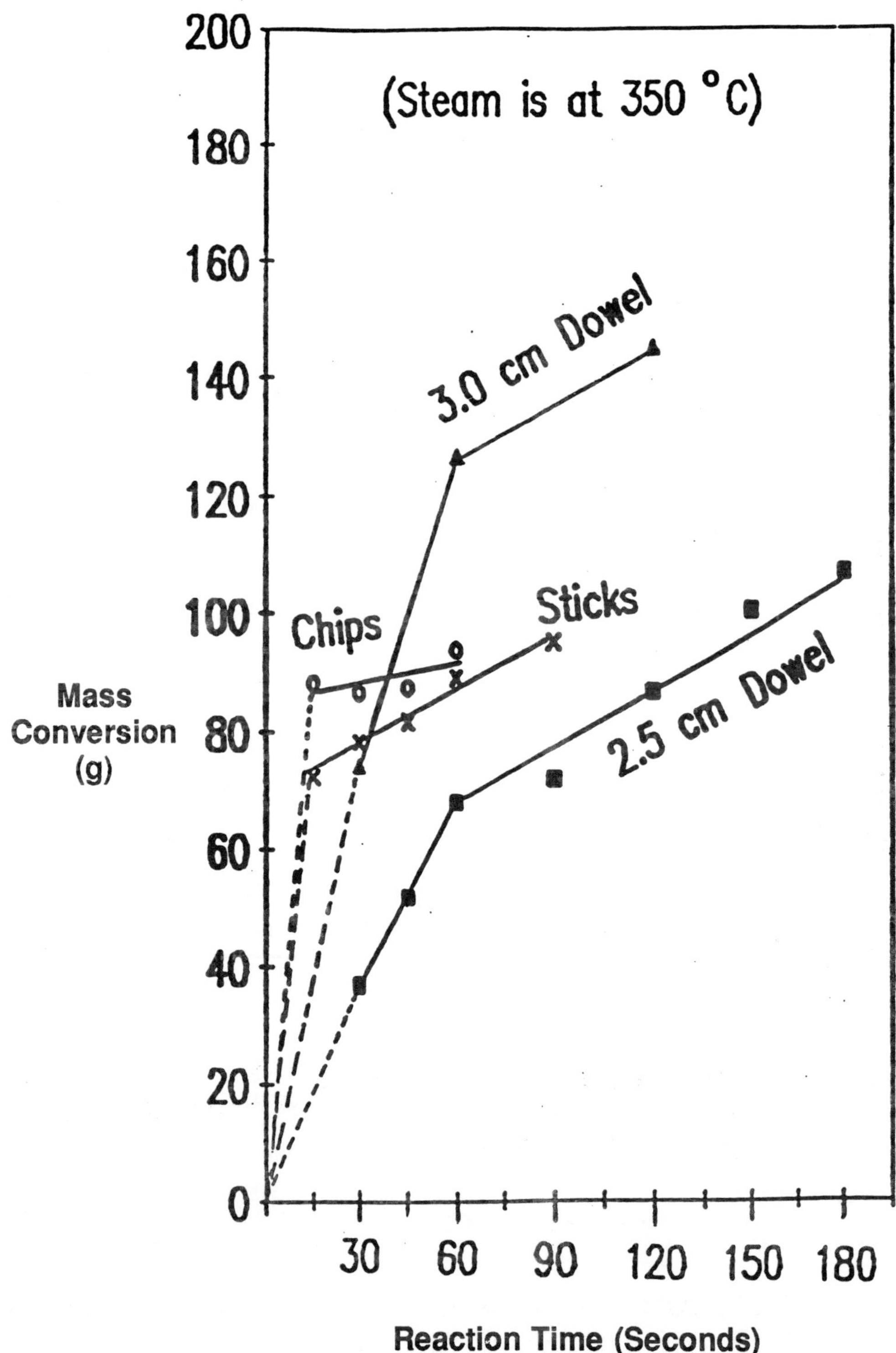

FIGURE 6. MASS CONVERSION OF POPLAR, CHIPS, STICKS AND DOWELS

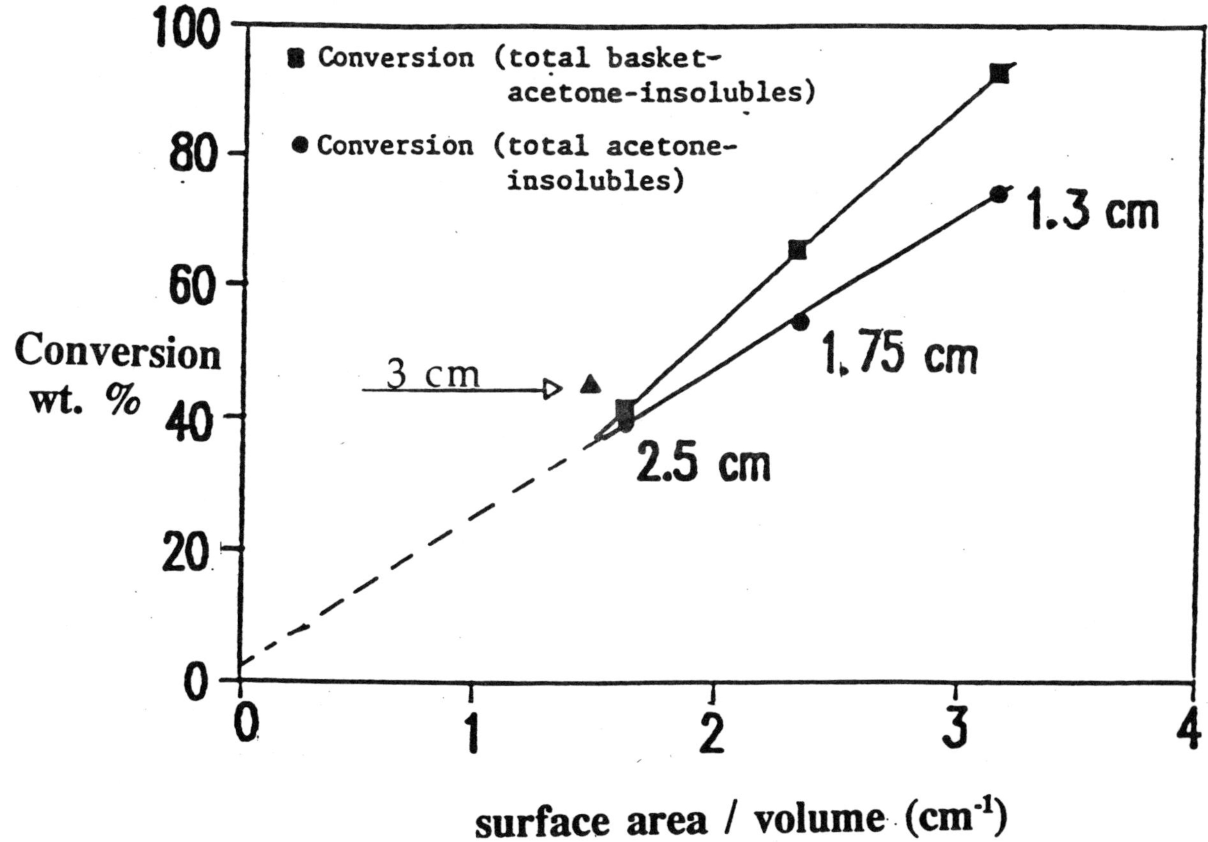

FIGURE 7. CONVERSION (WT.%) AS FUNCTION OF SURFACE ARE/VOLUME RATIO OF DOWELS (350°C STEAM, 30 S REACTION TIME)

Biomass Liquefaction at SERI

J. Diebold, J. Scahill, R. Bain, H. Chum, S. Black, T. Milne, R. Evans and B. Rajai
Solar Energy Research Institute, Golden, CO USA 80401

Current biomass liquefaction efforts at SERI are in two areas: direct liquefaction via fast pyrolysis to oxygenated oils and indirect liquefaction via gasification to a syngas for methanol synthesis. The fast pyrolysis of coarse sawdust in an externally heated, vortex reactor produces large yields of condensible, oxygenated, organic vapors and relatively small yields of char, water, and permanent gases. Interchem is currently scaling up the vortex reactor. The upgrading of the organic vapors is currently being studied with the use of a close-coupled primary catalytic reactor containing various zeolite catalysts, e.g., H-ZSM-5. A secondary catalytic reactor with different operating conditions converts byproduct benzene and gaseous olefins to less toxic gasoline components having higher octane values and lower volatility. The technoeconomics of this biomass-to-gasoline process are now being studied by the International Energy Agency Biomass Liquefaction Group. The fractionation of the condensed organic vapors to make a phenolic-rich extract for replacement of phenol in adhesive and plastic manufacture is of commercial interest at 1987 to 1990 market prices. An industrial consortium has been formed to exploit this attractive technology. A systematic analysis of the process to convert biomass to methanol (via syngas) has identified several areas for future R&D. These areas have the potential to significantly reduce the cost of production to be essentially competitive with natural gas feedstocks.

Les travaux actuellement réalisés au SERI sur la liquéfaction de biomasse portent sur deux grands domaines, soit la liquéfaction directe par pyrolyse, ce qui donne des huiles oxygénées, et la liquéfaction indirecte par gazéification, qui produit un gaz de synthèse destiné à servir à la fabrication de méthanol. La pyrolyse rapide de bran de scie grossier dans un réacteur vortex à une température très élevée produit de grandes quantités de vapeurs organiques oxygénées condensables et des quantités moindres de produits de carbonisation, d'eau et de gaz permanents. La société *Interchem* tente actuellement d'accroître l'échelle du réacteur vortex. On étudie présentement des façons d'améliorer les vapeurs organiques en utilisant un réacteur catalytique primaire à couplage direct contenant divers catalyseurs à base de zéolite, par exemple le catalyseur H-ZSM-5. Un réacteur catalytique secondaire fonctionnant dans différentes conditions permet de convertir le benzène et les oléfines gazeuses, des sous-produits, en constituants de l'essence moins toxiques et possédant un indice d'octane plus élevé et une volatilité plus faible. Le Groupe de la liquéfaction de la biomasse, de l'Agence internationale de l'énergie, étudie présentement les aspects techno-économiques de ce procédé permettant de produire de l'essence à partir de biomasse. Le fractionnement des vapeurs organiques condensées en vue d'obtenir un extrait riche en composés phénoliques destiné à remplacer le phénol dans la fabrication d'adhésifs et de plastiques est intéressant du point de vue commercial, compte tenu des prix de ce produit au cours de la période 1987 -1990. Un consortium industriel a été créé en vue d'exploiter cette technologie prometteuse. L'analyse systématique du procédé de fabrication de méthanol à partir de biomasse (par l'intermédiaire d'un gaz de synthèse) a permis de dégager plusieurs domaines où il y aurait lieu de faire des travaux de R. et D. L'exploitation de ces domaines permettrait peut-être de réduire le coût de production à un point tel qu'il serait essentiellement comparable au coût du procédé utilisant du gaz naturel comme charge d'alimentation.

Introduction

The thermal conversion of wood to about one-third each of char, gases, and organic liquids has its origin in antiquity. However, the application of modern science and engineering to shift the products to maximize the organic liquid yield has occurred in the last 20 years and is still a young technology. It is now widely accepted that the pyrolysis of biomass proceeds via a large number of competing pathways, which can be grossly simplified for the purpose of illustration to the global reactions shown in Figure 1 (**1**). In this global schematic, biomass undergoes both dehydration and depolymerization reactions. The dehydration reactions are more favored at low temperatures and form a plastic char and water, whereas the depolymerization reactions are favored at higher temperatures and rapidly form an oligomeric material. At low temperatures, the mixture of intermediate products is a highly viscous material that shrinks as the charring process proceeds through the further dehydration prior to the eventual polymerization of the residue to form an insoluble char. At higher temperatures, the oligomeric material continues to depolymerize to result in a low viscosity liquid, which eventually has a sufficiently low molecular weight to evaporate at atmospheric pressures to form organic pyrolysis vapors. The presence of small quantities of alkali metals catalyzes the dehydrating, char-forming reactions. These "primary" organic vapors are very reactive and readily undergo further cracking reactions to form permanent gases, as well as "secondary" organics, i.e., phenol and cresols. Under more severe cracking conditions, the condensible organics are reduced to a small yield of "tertiary" products, predominantly a mixture of mono and polycyclic aromatics (**2,3**). The removal of these small amounts of "tertiary" tars from pyrolysis gas streams is necessary for the use of the gases in internal combustion engines or for syngas. This tar removal is a technical challenge, because the tars tend to be very stable at high temperatures and form persistent aerosols at low temperatures.

For direct biomass liquefaction, it is usually desirable to quench the fast pyrolysis reactions before the "primary" pyrolysis products have been degraded to secondary gases and phenols. This has several distinct advantages: a maximum yield of organic condensates; a low viscosity, single phase condensate (if the biomass is dried prior to pyrolysis); a condensate having a demonstrated negligible carcinogenicity (on rodent skin)**(3)**; a highly oxygenated condensate that is very clean burning; and a minimum of energy input to the pyrolysis step, as well as a minimum reactor size resulting from the short residence time. The ease of biomass liquefaction to primary pyrolysis oils has generated a considerable interest in the production of these oils for boiler and gas turbine power applications. The energy density of the pyrolysis oil made from dried feed, but containing the water of pyrolysis, is only about half that of petroleum on a lower heating value basis. However, compared to the solid biomass feedstock, the pyrolysis oil is easier to use as a fuel, has a higher energy density, and better storability.

Fast Biomass Pyrolysis in a Vortex Reactor

One of several methods to achieve the fast pyrolysis of biomass to organic oils is the vortex reactor shown in Figure 2, which was pioneered by SERI **(1)**. In this reactor, the biomass particles are entrained tangentially at high velocities by a carrier gas into the vortex reactor. This causes the biomass particles to slide on the inside surface of the externally heated, cylindrical vortex reactor and to be preferentially heated relative to the carrier gas and the pyrolysis vapors. This type of physical contact can be easily shown with a hot wire to result in such a rapid heating of wood that ablative pyrolysis with rates as high as 3 cm/s is observed **(4)**. The underlying phenomena in the vortex reactor have been modeled based on first principles and ablative pyrolysis experimental data **(5)**. The resulting model shows that the rate of ablative pyrolysis is typically much slower in the vortex reactor than that demonstrated with the hot wire demonstration and explains why the solids recycle loop was necessary to achieve complete pyrolysis of the coarse sawdust particles **(6)**. The products recovered from the vortex pyrolysis reactor have been demonstrated to be 55% by weight organic condensates, 13% char, 12% water, and 14% gases for a 94% closure. The difficulty to achieve mass balance closure is thought to be due to the volatility of the aldehydes in the organic condensates. Note that these yields are on a solvent-free basis (because no solvent was used in their recovery, there is no possible ambiguity of residual solvent or solvent derivatives being confused for product oils).

Upgrading Pyrolysis Oil Vapors to Gasoline

Although the pyrolysis vapors can be readily cracked to form about 15 wt% C_2 + hydrocarbons, a very large amount of methane and free hydrogen is formed as byproducts **(7)**. This lack of specificity led to the abandonment of the thermal cracking approach and to the ongoing catalytic cracking effort using shape-selective catalysts, e.g., HZSM-5. This is the same catalyst that Mobil has used to convert methanol to gasoline in a commercial application in New Zealand. With hydrogen-rich methanol, the oxygen is rejected almost entirely as water and the products at atmospheric pressures are usually a mixture of aromatic liquids and gaseous paraffins. However, for the conversion of biomass pyrolysis vapors to gasoline without the addition of hydrogen, it is desirable to reject oxygen in the form of carbon dioxide, followed closely by carbon monoxide; this conserves the relatively scarce hydrogen for the desired hydrocarbon product. In fact, in this context, gaseous olefins are considered to be hydrogen donors as they oligomerize and then aromatize in the catalyst to release very reactive hydrogen. This reactive hydrogen is then transferred to hydrogen-lean hydrocarbons, such as other olefins (and it is speculated to polycyclic aromatics). Another source of reactive hydrogen is thought to be by the reaction of steam with coke on the catalyst to co-produce carbon oxides **(8)**.

Using extruded HZSM-5 catalyst supplied by Mobil R&D Corp. in a fixed-bed catalytic reactor, we have observed nearly steady-state catalyst operation with a very low rate of coke formation, i.e., as deduced from observing a temperature profile that moved very slowly down the reactor during one to two hour long experiments, using freshly made pyrolysis vapors in steam **(9)**. The beneficial effect of steam on catalyst life has also been reported in the literature for the conversion of canola oil to gasoline with ZSM-5 **(10)**.

The effects of zeolite catalyst variations are being studied with the molecular beam/mass spectrometer (MBMS) at SERI using catalysts made by Prof. Hanson at the University of Utah. These zeolite catalysts were made with varying ratios of silicon to aluminum, organic templates, and metal ions. It appears that there is a considerable effect of these variables on the product slate made from biomass pyrolysis vapors. This work will be reported at the Energy from Biomass and Solid Wastes XV Symposium sponsored by the Institute of Gas Technology.

With the ZSM-5 catalyst bed at the conditions required for the stable catalyst operation for cracking the pyrolysis vapors, the equilibrium product slate is shifted from alkylated aromatics, e.g., xylene, toward benzene, toluene, and olefins. To maintain high gasoline yields, a secondary upgrading reactor is operated with a catalyst and at conditions that favor benzene alkylation with gaseous olefins to form very high octane compounds, e.g., cumene. When properly operated, the catalyst in the secondary reactor is also quite resistant to degradation. Recent yields are shown in Table 1 for the two catalytic reactors in series. It is thought that with the recycling of hydrogen-rich olefinic oligomers, and high boiling polyalkyl aromatics, that the yield of coke and fuel oils can be significantly reduced with the concurrent increase in the yield of high octane gasoline. Although benzene is a valuable chemical ($1.50 to $2.00/gallon) and could be recovered, the recycling of benzene through the two catalytic reactors in series will result in a gasoline product having a negligible benzene content.

Table 1. Run 120 Summary

	WT% of Wood Fed	
C2+ Hydrocarbons		20.5
Primary Gasoline	6.3	
Primary Fuel Oil	6.2	
Distillation Losses	0.8	
Secondary Gasoline	3.4	
C2+ in Secondary Of-Gases	3.8	
Fuel Gases		25.0
Hydrogen	0.1	
Carbon Monoxide	15.8	
Carbon Dioxide	7.8	
Methane	1.3	
Coke		8.2
Char		14.8
Water (by difference)		31.5
		100

Beginning in 1989, the IEA Biomass Liquefaction Group selected this biomass to gasoline process for a technoeconomic evaluation. The process schematic is shown in Figure 3 and includes fast pyrolysis in a vortex reactor, followed by a catalytic cracking reactor and aromatic alkylation reactor. Both a "present" case and a "future" case will be analyzed. In the "present" case, there will be no distillation tower nor recycle of unwanted hydrocarbons, because the results of this hydrocarbon recycle have not yet been completely demonstrated. A light "crude" oil will be the product in the "present" case. This light crude oil would be sent to a central refinery for final processing. In the "future" case, a distillation column will be used to separate the product gasoline from the unwanted recycle streams. This will result in a higher projected yield of liquid hydrocarbons, predominantly a very high octane gasoline having both a low volatility and a low upper boiling point.

Fractionation of Pyrolysis Oils to a Phenolics-rich Extract

During the fast pyrolysis of biomass, the macropolymers of cellulose, hemicellulose, and lignin are depolymerized to monomers and monomer fragments, as well as undergoing mild dehydration reactions. Although chemical analysis has consistently revealed a very large number of compounds present, there are only a few generic types of compounds in the primary pyrolysis oils: acids, aldehydes, sugars, and furfurals derived from carbohydrates; and phenolics, aromatic acids, and aldehydes derived from lignin.

The goal of the fractionation of the pyrolysis oils was to concentrate the phenolic fraction sufficiently so that it could replace a portion of the phenol in phenol-formaldehyde resins. In this application, the chemical reactivity of the extract is of primary importance, rather than its purity **(11)**.

The extraction process developed at SERI involves the neutralization and solvent extraction of the whole pyrolysis oils made with the ablative, vortex pyrolysis reactor. A simplified process flow sheet is shown in Figure 4, in which sodium bicarbonate is used to neutralize the organic acids present in the oils. The phenolics and neutrals (PN) fraction is then extracted with ethyl acetate solvent. The ethyl acetate is evaporated from the PN fraction and recycled. The PN fraction is utilized without further purification to replace phenol in phenol-formaldehyde resins; quite good results have been shown with 50% of the phenol replaced with PN. The extraction process was originally demonstrated as a batch process, but has since been engineered into a continuous process including the solvent recovery **(12)**. The projected economics of the process to produce PN are very competitive at about half the price of phenol **(5)**, which could result in a simple payback of a year or so, depending on feedstock costs and plant size. This process was recognized as being one of the 100 most promising developments in 1990 by *Research and Development* magazine, which gave it the prestigious 1990 R&D 100 award **(13)**.

To commercialize the PN process, the title to the intellectual properties passed to a for-profit private company, Midwest Research Institute Ventures (MRIV). MRIV then advertised the pending formation of a consortium and eventually selected five U.S. companies having the appropriate expertise in the fields of pyrolysis, phenol production, and the use of phenol in resins. The Pyrolysis Materials Research Consortium was formed by MRIV in August 1989 with Interchem, Allied Chemical, Aristech, Georgia-Pacific, and Plastics Engineering **(14)**. SERI's role is to continue to develop intellectual property in this area, which is then passed to MRIV for license to the consortium. Patents developed by the member companies are retained by the originating company. Since the initial formation meeting, the consortium has met twice to discuss progress being made by the members and by SERI in the further research, development, and commercialization of the process. The scaleup of the vortex reactor to 1350 kg dry wood/h has progressed through construction and is currently in a shakedown phase by Interchem, as discussed elsewhere in these proceedings. This approach to technology transfer was awarded the 1990 Award for Excellence in Technology Transfer by the Federal Laboratory Consortium.

Indirect Liquefaction to Methanol

The conversion of biomass to methanol via gasification to syngas has been studied for well over 15 years by a variety of organizations, but always with the conclusion that biomass could not compete with inexpensive, hydrogen-rich natural gas as the feedstock. These previous studies were directed toward utilizing almost the same process flow sheet for biomass as was commercially used with natural gas feedstock. A different technoeconomic approach was initiated about 8 months ago, in which the senior chemical process engineer was directed to determine what advances in biomass gasification, syngas processing, and methanol catalysis would be needed in order for biomass to be a competitive feedstock with natural gas. In the resultant sensitivity analysis, several key areas for research and development were identified, which could result in commercially competitive methanol from biomass. These R+D needs are now being translated into a comprehensive program for DOE consideration.

Summary

The research and development of thermoconversion of biomass to liquid fuels in the program sponsored by the U.S. Department of Energy at SERI is primarily focussed on the fast pyrolysis in a vortex reactor to maximize the yield of organic vapors. Although the vapors can be condensed to a low viscosity, oxygenated liquid thought to be suitable for combustion purposes in furnaces and gas turbines, efforts are directed toward the deoxygenation of the oil vapors through the use of zeolite catalysis at low pressures. The oxygen would ideally be rejected in the form of carbon oxides, leaving the hydrogen in the gasoline product. A re-examination of the potential for gasification of biomass to syngas for methanol production, has revealed several areas that deserve research efforts in the future. Research into the conversion of biomass to chemicals has resulted in a patented process for the recovery of phenolics from pyrolysis oils for the replacement of phenol in certain applications. The Pyrolysis Materials Research Consortium was created to exploit this commercially attractive technology.

References

1. Diebold, J.P. and Scahill, J.W. (1988) "Production of Primary Pyrolysis Oils in a Vortex Reactor," *Pyrolysis Oils from Biomass*, E. Soltes and T. Milne, eds., ACS Symposium Series 376, pp 31-40.
2. Elliott, D.C. (1988) "Relation of Reaction Time and Temperature to Chemical Composition of Pyrolysis Oils," Ibid. pp. 55-66.
3. Evans, R.J. and Milne, T.A. (1987) "Molecular Characterization of the Pyrolysis of Biomass," *Energy and Fuels*, 1, pp. 127-137.
4. Diebold, J.P. (1980) "Ablative Pyrolysis of Macroparticles of Biomass," *Proceedings Specialists' Workshop on Fast Pyrolysis of Biomass*, J. Diebold, ed., Copper Mountain, CO, Oct. 19-22, Solar Energy Research Institute, Golden, CO, SERI/CP-622-1096, pp. 237-252.
5. Diebold, J.P. and Power, A.J. (1988) "Engineering Aspects of the Vortex Reactor to Produce Primary Pyrolysis Oil Vapors for Use in Resins and Adhesives," in *Research in Thermochemical Biomass Conversion*, A. Bridgwater and J. Kuester, eds., Elsevier Applied Science, pp. 609-628.
6. Diebold, J.P. and Scahill, J.W. (1985) "Ablative Pyrolysis of Biomass in Solid-Convective Heat Transfer Environments," in *Fundamentals of Thermochemical Biomass Conversion*, R. Overend, T. Milne, and L. Mudge eds., Elsevier Applied Science, pp. 539-556.
7. Diebold, J.P. (1985) "The Cracking Kinetics of Depolymerized Biomass Vapors in a Continuous, Tubular Reactor", Thesis T-3007, Colorado School of Mines, Golden, CO.
8. Diebold, J.P. and Scahill, J.W. "Engineering Aspects of Upgrading Pyrolysis Oil using Zeolites," in *Research in Thermochemical Biomass Conversion*, loc. cit., pp. 927-940.
9. Diebold, J.P. and Scahill, J.W. (1988) "Zeolite Catalysts for Producing Hydrocarbon Fuels from Biomass", *Thermochemical Conversion Program. Annual Meeting. June 21-22, 1988*, Solar Energy Research Institute, Golden, CO, pp. 21-32.

10. Prasad, Y.S. and Bakhshi, N.N. (1986) "Catalytic Conversion of Canola Oil to Fuels and Chemical Feedstocks. Part II. Effect of Co-feeding Steam on the Performance of HZSM-5 Catalyst," *Can. J. Chem. E.*, 64, pp.285-292.

11. Chum, H.L., Diebold, J.P., Scahill, J.W., Johnson, D.K., Black, S.K., Schroeder, H., and Kreibich, R. (1989) "Biomass Pyrolysis Oil Feedstocks for Phenolic Adhesives," in *Adhesives from Renewable Resources*, R. Hemingway and A. Conner eds., ACS Symposium Series 385, pp. 135-154.

12. Chum, H.L. and Black, S.K. (1990) "Process for Fractionating Fast-Pyrolysis Oils, and Products Derived Therefrom", U.S. Patent 4,942,269.

13. Anon. (1990) "Year's Best High-Tech Products Earn R&D 100 Recognition", *R&D Magazine*, October, pp. 48-101.

14. Anon. (1989) "Biomass-to-Phenolics Route Moves Closer to Market", *Chemical Engineering*, September, p. 25.

Discussion:

N. Bakhshi (University of Saskatchewan): Jim, I saw in that floor sheet that the gases which you produce from your paralyzer projectly going to the Ensyn 5 Catalyst. If I understand, the temperature is around 525, 530 degrees centigrade, something like that?

J. Diebold (Speaker): That is correct.

N. Bakhshi: Now, what is the deactivation characteristics of our catalyst. I think at that temperature you will have a very severe deactivation of the catalyst.

Speaker: At those temperatures the catalyst maintains its activity very well. In fact, in order to maintain activity you must operate at those temperatures.

N. Bakhshi: Can you run it for quite a long time, for example?

Speaker: Yes.

J. Piskorz (University of Waterloo): A few years ago an International Energy Agency evaluated different pathways from biomass to gasoline. One of the conclusions of this research was that fast pyrolysis plus high percent liquefaction using catalysts, is the best way to go. Now you are presenting this zeoloite upgrading. Is this pathway now better than the previous one, or has the evaluation changed in this aspect?

Speaker: I think the conclusion that the only way to reduce gasoline via fast pyrolysis and hydrogenization was not made by the IEA Committee. What they said was, that if you are going to make a pyrolysis liquid, that fast pyrolysis is superior to high pressure liquefaction processes; that is the Albany Process. The first method that they used to upgrade those pyrolytic liquids was with hydrogenization. At that time the Committee felt that there was not enough information known about the zeolite approach to make a meaningful techno-economic study. In the meantime, we have continued to work in this area and the Committee now apparently feels that there is enough information to make this comparison. I don't think that there is, at this point, any conclusion that hydrogenization is better or worse than the zeolite approach. I think that will come out of the study which will be out in a year or so.

J. Piskorz: Correct me Jim, am I right or wrong? By nature of a fast pyrolysis, the oil that is produced, contains a lot of phenolics? It seems to me it does not matter what kind of oil it is, Ensyn oil, water oil or vacuum oil and it seems to me it does not matter what kind of separation scheme exists, those phenolics can be replaced and used in formulation of phenoformaldehyde resins. Am I right?

Speaker: I am not sure on that. You have evidence that says that your water oils can be used in adhesives. The nature of the fast pyrolysis of oils is a function of the operating conditions in the reactor, as well as the feedstock, which will effect the quality of the phenolics.

So, the thing that we have patented is not the production of the pyrolysis oil, but rather the conversion of the oils to the phenolic rich extract, which we feel is a superior product to a simpler or, in some cases, more complex ways to make the extract. Does that answer your question?

J. Piskorz: Yes.

Speaker: I think that fast pyrolysis, by its very nature, makes similar products.

J. Piskorz: Yes, we already covered two of your topics. I have a question about the third one. May I?

Speaker: I want to inject one more comment.

Some of you may be familiar with the magazine "Research & Development", "R&D". The process to make the phenolic rich extract was awarded the very prestigious "R&D 100 Award" this year for one of the top 100 innovations and we feel very proud about that.

J. Piskorz: My question will be scientific. Knowing chemistry, what is going on inside zeolites? In most of the reactions, for example, methanol to gasoline using zeolite dehydration, it is possible to obtain hydrocarbons because the hydrogen to oxygen ratio is much higher than in bio-oils. If you use bio-oils using zeolite, you will obtain mostly carbons. How are you going to visualize or proceed to improve your hydrogen/oxygen ratio in your process, using your zeolite and bio-oil?

J. Diebold: It turns out that as you inject oxygen as water, you can very readily see that you will only make coke on the catalyst. So, you must reject the oxygen as oxides of carbon, carbon monoxide and carbon dioxide in order to make hydrocarbons. That is what we are trying to do, to maximize the rejection of oxygen as carbon monoxides and it does it. The water is the preferred means of rejection of oxygen, only if you have a lot of hydrogen there as you do with methanol. Methanol is so rich in hydrogen it makes me jealous because biomass is very poor in hydrogen.

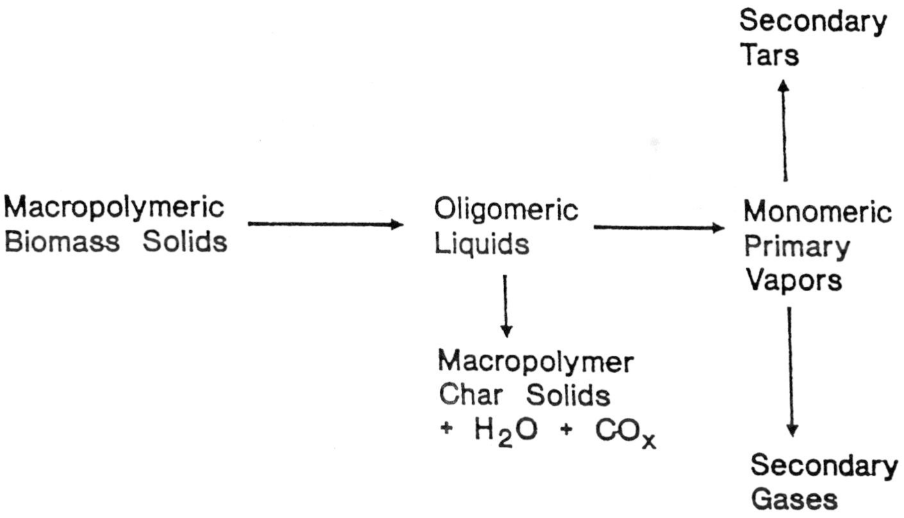
GLOBAL REACTIONS IN FAST PYROLYSIS
Secondary Tars
Macropolymeric Biomass Solids
Oligomeric Liquids
Monomeric Primary Vapors
Macropolymer Char Solids + H_2O + CO_x
Secondary Gases

FIG. 1

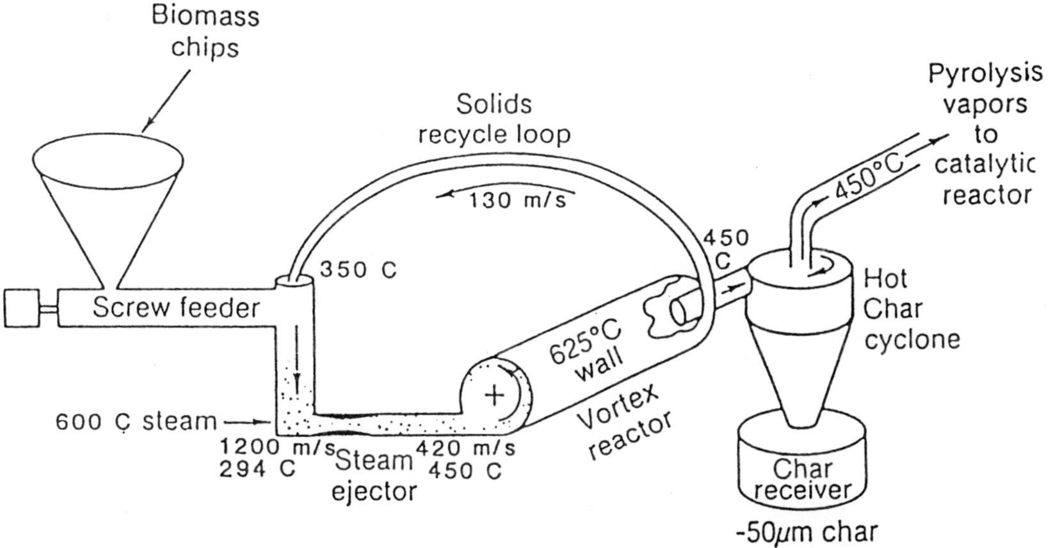
Vortex Reactor Schematic
Biomass chips
Solids recycle loop
130 m/s
Pyrolysis vapors to catalytic reactor
450°C
450 C
350 C
Screw feeder
Hot Char cyclone
625°C wall
Vortex reactor
600 C steam
1200 m/s 294 C
Steam ejector
420 m/s 450 C
Char receiver
-50μm char

FIG. 2

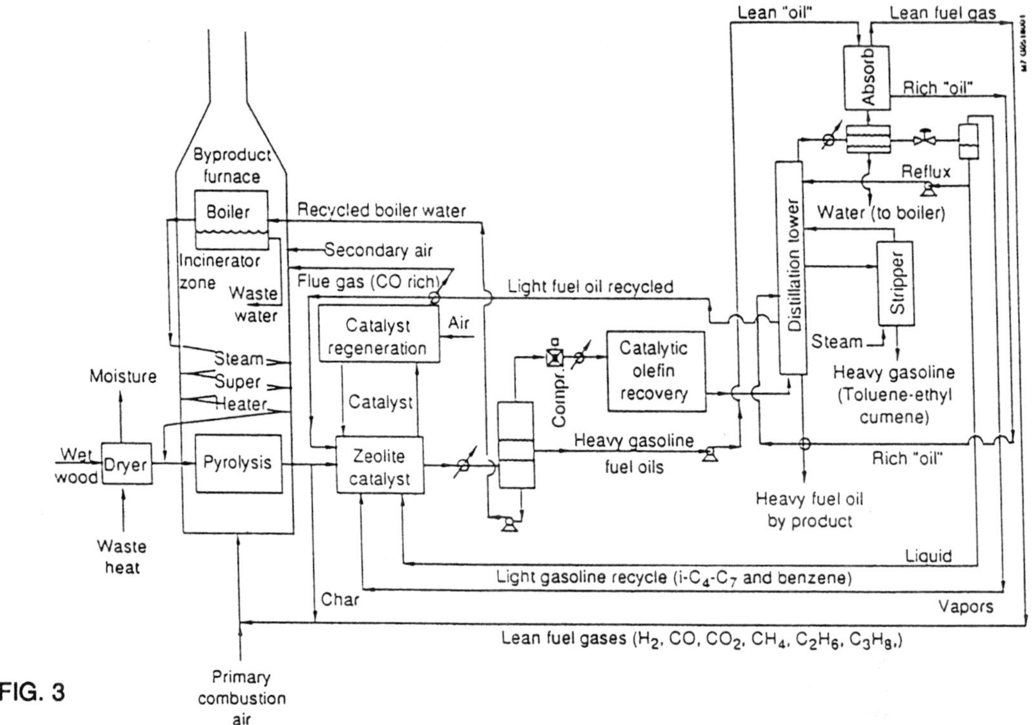

FIG. 3

Production of Phenolic/Neutral Fraction

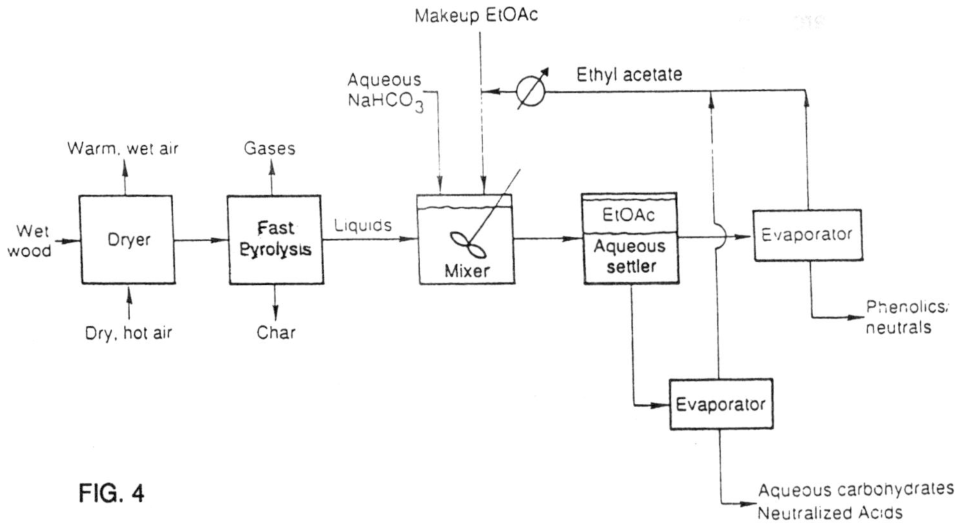

FIG. 4

Vacuum Pyrolysis of Used Tires, Petroleum Sludges and Forestry Wastes: Technological Development and Implementation Perspectives

C. Roy, B. de Caumia, H. Pakdel, P. Plante, D. Blanchette and B. Labrecque[1]
Université Laval, Department of Chemical Engineering, Sainte-Foy (Québec) J3X 1R3
[1]Energie, Mines et Ressources, CANMET/LRDE
2082, Marie-Victorin, suite 210, Varennes (Québec) J3X 1R3

Introduction

Vacuum pyrolysis is a process under development in Quebec since the early 1980's. Approximately 5M$ have been spent so far for the establishment of a solid background basis of the reaction mechanisms, pyrolysis product characterization and design and operation of Process Development Units with capacities ranging from batch up to 200 kg/h continuous pilot reactor.

At the beginning, the research program was focused on wood and related lignocellulosic materials. Between 1981 and 1985, the main objective was to produce liquid fuels from wood. It quickly appeared that the process economics would greatly benefit from the recovery of chemicals, and especially fine chemicals, from the pyrolysis oils. A large research program began in this area in 1984. On the other hand, it also appeared to the authors back in 1985 that vacuum pyrolysis by enabling the production of high yields of pyrolysis oils could represent an attractive solution for the recycling of several industrial and municipal waste materials. Additionally, it was anticipated that methods like vacuum pyrolysis would represent alternative solutions to the more conventional elimination methods like incineration and landfilling.

Consequently, a research program was established back in 1986 in collaboration with Petro Sun International Inc. for the recycling of used tires. The pilot plant unit was built and tested during the fall of 1987 in Saint-Amable near Montreal. Unfortunately, the project was momentarily stopped during early 1988 because of the bankruptcy of the licensed company (Petro-Sun International Inc.). The failure was totally unrelated to the pyrolysis project. With the financial support of both Université Laval and Energy, Mines and Resources Canada (Bioenergy Development Program), the project was continued at the PDU level until now.

A second industrial project was started back in 1987 with a major Canadian oil refinery: Ultramar Canada Inc. This project is now ready for scale-up demonstration. Some of the experimental results obtained so far with the Process Development Unit (40 kg/h) will be disclosed in this paper.

The conversion of forestry wastes to heating oil and activated charcoal is stirring the interest of a Canadian corporation based in British Columbia. The commercial prospects of this third application of the technology will also be discussed in the paper.

Used Tire Project

Tire recycling has become a necessity because of the accumulation of discarded tires that are a potential environmental risk. Each year 24 million tires (0.22 Mt) are disposed of in Canada and about 250 million tires (2.3 Mt) in the U.S. Table 1 reports estimates of used tire arisings for the Economic European Communities and other countries. While some of these tires are recapped or ground for special uses, most are dumped in rural areas or landfills. When buried in landfills, they eventualy float to the surface. In piles, the non-biodegradable rubber can cause serious harm if ignited. Tires infested with mosquitos is a subject of increasing concern **(1)**.

Tires represent a source of energy and chemicals. By thermal decomposition, it is possible to recover useful products. There have been numerous attempts to pyrolyze tires. It is beyond the scope of this paper to describe the various ventures and adaptions of technology. Literature reviews have been published **(2,4-6)**.

There appears to be only a few scrap tire commercial pyrolysis plants in the world, most of which operate in Japan (Table 2). Of these, three operate on a continuous feed and one batch. Typical yields from the Kobe plant are 31% oil, 29% carbon black, 15% gas, 10% steel, 5% sludge and 2% water (8% loss in material balance). The oil is sold to a cement kiln company. The economics of the plant were reported to be marginal **(2)**.

The Kleenair process was developed by Conrad Industries in 1986 for the recycling of used rubber to gas, oil and carbon black. The plant capacity was 1000 kg/h, similar to the Kobe plant. In both cases, the feedstock consisted of shredded tires. The plant is not operating.

The distinct feature of the Onahama plant is that whole tires are used as feedstock to the reactor. The major saleable products are heating oil (25-30%), carbon black (35-40%) and steel (10%). The carbon black is used in an adjacent copper smelting plant **(7)**. The economics of the process is good, although it has been reported to be less profitable than burning the tires at the plant site.

Pilot Plant Study

Vacuum pyrolysis of used tires has been successfully performed in our laboratories at the bench scale and the Process Development scale **(6)**. Based on this background information, the process was then tested as mentioned before at the pilot plant scale, using a horizontal sliding blade reactor. The system was designed to continuously decompose 200 kg of steel belt tires at a pressure below 1.3 kPa. Large pieces of rubber were continuously fed across a column of water which connected a tank on the ground and the top of a horizontal reactor which was elevated 14 m in the air. The reactor was externally fired with gas and a small portion of the pyrolysis oil. The vapors were quenched by two scrubbers set in series. The carbon black was recovered at the bottom of another water head which connected the bottom of the reactor and the ground. A sharp separation of the fiber, steel and carbon black was made in the water phase. The process schematic is given in Figure 1.

The objective of the pilot plant study was to demonstrate the feasibillity of the vacuum pyrolsis process using semi-industrial scale equipment under continuous operation over several hours. The reactor configuration and the reliability of the downstream equipment, including the scrubbers, gas cleaning system and pumps, were evaluated. The results obtained were:

- Continous feeding of tires in vacuum is practicable.
- Separation of steel, fiber and carbon black at the reactor outlet is feasible.
- The equipment designed for the condensation and recovery of the pyrolysis oils performed well.
- The overall thermal efficiency of the process is good.
- The optimum temperature and pressure conditions to produce large yields of oils have been found.
- The three major products, oil, carbon black and steel, are saleable and marketable.
- No major problem is expected with the control of emissions.

The main limitation with this system was the low rate of heat transferred to the rubber per unit area of reactor surface. The reactor operated at less than half of its throughput capacity due to inefficient heat exchange in the chamber. Another improvement in progress is a better system for handling the rubber material inside the reactor by using large rubber pieces (quarters) as feedstock.

Process Feasibility

Based on the reported data, a preliminary feasibility study of the process was performed. The assumptions used are summarized in Table 3. The "improved grade" for the oil and carbon black refers to upgraded product quality as a consequence of additional R and D. The price of $220/ton of carbon black for year three represents 40% of today's price for the bottom line blacks from major producers (all prices are in Canadian dollars). The assumed tipping fee of $1/tire fits inside the current range of fees charged to dispose of tires near large North American cities. More and more states and provinces are enacting or considering legislation to clean up tire piles, which seems to be necessary to prevent speculation when there will be a market for tire recycling **(8)**. Table 4 shows that the profitability of a 3 tons per hour plant (20 000 tons per year), with a capital investment of 7 M Canadian dollars, is attractive if a tipping fee can be collected for recycling tires. The return on invested capital shown in Table 4 represents the invested capital valued at net book value of assets.

Petroleum Sludge Project

The objective of this study which started during 1987 was to evaluate the recycling of petroleum residues by vacuum pyrolysis. Petroleum wastes tested during this study were a complex blend of tar, oils, water, sand and other inorganic materials and were located on site of the Ultramar Canada Ltd. refinery in Saint-Romuald, Quebec. Approximately 60 000 barrels of residues are accumulated and 6 000 more (1 000 t) are generated each year at the refinery plant. The Ministry of Environment of Quebec is progressively putting in place a new regulation that renders on site storage of petroleum residues prohibited. Refineries must find new ways of safely disposing of or recycling such wastes.

Bench Scale Study

A random on site sampling was made which served to prepare two composite samples that were likely representative of the waste materials. The first composite sample was mainly composed of a hard, solid material which represented a small portion of the dump. It was composed of 85.1% inorganic materials (mainly sand and soil), 12.7% water and 2.2% organic matters. The second composite sample was a sort of semi-liquid waste material which represented the bulk of the waste storage area. Its average composition was 54.5% water, 36.0% organics and 9.5% inorganics.

The vacuum pyrolysis experiments were conducted in a batch reactor using 1 kg each of the two composite samples. The maximum temperature reached in the reactor was 425°C under a total pressure lower than 4kPa. Pyrolysis of the solid sample produced 86.5% solid residues, 13.2% water, 0.2% gas and 0.2% oils. In this case a large portion of the organics in the raw materials was converted to carbon which found its way in the pyrolysis solid residues. Pyrolysis of the semi-liquid sample generated 55.9% water, 30.5% oils, 12.3% solid residues and 1.3% gas. The high conversion (85%) of the waste organics to oils is noticeable.

The pyrolysis oils contained less aromatics, asphaltenes and resins than the oils from the raw petroleum sludges. The pyrolysis oils split in two different phases: a light oil fraction and a heavy oil fraction. Both can be recycled by the refinery. However, the pour point of the heaviest oil fraction is high (> 50°C) and the materials will have to be warmed up during its transportation in the pipelines. Alternatively this fraction will be burned, providing heat the pyrolysis process.

The solid residues left after pyrolysis, have been leached according to the procedure described by the Ministry of Environment of Quebec. The leachates were examined for toxic metals, phenols and PAH contents. All tested samples were well below the maximum allowed concentrations. A leaching test was also performed on the solid residues from the raw solid composite sample. Again, no trace of significant contamination was detected in the leachates.

Process Development Unit Study

This phase of the project which just ended a few months ago was to demonstrate the process technical feasibility by constructing and operating a 40 kg/h Process Development Unit (P.D.U.).

Other aspects of the mandate were to prepare the engineering plans for a demonstration plant (1 to 2 t/h), and to evaluate the process feasibility.

As a first step, the preliminary process flow sheet prepared during Phase 1 was revised and the engineering calculations for each unit and component of the process were done. The P.D.U. was installed at the Université Laval pilot plant room in C.R.I.Q., Sainte-Foy. A sketch of the system tested is shown in Figure 2 of the paper.

Samples of petroleum sludges were supplied by Ultramar which included a total of eleven 45-gal. barrels. Six barrels contained hydrocarbon liquid fuels. Five barrels contained solid residual wastes left after centrifugating petroleum sludges on a large scale. All this waste material plus sludges used during Phase 1 of the project were used to prepare a representative feedstock for the P.D.U. demonstration experiments.

At first, small reactor prototypes were empirically tested, especially an evaporation unit and a screw type reactor which was designed to treat the more solid portion of the petroleum sludges. In the evaporation unit which was operated under vacuum at 70-75°C, the sludge was dried so that the water content of the exit materials was less than 1%.

Vacuum pyrolysis of the predried sludge was done at a large scale in a special type of rotating drum reactor. Some design problems were encountered with this reactor principally with the O-ring gaskets between the mobile parts of the apparatus, but these problems were indirectly solved and the reactor concept was found positive.

The heavy and light hydrocarbon vapors generated during the reactions were cooled and collected in two scrubbers, while the non-condensable gases were burnt in a gas burner. Light and heavy oils from the process were tested by the quality control laboratory of the refinery. These tests showed that there would be no problem at all retreating this oil in the refinery.

The pyrolysis solid residue was principally composed of sand, minerals and a small amount of coke. The leaching test for those residues again indicated that the heavy metals content is meeting the MENVIQ regulations. The rotating reactor was best operated under the following condition: drum surface temperature, 540°C-560°C; total pressure, 6.5-10.5 kPa; drum rotation, 0.3 rpm; feeding rate, 15-18 kg/h. For the condensation system temperatures of 40° -50° C and 10° -15°C for the 1st and the 2nd scrubbers, respectively, with recirculation flowrates of 20-25 l/min, have been found optimum.

A typical distribution of the predried petroleum sludge pyrolysis products was as follows: solid residues, 37%; heavy oil, 44%; light oil, 15%; gas, 4%. the predried sludge feedstock (0.5% moisture) was composed of 70% organics and 30% inorganics.

Assuming a 50% water content in the raw petroleum sludge, the required evaporation energy was calculated to be 1250 MJ/t, which is twice that of the pyrolysis energy (600 MJ/t). With respect to the total calorific value contained in the heavy and light oils and the non-condensable gases, the energy required for the process is only 13% of the energy value of the products.

A feasibility study for an industrial demonstration unit, with throughput capacities of 0.5, 1 and 2 t/h, revealed that the scale factor favorably contributes to the process profitability. The operating costs, including incomes for the product oil, were respectively, 60$/t, 19$/t and 4$/t for the three plant capacities considered. Profitability turned to breakeven for a plant capacity of 1.7 t/h. The pyrolysis process also appeared to be rather advantageous when compared with centrifugation (> $125/t) and incineration (> $350/t) methods.

Forestry Waste Project

In Canada the forest industry generates large quantities of wastes which could be recovered and reused. The thermal conversion of wood wastes by vacuum pyrolysis mainly produces pyrolytic oils, and to a lesser extent, wood charcoal. Typical yields are presented in Table 5.

Vacuum pyrolysis of wood and forestry wastes was tested in a multiple hearth furnace Process Development Unit (PDU). The results have been compiled in a report submitted to EMR in 1989 **(9)**. The feedstock typically consisted of air dried wood chip residues composed of 60% wood and 40% bark. These were continuously fed in the reactor at constant flowrate ranging from 4 to 23 kg/h. The optimum contact time of the solids inside the reactor was varied as a function of the reactor temperature and pressure in order to achieve maximum conversion to oils. In this system, the organic vapor and pyrolytic water generated during the pyrolysis are condensed in two quenching units in series. The charcoal residues are collected by a screw and recovered in a bin at the bottom of the pyrolysis unit. The non-condensable pyrolysis gas was flared after passing through the vacuum pump.

A second series of tests recently reported to EMR **(10)** were performed in the PDU to evaluate the capability of the pyrolysis unit to handle large feeding rates. Contact time of the solids inside the reactor was set at 30 minutes. Temperature of the last hearth was maintained between 550 and 570°C while temperature of the top hearth was around 250°C. Absolute pressure inside the reactor ranged between 9.3 and 16.7 kPa. Some technical problems occurred when feeding the PDU at a high flowrate (close to 30 kg h^{-1}): it was sometimes difficult to maintain a stable throughput of wood residues in the reactor over a long period of time.

Results of the proximate analyses of residual charcoals showed that for the same temperature wood conversion was better achieved at lower pressure. Measurements revealed that the power required to maintain vacuum in the system represented only 1/10 of the brake power of the mechanical pump. The process thermal efficiency of the process is high: around 83%. Table 6 summarizes some properties of the pyrolytic oil and charcoal obtained during these tests.

The experimental results with the PDU were used to prepare a flowsheet for a 4 t/h capacity pyrolysis plant (Figure 3). In this process, the moisture content of the forestry residue is lowered from 50% down to 10% with a rotary dryer. The dryer uses the heat generated by burning a small amount of wood flour feedstock. The heat content of the burned flour is equivalent to 1830 kWh. A mill pulverizes the required amount of dry wood and supplies it to the burner. Some energy is also obtained by combustion of the pyrolytic gases (690 kWh) in the gas burner. Four tons per hour of wood chips (10% moisture) are fed with an airtight screw feeder in a battery of three multiple hearth furnaces in parallel (1.3 t/h capacity each). A total energy of 1300 kWh (25% heat loss has been accounted) is required to heat the three electrical pyrolysis furnaces. During pyrolysis, organic vapors, steam and non-condensable gases are formed.

Two quenching towers in series condense the organic vapors and steam. Temperature of the first and second quenchers was adjusted so that the heavier oil phase collected at the first stage contained almost no water. Most of the water as well as the light oil phase were condensed in the second quenching unit. The two liquids recovered at the secondary stage are miscible and constitute an homogeneous aqueous phase with a 2:1, water to light oil ratio. No special water treatment unit was included during the design since the aqueous phase would be disposed of by burning it at the dryer stage. The calorific value of the light oil present in the aqueous phase is recovered to evaporate the water contained in the aqueous phase.

The non-condensable gas is removed by the vacuum pump and is expelled at a pressure slightly higher than atmospheric pressure towards the burner for the wood dryer. The charcoal is collected at the bottom of the three reactors using screw devices externally cooled with tap water. Water is also sprayed over the charcoal to help in cooling it. The resulting slurry is removed from the vacuum system by a slurry pump. Finally the water is removed from the slurry by centrifugation leaving a moistened solid charcoal which can be used as a fuel or as a metallurgical coke when dried.

Upgrading the products is one way of improving the process profitability. The heavy oil has a very low sulphur content and can be burned after being mixed under certain conditions with a lower grade petroleum oil (a high sulphur content oil for example) in power plants. The water/light oil fraction contains valuable chemicals such as carboxylic acids, phenols and fine chemicals such as food aromas (e.g., vanilline) and pharmaceutical compounds. These can eventually be extracted thus providing a better pay off for the process. Charcoal can alternatively be used as a base material for the production of activated carbon. High quality activated charcoals have a pretty good value on the market right now.

A profitability study was done based on six different scenarios for a 4 t/h capacity plant (Table 7). The fixed capital investment for this plant was estimated to be $ CAN 12 M. It was assumed that the heavy oil product would be sold at $20/bbl. Wood charcoal used as a solid fuel is worth $30/t; used as metallurgical coke, $75/t. The price of charcoal upgraded to activated carbon was estimated to be not less than $1000/t. It was assumed however that the activation process would yield only 30% of activated carbon from the initial charcoal feedstock. In scenarios 4, 5 and 6 a tipping fee of $30/t of wood waste treated was included as an income.

The study showed that without collecting a tipping fee, the process generated substantial profits only when the charcoal is upgraded to activated carbon. The collection of $30/t tipping fee makes the process much more attractive.

At this stage of our study it wil become important to more precisely assess the yield and the quality of the activated carbon which can be obtained. Alternatively, the development of methods to extract and purify fine chemicals from the oil phase is also recommendable.

Acknowledgement

The following agencies and corporation have made these projects possible: Energy, Mines and Resources Canada (Bienergy Development Program) and CANMET; Energie ressources Québec; the Natural Science and Engineering Research Council of Canada; the Fonds pour la Formation de Chercheurs et l'Aide à la Recherche; the Université Laval and Ultramar Canada Ltd.

References

1. Craven, R.B., Eliason, D.A., Francy, D.B., Reiter, P., Campos, E.G., Jakob, W.L., Smith, G.C., Bozzi, C.J., Moore, C.G., Maupin, G.O. and Monath, T.P., Importation of *Aedes albopictus* and other exotic mosquito species into the United States in used tires from Asia. J. Am. Mosquito Control Assoc., (1988) 4: 138-142.
2. Dufton, P.W., The Value and Use of Scrap Tyres. Rapra Technology Ltd., England (1987).
3. The Rubber Association of Canada, Mississauga, Ont. (1989)
4. Dodds, J., Domenico, W.F. and Evans, D.R., Scrap tires: A resources and technology evaluation of tire pyrolysis and other selected alternate technologies. N.T.I.S. Report # EGG-2241, presented to the U.S. Dept. of Energy (November 1983).
5. Rivard, P., Roy, C. and Vendette, A., *Les possibilités de réutilisation commerciale des pneumatiques usagés.* L'ingénieur, (1985) 366: 25-30.
6. Labrecque, B., *Etude du transfert de chaleur par radiation thermique dans un réacteur de pyrolyse sous vide des vieux pneumatiques.* M.Sc.A. Thesis. Université de Sherbrooke, Sherbrooke, Qué. (1987).
7. Kono, H., 1987. Onahama Smelting and Refining Co. Ltd., Iwaki City, Fukushima, Japan, private communication.
8. Anon. Scrap Tire News, (1990) 4(1):1-13.
9. Roy, C. et al, *Etude de performance d'une Unité de Développement de Procédé pour la pyrolyse sous vide de la biomasse.* Rapport présenté à Energie, Mines et Ressources Canada /Programme du développement de la bioénergie. Contrat A.S.C. # 23283-7-6334. Janvier 1989.
10. Roy, C., Plante, P., de Caumia, B., Blanchette, D., Pakdel, H., Zhang, H.-G. and Labrecque, B., *Development of a Reactor for the Vacuum Pyrolysis of Biomass,* SSC Contract # 051SZ-23283-8-6100, Energy, Mines and Resources Canada, Ottawa, Canada, 1991.

Discussion:

J. Piskorz (University of Waterloo): It is a general comment not only about your presentation but other developers of commercial scale units. When do you think we should stop involving public funds and public money into this commercial, successful development? At which stage?

C.Roy (Speaker): I think the public is almost ready to get involved in the environment issue. I think many people now want to do something practical, because they are tired of listening to the politicians talking and not doing anything. I think there is a possibility with good technology, to involve people in a true stock exchange. I also think the environment industry is going to require massive capital investment.

I don't think the private sector can even imagine finding the money by themselves for doing it, because it is a big, big industry coming in the next years.

So, once a plant is proven, you have got to find the money to build another plant and another plant and yet another plant, in many different places. That is going to require capital investment and one way of raising funds is through the public. I think it is a good way to involve people and this is what they want. That is how I feel about the future.

P. Fransham (Encon Enterprises): Your last slide, I think, unfairly distorts the economics. In four states now hydrocarbons are considered a hazardous waste. It must be incinerated at somewhere around $900 a barrel. Do your economics take into effect that you could, when you set up a plant, actually end up getting a tipping fee somewhere around $100 to $200 a barrel - you would have discount at a certain amount.

Speaker: Okay, understand me. This is a petroleum sludge oil. It is not meant to be burned; it is meant to be recycled by the refinery itself. So, the pyrolysis plant is going to be attached to the refinery. It is going to be a body, a party of the refinery. Taking that the oil, which is valuable material, but for some reasons due to the presence of toxic metals in this bitumen oil, it is classified as arduous. Now, pyrolysis is going to separate the sand, the toxic metals with some forming coke. I could show here the numbers for the metal, but they are, you know extremely low. The oil is meant to be sold to the refinery - because this is oil coming from the refinery, which is mixed with sand and with heavy bottoms.

So, if you correct the numbers here, not including the value of the oil, the cost here of $60 becomes $100 per ton, as if the oil had no value. But actually, it will have a value, because it is going to increase and add to the total amount of crude oils for the refinery instead of burning it and shipping it to an incinerator which would cost more than $350 per ton. This oil would add as another line, to the crude oil for the refinery.

P. Fransham: That is why I think you have sort of underestimated your case. That it is far better, if you were to take the cost of incineration and put that into your graph, you will find that pyrolysis is very much a net gain because there is a cash outlow that is required for them now to dispose of that oil. As you have said, it is over $300 a barrel.

Speaker: Yes, this is just an extract from our report, but our final report, which was submitted a few days ago to Ultramar, which is happy to share this information with the community here because they believe that pyrolysis has a niche in the industry. So, of course, we are stressing this point and not talking about all the transportation problems and the permits for transporting this sludge.

Instead of transporting this lagoon (60,000 barrels of sludge) 500 kilometres away, it is better to install a small pyrolysis plant for two or three years there, which can be moved and transported and then used elsewhere. It is much more logical this way.

J. Diebold (SERI): I don't quite understand where in the refinery the petroleum sludge is coming. Is this material that would normally go to a coking unit? And is the refinery too small to have their own coking unit or how does that fit in?

Speaker: The petroleum sludge is produced in all refineries. They are bottom residues from the distilling towers: they are, you know, these big, large, wide tanks which are filled with crude oil. When it is empty there is a sedimentation and heavy stuff - toxic metals find their way there. It is skimmed off with steam and they remove all that they can, but there is already this bitumen stuff at the bottom. It is that as well. It is spills, spillage and since you take sand and putting it down, it becomes a mixture of all sorts and from all sources. More or less, the number you have got to keep in mind is in the order of 1 per cent of the production of the refinery, which is a residue. Would you agree with this number?

J. Diebold: I am not sure.

Speaker: Take a refinery like Ultramar, when this study was done, 100,000 barrels per day produces 1,000 a year of waste. So, it is a stream, a continuous stream of residues. Of all the organic waste and we have been talking a lot about PCBs and chemicals like them, but organic, petroleum sludge counts for 2.5 per cent of arduous organic materials in North America. This problem has not been addressed. I think it is a great opportunity for us (pyrolysis 'guys') to get involved in this area as well as biomass.

Perhaps, my message here is, that this nice work on wood is interesting, but I think we have got to also divert a little bit of our expertise in those very promising and demanding technologies - opportunities. It is a challenge for us.

G. Grassi (European Community): Two questions - The first one, we agree with you that the two-ton per hour is the minimum capacity needed to make a profit. In the case of a pyrolysis flash or thermal pyrolysis, the specific estimate is around $7 or $8 a barrel. So, will you see more of a magnitude for drilling oil fields. The cost in future will be around perhaps $10, $15 a barrel for your technology, which is the anticipated, specific investment for a barrel, assuming you utilize it as a fuel?

Speaker: I can comment on two figures, perhaps which have been compared with the Japanese on the tire industry. For the tire, the numbers we came up with was for a two million tires per year plant, which is in the range of a three to four tons per hour plant. We estimated $6 million capital investment.

Compare that with the incineration, with $100 US million - the first number was Canadian - for $10 million - so, there is a factor of five, but you would come up with four times more expensive than incineration.

We have also compared that number with the Japanese, the Iwaki Plant, the Onahama Smelting. They invested $10 million Canadian for their 4.5 ton an hour. So, I think we are in the range.

As for the petroleum sludge, the numbers we estimated were $2 million capital investment for a two-ton per hour plant.

G. Grassi: The second question. Assuming that your product is envisaged for electricity production applied to gas to have a very high efficiency and no specific investment, the cost of a product - of a fuel - is not to be taken as a comparison of the barrel of crude oil, because there are several questions.

First of all, for new limitations, at least in Europe - we have a particular case in Italy by ANAR - you are obliged to utilize very low sulphur content oil.

We know, in general, in the refinery they have a residue with a very high sulphur content, up to 3%. The sulphurization of the oil is much more difficult than coal and the cost is from 3% to 0.1% - $14 a barrel. You then have to take into account the anticipated taxation on this kind of fuel. We evaluated several cases, not experimental plants, but a project of a larger dimension. It is around $5 a barrel. When you utilize the power station or refinery, the main power, the internal power to produce the fuel is very high and the valuation is around $5.

Consider the case where you want or you are obliged to produce a cleaner electricity, you have a difference of at least $15 more or $20 a barrel in total, although it is in a system where you are going from production, harvesting and conversion of electricity - the same side. Now, the economical evaluation indicated the utilization of these kinds of fuels could be competitive.

Now, there is the new carbon dioxide taxation. I am participating in the meeting of the Commission in two or three years time and we will have a carbon dioxide tax. For oil it would be perhaps $10, $15 a barrel which you would have to take into account - and this is not far away in fact in a few years time. In any case, we cannot produce large quantities before that happens.

I think the analyses seems very, very promising considering $25 a barrel and not taking into consideration you may have to have 38 or 37.

Speaker: The arguments you are using will be stronger and stronger in the near future in the environment context. That is the basis of the argument we are using in B.C.; whereby the government is formally forfeiting combustion in the opening of wood waste in B.C.

This is the reason why this project is actually flying in B.C., because the shredders, the forestry people, cannot combust or incinerate in the open any more wood waste. It is forbidden by law.

The only option they have is to pile the wood on their land. If they don't have land, they have got to lend to use - a recourse landfill which is a big problem in B.C. Currently the government is giving $40 per ton as a tipping fee to the shredders, to do something with the wood waste. This is an excellent incentive for the project and this was not calculated in our previous scenario.

We are documenting this aspect of the research with wood based on this preliminary engineering and feasibility of the process over the next weeks. I will be pleased to keep you informed on this particular project with wood.

Table 1. Total Arisings of Used Tires in Different Countries

Country	Mg/Year	Reference
EEC	1 500 000	(2)
USA	2 300 000	(3)
JAPAN	579 000	(2)
CANADA	220 000	(3)

Table 2. Existing Commercial Scrap Tire Pyrolysis Plants

Process or Company	Capacity	Location	Start-up Date
Hyben Recycler	Batch	Britain and Japan	1977
Kobe Steel	1 t/h	Aioi, Japan	1978
Onahama Smelting and Refining	4 t/h	Iwaki, Japan	1981
Kleenair	1 t/h	Centralia, Washington	1986 (now closed)

Table 3. Assumptions used for the process feasibility study

Product	Yield	Price
Oil	55%	Regular grade: 16¢/kg Improved grade: 18¢/kg
Carbon black	25%	Regular grade: 100$/t Improved grade: 220$/t
Steel	9%	120$/t
Fiber/Kevlar	5%	2 $/t
Gas	6%	Used for make-up heat
Prices are in Can. $.		
Other assumptions:		
Tipping fee	1 $/tire	
Tires per ton	110	
Rate of inflation	5%/year	
Re-usable tires	2% of tires @ 6 $/unit	
Re-caps	10% of tires @ 2 $/unit	

Table 4. Profitability of a 20 000 t/year plant for vacuum pyrolysis of used tires

Sales/Revenues (Can. $)	Year 1	Year 2	Year 3
Oil, regular	645 000	1 354 000	
improved			1 422 000
Carbon black, regular	220 000	462 000	
improved			968 000
Steel	95 000	200 000	210 000
Fiber	1 000	2 000	2 000
Tipping fees/Cleanup	1 100 000	2 200 000	2 200 000
Re-usable tire	132 000	277 000	291 000
Re-caps	220 000	462 000	485 000
Total Sales/Revenues	2 430 000	4 957 000	5 578 000
Cost of sales	651 000	1 509 000	1 657 000
Gross margin	1 779 000	3 448 000	3 921 000
Adm. and comm. expenses	177 000	263 000	287 000
Depreciation	675 000	675 000	675 000
Final expenses	473 000	473 000	473 000
Taxes	145 000	785 000	964 000
Profits (loss) of the year	309 000	1 252 000	1 522 000
Undivided profits (deficit) at the commencement	NIL	309 000	1 561 000
Undivided profits (deficit) at the end	309 000	1 561 000	3 083 000
Return on invested capital	5%	22%	31%

Table 5

Yields on pyrolysis of wood containing
10% moisture (wt. %, as-received basis)

PRODUCT	YIELD (%)
Oil	42
Char	24
Water	25
Gas	9

Table 6

General Characteristics of Pyrolysis Oils and Charcoal

	OILS	CHARCOAL
Water content (wt. %)	4.4	<1
pH	2.5	
*Pour point (°C)	15	
*Autoignition point (°C)	440	
*Density at 20°C (g/ml)	1.17	
*Viscosity at 40°C (cSt)	250	
*Viscosity at 100°C (cSt)	16.5	
Gross calorific value (kJ/kg, anhydrous basis)	22 500	32 000
Elemental analysis (%)		
C	55.6	84.4
H	5.8	3.5
N	0.7	0.2
S	0.6	0.8
O (by diff.)	37.3	11.1
Proximate analysis (%)		
Ash	0.13	2.8
Volatile Matter		18.5
Fixed carbon		78.7
Specific suface(BET, m^2/g)		≈3

* Test conducted with an oil sample which contained approximately 10.7% water

Table 7

Feasibility of a 4 t/h Capacity Pyrolysis Plant for Forestry Wastes*
(in M $ Can.)

	SCENARIO #					
	I	II	III	IV	V	VI
SALES						
Charcoal: Fuel (@ $30/ton)	0.2	-	-	0.2	-	-
Charcoal: Coke (@ $75/ton)	-	0.5	-	-	0.5	-
Charcoal: Activated Carbon (@ $1000/ton)**	-	-	2.3	-	-	2.3
OIL (@ $20/bbl)	1.1	1.1	1.1	1.1	1.1	1.1
TIPPING FEE (@ $30/ton)	-	-	-	1.7	1.7	1.7
TOTAL INCOMES	1.3	1.6	3.4	3.0	3.3	5.1
TOTAL EXPENSES	2.9	2.9	2.9	3.0	3.0	3.0
PROFIT (before tax)	(1.6)	(1.3)	0.5	0.0	0.3	2.1
RETURN ON CAPITAL INVESTMENT (before tax), %	(13)	(11)	4	0	3	18

* Investment: 12 M $ CAN.

** A 30% yield of activated charcoal from wood charcoal was assumed

( ) Indicate negative values

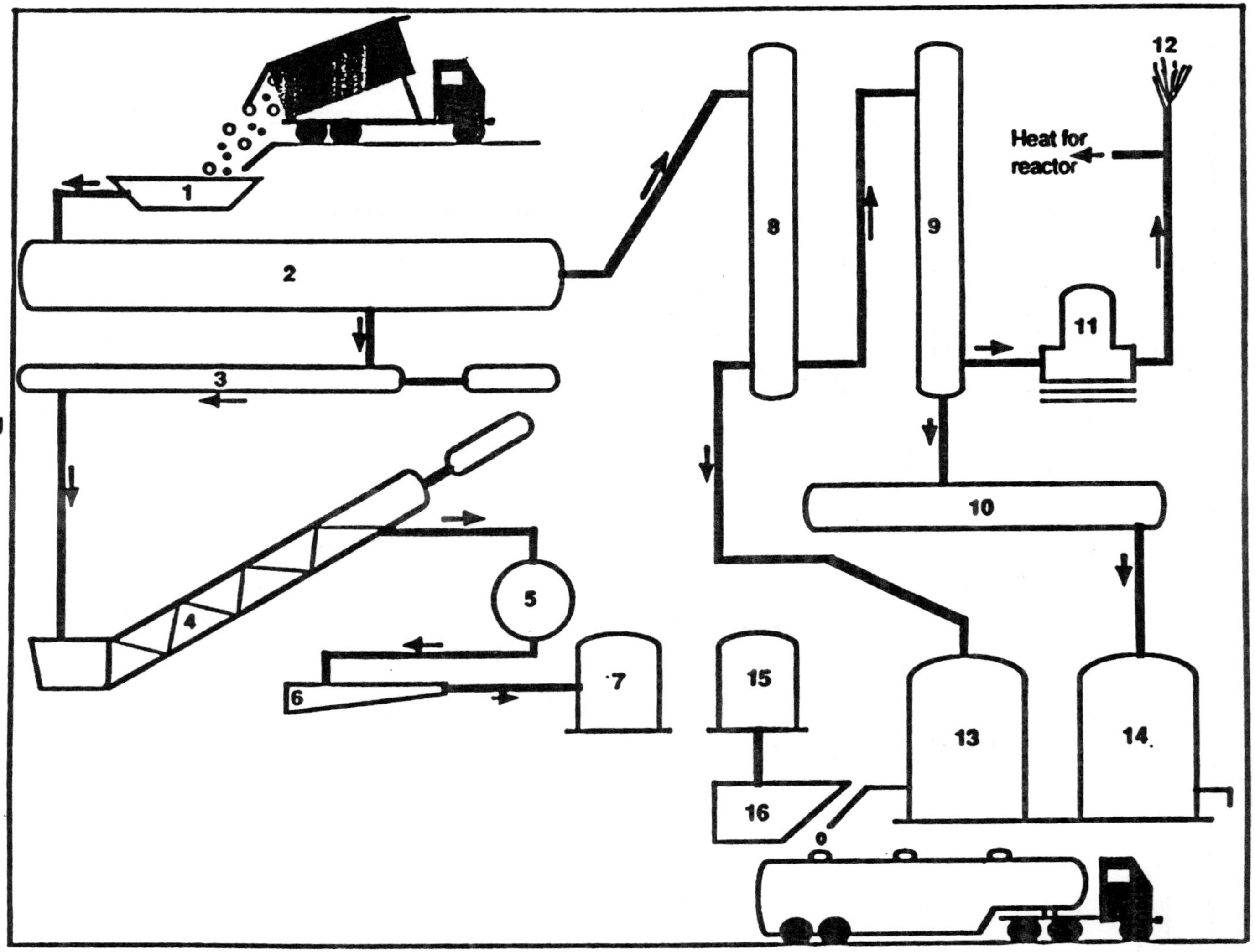

Heat for reactor

Legend for Figure 1:

1. Feed conveyor
2. Vacuum reactor
3. Cooling screw
4. Discharge screw
5. Crusher
6. Vibratory screen
7. Carbon black handling system
8. Heavy oil quencher
9. Light oil quencher
10. Decanter
11. Vacuum pump
12. Flare stack
13. Heavy oil storage
14. Light oil storage
15. Magnetic separator
16. Steel recovery bin

Figure 1. Schematic of the St-Amable Pilot Plant for Vacuum Pyrolysis of Scrap Tires

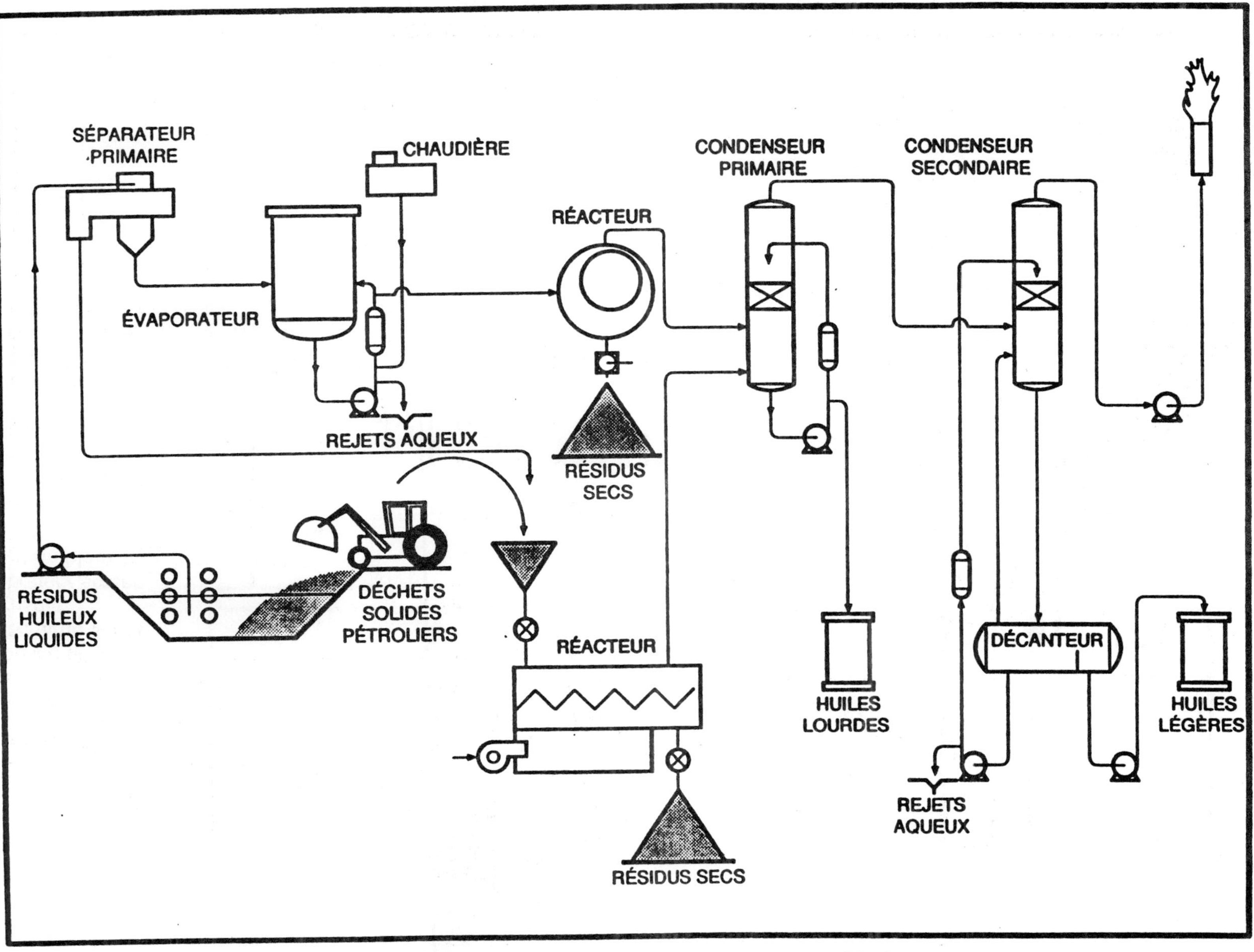

Figure 2. Schematic of the Ultramar/U. Laval Petroleum Sludges Vacuum Pyrolysis Process Development Unit

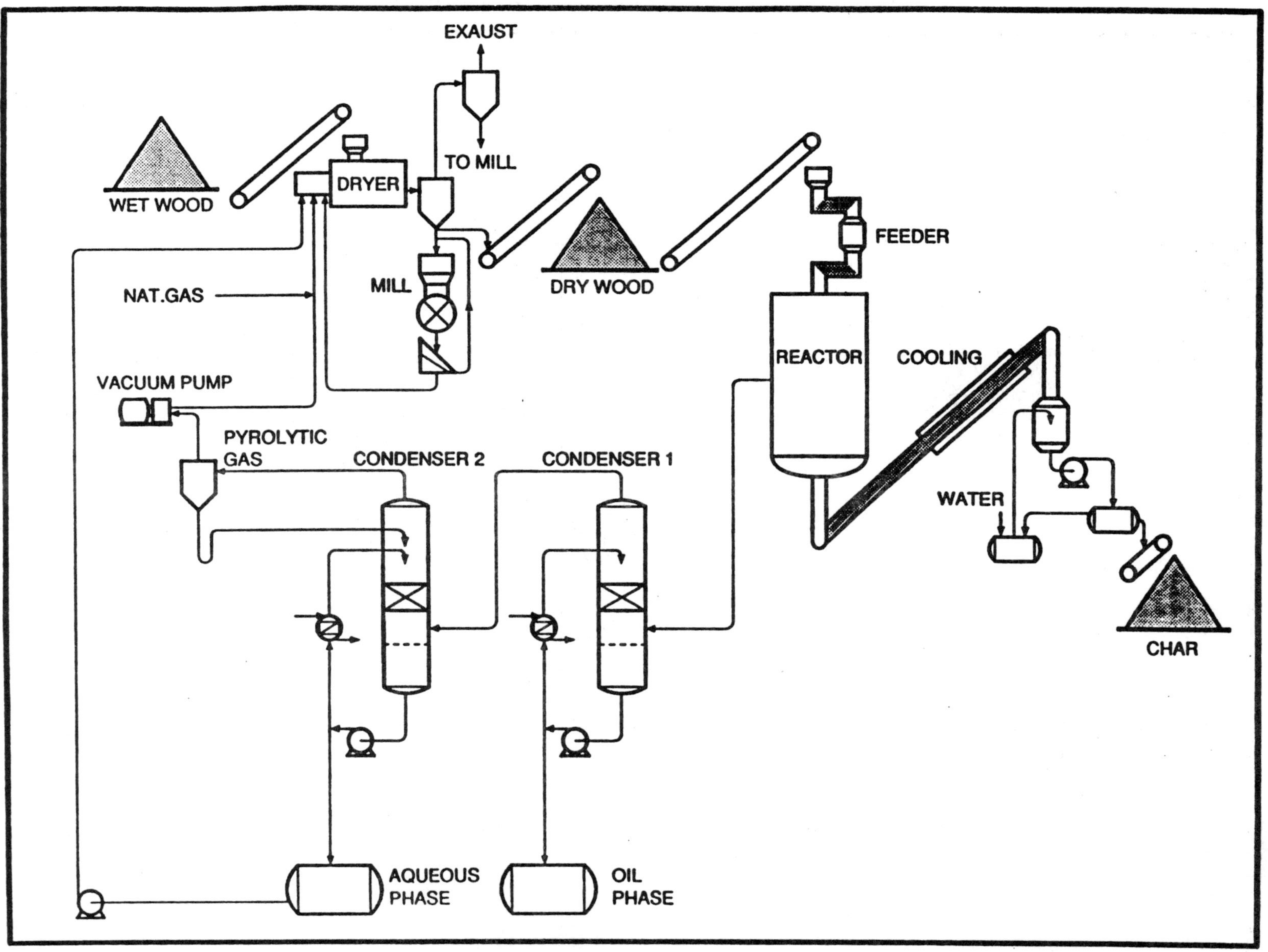

Figure 3. Flowsheet of the Contemplated 4000 kg/h Biomass

Preliminary Mass Balance Testing of the Continuous Ablation Reactor

J.W. Black and D.B. Brown
BBC Engineering Ltd.

Introduction

The complete disposal of rubber tires continues to be one of the most visible and hazardous problems facing our society today. The Rubber Association of Canada reports that some 26 million tires are discarded annually, with 85% of this amount ending up in municipal landfills or privately owned dumps.

Tires are a petroleum-based product and represent a significant source of hydrocarbon chemicals and energy. We believe that recovery of value-added liquids and gases from this source, provides a greater opportunity for a commercially viable facility, than simply combusting the whole tire. In addition, the recovery of carbon black, along with the steel present in the bead and the belt, are also contributors to profitability. The residual liquids, after extraction of the petro-chemicals, are rich in aromatics and hence have value as a raw material for carbon black manufacture.

Pyrolysis is the key to recovery of these chemicals and considerable work has been done on this technique for the production fuel oils. These efforts have resulted in marginal economics. Greater emphasis on the production of chemicals will sufficiently improve the economics to ensure sound profitability.

It is known from studies on wood, oil shale and coal, that short vapour residence time reactors generally produce simpler organic moieties than long residence time systems. These low molecular weight materials tend to be easier to separate and be more valuable. A number of different fast reactor designs are available, the Rapid Thermal Processing (RTP) unit of Ensyn Engineering, the fast pyrolysis fluidized system from the University of Waterloo and the vortex reactor of SERI.

Ireton International Waste Management, a newly formed company, have just recently developed another fast reactor design involving a simple continuous ablation reactor, C.A.R., which provides a continuous steady force for ablation, eliminates inert solids recycle and yet maintains the very high heat transfer rates associated with such reactors. Reactor residence times are in the range of the ultra fast reactor design, 50 to 3000 msec. Tests with a large pilot reactor at rates up to 50 kg/hr have successfully demonstrated efficient heat and mass transfer. The CAR can also be scaled-up readily and can be used with a wide variety of feedstocks having particle sizes larger than is normally encountered in ultra fast reactors.

We propose to evaluate this simple, ultra fast reactor as a means of maximising the recovery of primary pyrolysis (highest value) products from rubber tires.

Objectives

The primary goal of the program, at this stage of development, is to evaluate the continuous ablation reactor (CAR) as a rapid heat transfer vehicle for ultra fast pyrolysis of rubber.

The objectives are:

- Process tire particles at short residence times (500ms)
- Evaluate the products
- Confirm that complete pyrolytic decomposition is occurring within the reactor

Equipment Description

The equipment is shown in figure 1 and consists of a feed system mounted on load cells, a continuous ablation reactor (CAR), a cyclone, a condensation train, various instruments and analytical devices and a data acquisition system. The capacity of the small test unit is 10 to 25 kg/h. A larger reactor with a capacity of four times the small unit is also available for small scale production tests.

The feeding system comprises a pressurized lock-hopper with a variable speed discharge screw which in turn feeds material into the reactor. A purge gas is added to the feed to assist in removing the vapours from the reactor, thus controlling the vapour residence time. For simplicity in the early stages of the program, nitrogen will be used as the carrier gas. Later, the gas produced by pyrolysis will be recycled for use as the extraction gas. The short residence time CAR has staged external heating to provide zonal temperature control. Gases and vapours leave the reaction zone and pass through a cyclone to remove the fine carbon black and steel, which is blown out of the reactor by the purge gas. The residues are collected in a sealed bin for later weighing and analysis. Clean steel can be magnetically separated from the carbon. A fixed charge of catalyst can also be added to the reactor, if desired.

The vapours are condensed in two stages of direct contact quenching followed by a final indirect condenser. The first stage condenser is a baffle plate column, which has been designed to eliminate plugging. If necessary, a final filter can be installed to collect any mist that may be formed.

The pilot plant is well instrumented for temperature, pressure and flow measurement. Gas analyses are provided on-line by a combination TCD/FID gas chromatograph. Rubber feed rate is determined by a load cell system which supports the feed hopper. Liquid product and carbon black are weighed at the end of each test and samples sent for analysis. All of the instruments are hooked up to a Hewlett Packard data acquisition system and the data is sent to computer for instantaneous readout and storage.

Results

The reactor has been tested over the period December, 90 to March, 91. Some minor modifications have been made to improve sample collection.

Typical reactor operating conditions to date are:

Vapour residence time:	350 - 650 ms
Reactor pressure drop:	10 - 15 kPa
Operating temperature:	500 - 550~ ^oC
Solids flow rate:	2.5 - 16 kg/h
Carbon black yield:	29 - 33 %
Particle size:	250 - 3000 microns

The results of the most recent test in which accurate measurements of solids and gas flows were as follows:

Vapour residence time:	0.41 s
Reactor pressure drop:	13.5 kPa
Operating temperature:	550~ ^oC
Rubber flow rate:	2.5 kg/h
Particle size:	1 - 3 mm

Mass Balance

Raw Material	Products		Yield (%)
Rubber: 2.454g	Carbon Black:	0.821g	33.5
	Gas:	0.416g	16.9
	Liquids: (by difference)	1.217g	49.6

A typical gas analysis is presented in Table 1.

Table 1 - Gas Analysis

HYDROGEN	12.59
CARBON MONOXIDE	4.30
CARBON DIOXIDE	3.68
METHANE	22.46
ETHANE	2.69
ETHYLENE	15.15
PROPANE	0.88
ACETYLENE	0.67
PROPYLENE	13.77
iso-BUTANE	0.33
UNKNOWN	0.31
n-BUTANE	0.09
1-BUTENE	1.19
iso-BUTYLENE	9.67
trans-2-BUTENE	1.11
PROPYNE	0.44
cis-2-BUTENE	0.99
1-3 BUTADIENE	9.35
PENTANE	0.32
	100.00

Analysis of the solids and liquids is not yet complete. Subjectively, however, the carbon black is very fine, similar in texture to copier toner and does not resemble the char described by other investigators.

Conclusions

The gas analysis shown in Table 1 is very high in olefin content, more than 50%. These levels are higher than those described by other investigators whose results tend to be in the range 20 - 30%. This difference strongly suggests that, as originally purported, the CAR is producing primary pyrolysis products, otherwise the olefin content would have been lower, due to addition and hydrogenation reactions.

Carbon black yield, at around 30%, is the normal value found for tire pyrolysis. The material also subjectively resembles the carbon black which is initially blended into the rubber, in its particulate nature. From the nature of the carbon black and its yield, it can be surmised that pyrolysis is essentially complete.

It can be concluded, therefore, that the CAR is performing as postulated, i.e., that high heating rates can be obtained without the use of inert solids and that ultra fast pyrolysis can be attained within the residence time available in the reactor.

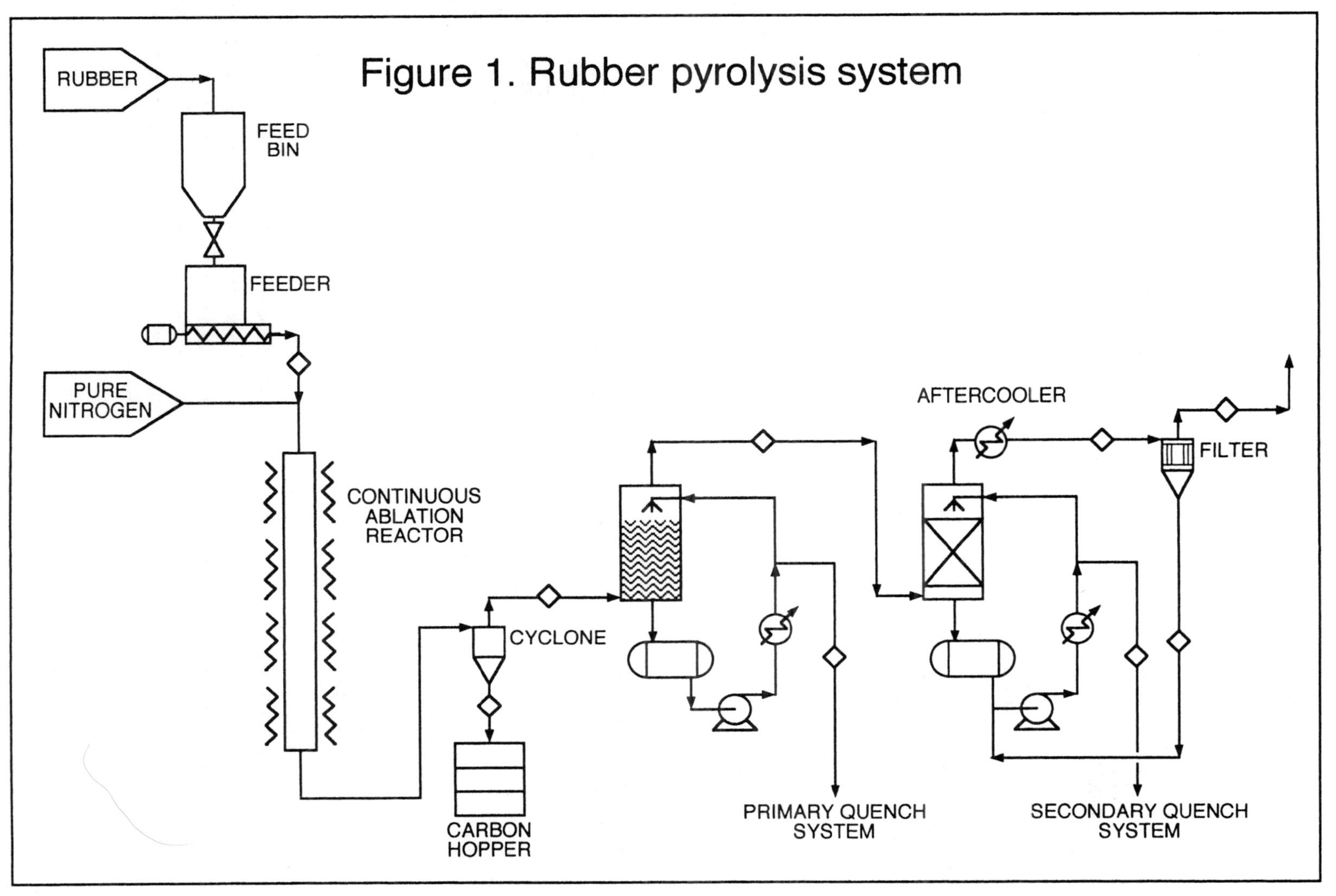

Figure 1. Rubber pyrolysis system

Biomass Liquefaction: Centralized Analysis

C.S. Alleyne, E.S. Skelton, and J. McKinley
British Columbia Research Corporation, Vancouver, B.C.

B.C. Research serves as a centralized analytical facility for the characterization of biomass liquefaction products. It provides for the needs of the biomass liquefaction community through its two designated objectives: (1) To provide a centralized analytical service, and (2) To undertake special projects requiring analytical expertise.

Analytical services requested most frequently are those required for mass balance calculations and characterization of the chemical compositions of liquefaction products. They include elemental analyses, determinations of water content and density, and chemical characterization of biomass pyrolysis materials requiring techniques such as high resolution gas chromatography (HRGC), gas chromatography-mass spectrometry (GC-MS), high performance liquid chromatography (HPLC), and carbon-13 nuclear magnetic resonance (^{13}C-NMR). Other analyses include determinations of calorific value, particle size distributions, and benzene, toluene and xylene content.

B.C. Research played a central role in the first Thermochemical Round Robin to determine the interlaboratory precision of carbon, hydrogen, nitrogen, oxygen, and water measurements in biomass oils. We are currently validating the protocol of Dr. Olof Theander (Swedish University of Agricultural Sciences) for analysis of extractives, polysaccharides and lignin, as well as a colorimetric method for uronic acids. These procedures will be documented in a suitable format for the forthcoming third IEA Analytical Round Robin for whole feedstocks.

La *B.C. Research* sert d'installation centralisée d'analyse où l'on procède à la caractérisation des produits de liquéfaction de biomasse. En réalisant ses deux objectifs désignés, soit assurer la prestation de services centralisés d'analyse et entreprendre des projets spéciaux exigeant des compétences dans le domaine analytique, la *B.C. Research* repond aux besoins des personnes s'occupant de liquéfaction de biomasse.

Le calcul du bilan massique et la détermination de la composition chimique des produits de liquéfaction sont les services d'analyse les plus fréquemment demandés. Ces services comprennent l'analyse élémentaire, la détermination de la teneur en eau et de la masse volumique et la caractérisation chimique des produits de pyrolyse de la biomasse, à l'aide de techniques comme la chromatographie en phase gazeuse à haute résolution, la spectrométrie de masse couplée à la chromatographie en phase gazeuse, la chromatographie en phase liquide à haute performance et la résonance magnétique nucléaire du carbone-13. Parmi les autres analyses, on compte la détermination du pouvoir calorifique, l'analyse granulométrique et le dosage du benzène, du toluène et du xylène.

La *B.C. Research* a joué un rôle crucial au cours de la première série d'essais comparatifs sur la thermochimie, dont le but était de déterminer la précision interlaboratoire des méthodes de dosage du carbone, de l'hydrogène, de l'azote, de l'oxygène et de l'eau dans les huiles tirées de la biomasse. Nous procédons actuellement à la validation du protocole du Dr Olof Theander (Université suédoise des sciences agricoles) qui permet d'analyser les produits d'extraction, les polysaccharides et la lignine, ainsi que d'une méthode colorimétrique applicable aux acides uroniques. Ces méthodes seront présentées, pour des charges d'alimentation entières, dans un cadre convenable en prévision de la prochaine série d'essais comparatifs sur les méthodes d'analyse de l'AIÉ.

Characterization of Biomass Materials

An initial objective of the Centralized Analysis project was to serve as a central analytical facility for acquiring experimental data for characterization of biomass liquefaction products derived from different feedstocks and different process conditions. However, this project has been diverted from that original objective into providing an analytical service for characterization of biomass materials.

Several members of the biomass liquefaction community have taken advantage of this service and use it on a routine basis. Some of the analytical instruments that are available at B.C. Research for this service are listed below.

- High Resolution Gas Chromatograph (HRGC) - FID, FPD, TCD, ECD, NPD
- High Resolution Gas Chromatograph-Low Resolution Mass Spectrometer (HRGC-LRMS)
- Pyrolysis Gas Chromatograph
- High Performance Liquid Chromatograph (HPLC)
- Ion Chromatograph
- Thermogravimetric Analyzer
- ^{1}H and ^{13}C-Nuclear Magnetic Resonance (NMR) Spectrometers (University of British Columbia)
- Sequential Elution Solvent Chromatography (SESC)
- Infrared Spectrometer

Other analytical techniques may also be available for biomass materials characterization through special arrangements with other laboratories. They include elemental analysis, tandem mass spectrometry (MS/MS) and gel permeation chromatography.

Analyses carried out on biomass liquefaction products include identification and quantification of selected organic components, such as aldehydes, phenolics, carboxylic acids, acid esters, and organic solvents. Determinations of elemental composition (C, H, N, O, S, halogens), ash, Karl Fischer water content, density, ^{13}C-NMR analysis, and particle size distribution on feedstocks are also carried out. Materials analyzed include feedstocks, residues and pyrolysis oils from hardwood, softwood, lignin, cellulose, scrap tires, peat, and sewage sludge.

Special Projects

The second task of the Centralized Analysis project involves collaborating with members of the biomass liquefaction community in special projects. B.C. Research played a central role in coordinating the first IEA Thermochemical Round Robin in 1988 to determine the variability associated with measurement of carbon, hydrogen, nitrogen, oxygen, and water in biomass oils. Fifteen laboratories participated in that study.

A project is currently underway in support of the forthcoming third IEA Analytical Round Robin test of woody and herbaceous feedstocks. B.C. Research is validating the procedure for analysis of extractives, polysaccharides and lignin in lignocellulosic feedstocks which was developed by Dr. Olof Theander, of the University of Agricultural Science in Uppsala, Sweden. A method for the colorimetric determination of uronic acids is also being tested. The Theander and uronic acid procedures have been validated with the NRC Standard Poplar and further tests with alfalfa are also planned.

The Theander method for analysis of feedstocks involves removal of low molecular weight extractives with aqueous ethanol in the first step. The extracted feedstock is prepared for further analysis by removing starch enzymatically with amyloglucosidase and α-amylase, followed by acidic hydrolysis with sulphuric acid. Quantitative measurement of acid-insoluble material gives the Klason lignin content. Neutral polysaccharide residues in the hydrolysate are quantified by gas liquid chromatography as alditol acetates after reduction with potassium borohydride and 1-methylimidazole as catalyst, followed by acetylation with acetic anhydride.

Uronic acids are determined in a separate procedure by first incubating the feedstock with concentrated sulphuric acid, then reacting with 3,5-dimethylphenol. Intensities of the colour developed are determined at 450 and 400 nm. The 450-400 nm absorbance difference is used instead of a single absorbance measurement at 450 nm in order to reduce the interference from neutral sugars and also absorbance variations over time. Background absorption is corrected with reagent-less blanks. Standard solutions of glucuronic acid are used for calibration.

Acknowledgement

We thank Energy, Mines and Resources, Bioenergy Development Program for providing funding for this project.

Discussion:

G. Grassi (Commission of European Communities): We have a particular problem for biomass feedstock analysis. Assuming tomorrow you have several feedstocks coming to a central conversion unit, you had to pave the feedstocking function of these characteristics. You had to do it quickly and in an empirical way.

In your evaluation, do you see any possibility to have a practical instrument to give economic value to the different kind of sources say three or four lorries transport one feedstock and one another and we had to mix it to convert into bio-oil. Which value could we give quickly - do you envisage any possibility to have a plant - to give economic value to?

C. Alleyne (Speaker): I am not sure if I quite understand what you are asking.

If you have biomass materials from very many locations coming in, you're asking and you have to determine the quality of the feedstock very quickly and you want to know is there any key property.

G. Grassi: Yes, the value of how much we had to pay. You see a growing crop in Europe - the average cost is $60 a ton of a dry material. If you have a different source, one that could be 65, one 55, how do you assess quickly the economic value coming to your conversion installation?

Speaker: This would be very difficult to do unless your biomass liquefaction, you know, a conversion process is all standardized to use the same type of process, because the economics of conversions would be so different.

So, I find it difficult to think that we can find some magic, like central property that would allow us to do that sort of economic optimization.

G. Grassi: The importance of the water content, we can do quickly. Every second is valuable, so, I don't know if there is a relation to a carbon content.

Speaker: Yes, it depends on what the main focus of your process is as well, whether you are interested in upgraded material. If you want to take it all the way to, say, hydroxyacetaldehyde that some of the Canadian producers are interested in. You monitor the final product, the yield of the final product.

G. Grassi: But you have to pay in advance before you decide if it's worth it to buy. So, I think it is a practical problem to be solved in some way just for a large-scale exploitation, not just some pilot plant. But when you want economical exploitation on a large-scale you need to solve the problem how to value the feedstock.

Speaker: Yes, it is a difficult problem. I'm afraid I don't have a ready answer for that.

Molecular Weight Determination of Lignins and Celluloses Derived from Liquefaction/Fractionation Processes

P.F. Vidal, R.P. Overend and E. Chornet
Laboratoire des procédés et technologies de conversion, Department of Chemical Engineering, University of Sherbrooke, Sherbrooke, Québec J1K 2R1

Introduction

This paper presents the results concerning size exclusion chromatography, a method which is part of the characterization protocol for biomass-derived products developed at the University of Sherbrooke.

I would like to focus on why it is important to precisely determine the molecular weight distribution of a polymer, especially if this polymer is derived from biomass via a chemical process.

Professor Chornet explained this morning his conception - shared by many other researchers - of the so-called fractionation of biomass, once named the fractional liquefaction. Given the fact that biomass in the form of lignocellulosics is now considered by most of the people interested and involved in its upgrading as a complex edifice in which several polymers, mainly cellulose, lignin and hemicellulose, are intimately coexisting and interacting together, any given transformation cannot be carried out without considering the complexity of the starting material as each fraction will have a specific behavior and will lead to different products.

Of course, we are able to completely liquefy a given biomass but the oxygenated product obtained has limited interest. It seems that, nowadays, the philosophy concerning biomass upgrading aims at the production of valuable products which in some way are related to the heterogeneity, the complexity, and the specificity of the starting material. For example, it now appears worthwhile to take advantage of the intrinsic properties of lignin as a plastifier instead of willing at the production of phenols by depolymerizing lignin for their subsequent use as plastic precursors.

Since significant improvements have been obtained during the past decade, particularly for the processing of high consistency suspensions, the technologies aiming at the fractionation of biomass are today available and the scale-up of process development units is only waiting for an adequate and economic utilization of the polymeric fractions. Let us remember that Thermo-Mechano-Chemical (TMC) processes induce (a) destructuring, (b) disaggregation, and c) depolymerization of the material. Moreover these modifications are most of the time a function of the treatment severity. As lignocellulosics are, from a polymer valorization perspective, fragile compounds this notion of severity is of importance: a too elevated severity will conduct to denaturation, in other words irreversible modification, of one of the constitutive polymers, while a TMC treatment with at weak severity will not induce sufficient forces to accomplish the expected fractionation of the complex-matrix.

To summarize, in order to valorize one or more of the produced fractions, their physical and chemical characterization is a condition *sine qua none* since TMC processes could induce on a given polymeric fraction (a) an incomplete recovery (too low severity), (b) an irreversible modification (too high severity), or (c) a depolymerization whose extent varies according to the severity.

Assuming that the integrity of the product under study has been confirmed using an appropriate method (i.e., FTIR, ^{13}C RMN), the effect of the treatment upon the molecular weight can be adequately followed by size exclusion chromatography since this analytical method is the only one that gives the molecular weight distribution.

Size Exclusion Chromatography

Theory of Operation

Gel Permeation Chromatography (GPC), more appropriately named nowadays Size Exclusion Chromatography (SEC), given that most of the newly developed packings used are no longer gels, presents the unique advantage of giving the molecular weight distribution (MWD) of a given polymer in a relatively short time.

The method is based on the ability of a given packing to fractionate the eluted polymer according the molecular weight of its constitutive fractions. This capability can be schematized as represented in Figure 1. The biggest molecules are unable to enter the pores of the packing and will then elute in the volume corresponding to the intersticial volume of the packing (Vo), while the total volume of the column (Vt) is available to the smallest molecules which will elute to this corresponding volume. Molecules having molecular weight within the fractionation range will elute between Vo and Vt. Of course this model is purely physical and leaves aside the probable chemical interactions between the eluting polymer and the chromatographic packing.

Molecular Weight Averages

Once the crude chromatogram has been transformed in a molecular weight distribution curve, several averages can be defined among which the number-average MW and the weight-average MW. These are of importance since mechanistic models for the depolymerization can be developed when considering their relative variations. The viscosity-average MW allows direct comparison between viscometry and SEC results. The definitions used are the following:

- the viscosity-average molecular weight $MW_{v,SEC,\alpha}$:

$$MW_{v,SEC,\alpha} = \left(\frac{\Sigma n_i MW_i^{\alpha+1}}{\Sigma n_i MW_i} \right)^{\frac{1}{\alpha}} \qquad (1)$$

- where α is the exponent in the Mark-Houwink equation $[\eta] = K.MW^{\alpha}$ and n_i is the number of molecules having a molecular weight equal to MW_i.

- the number-average MW_n (j=0), the weight-average MW_w (j=1) and the Z-average molecular weight MW_z (j=2):

$$MW_j = \left(\frac{\Sigma n_i MW_i^{j+1}}{\Sigma n_i MW_i^{j}} \right) \qquad (2)$$

- where n_i is the number of molecules having a molecular weight equal to MW_i.

Concerning the Maxima 820 software, the number-average MW_n (k=0), the weight-average MW_w (k=1) and the Z-average molecular weight MW_z (k=2) were calculated as shown below, once the integration parameters - mainly the baseline definition and the number of slices to be used - had been specified.

$$MW_k = \frac{\Sigma A_i.MW_i^{k}}{\Sigma A_i.MW_i^{k-1}} \qquad (3)$$

- where A_i is the area under the chromatogram curve of the ith slice whose corresponding molecular weight at the midpoint elution volume - determined from an input calibration curve - is MW_i.

The corresponding degrees of polymerization (DP) are obtained by dividing the calculated molecular weight by the monomer weight, which is 162 for the anhydroglucose unit and 519 for the anhydroglucose tricarbanilate unit. For lignin, depending on the species, the monomer weight varies around 200.

High Performance Size Exclusion Chromatography

The system used consists of an automatic injector (Waters WISP), a Waters 590 pump, a column heater module controlled with a refractive index detector (Waters 410), and a variable wavelength UV detector (Perkin Elmer). Four columns in series thermostated at 40°C (PL Gel, 300 x 7.5 mm, particle sizes of 5 and 10 µm, Polymer Laboratories) were used; the first having a porosity of 10^4 Å, the second 10^5Å , the third of 10^5Å , and the fourth of 10^5 Å. The eluent, degassed and filtered (0.45 µm) THF (Anachemia, UV grade), was kept peroxide-free by bubbling inert gas through it. The samples were dissolved in THF at a concentration of 1.0 $mg.ml^{-1}$, filtered (0.45 µm), injected via a 10 µl loop, and eluted at a flow rate of 1.0 $ml.min^{-1}$. Detection was carried out using UV at 225 nm for the narrow distribution (polydispersity <1.1) polystyrene standards (Polymer Laboratories), 254 nm for the cellulose samples, and 280 nm for the lignin samples.

All data was acquired, stored, and processed on a microcomputer with the aid of a Waters software program (Maxima 820).

A Case Study Based on Cellulose Depolymerization

Depolymerized Cellulose Production

The production of the thermo-mechano-solvolytic (TMS) celluloses has been described in a preceding paper. To summarize the approach, Sigmacell α-cellulose (Sigma, C-8002, lot 104F-027C) was prepared as an ethylene glycol suspension (7% w/w) and treated in a cascade reactor system. The temperature varied between 220 and 340 °C and the reaction time between 2 and 10 minutes. The system was equipped with a 2 mm ID by 12 cm long capillary which acted as a shear device when a differential pressure of 10.3 MPa was applied.

Tricarbanilation

The tricarbanilation of the cellulose samples was carried out using a modified version of the Hall and Horne method. The degree of substitution was controlled through the percentage of nitrogen as determined by elemental analysis.

Evidence for cellulose depolymerization

When cellulose suspension in ethylene glycol is treated in the UDES-S unit at temperature ranging from 250 to 350°C for 2 to 10 minutes, a certain solubilization occurs while the remaining solid is depolymerized. Furthermore, we proved that this depolymerization can be monitored by adjusting the severity: the more severe the treatment the more depolymerized the remaining solid. The protective role of ethylene glycol - cellulose is treated well above its transition temperature without being pyrolyzed - is a striking point of this process.

The possibility of producing microcrystalline cellulose in high yield is an interesting route to consider in the perspective of a complete utilization of the different fractions.

Figure 1 presents the chromatograms of the starting material and of two treated samples: a medium (DP=133) and a low (DP=90) degree of polymerization cellulose.

Universal Calibration

The principle for the separation of polymeric families by size exclusion chromatography induces, as indicated by its name, a fractionation according to the size and, as a consequence but under certain restrictions, according to the molecular weight. These restrictions concern the shape a given molecule will adopt in solution: molecules having the same MW but with different shapes (i.e., rod-like, sphere, random coil) will elute at different volumes. From this principle, that governs SEC, emerges the theory of universal calibration proposed by Benoit. Instead of calibrating the system in a MW scale, the universal calibration curve represents the elution volume as a function of the hydrodynamic volume which is the product of the molecular weight by the intrinsic viscosity. This calibration procedure presents the advantage of being valid for all the polymers, providing that there is no chemical interaction with the column packing.

The first step in any SEC analysis is to construct a calibration curve, preferably with narrow standards of the substance to analyze itself (i.e., cellulose standards for cellulose analysis). However most of the time these standards are not commercially available or difficult to isolate; that is why polystyrene standards are so widely used. Figure 2 presents the calibration curve obtained using such standards.

In order to employ the principle of universal calibration, we had to adapt the Maxima 820 calculation procedure since, for the earlier versions of the software, it is impossible to incorporate the hydrodynamic volume values, either for the polystyrene standards (i.e., to construct the universal calibration curve) or for the cellulose samples so as to determine their average molecular weights. Thus, we have constructed a pseudo-universal calibration curve which is in fact a calibration curve based on the hydrodynamic volume principle but valid only for the CTC samples. This calibration curve has been obtained by substituting a corrected MW ($MW_{PS,CTC}$) for the actual MW of the polystyrene standards MW_{PS}. The corrected value is expressed as:

$$MW_{PS,CTC} = \left([\eta]_{PS} \cdot \frac{MW_{PS}}{K_{CTC}} \right)^{\frac{1}{\alpha_{CTC}+1}} \qquad (4)$$

where

- $[\eta]_{PS}$ is the intrinsic viscosity of the polystyrene standard having the molecular weight MW_{PS}. We used the following Mark-Houwink relation for the polystyrene in THF at 25 °C:

$$[\eta]_{PS} = K_{PS} MW_{PS}^{\alpha_{PS}} = 0.0112 \cdot MW_{PS}^{0.72}$$

- K_{CTC} and α_{CTC} are the Mark-Houwink constants for CTC in THF at 25 °C: $K_{CTC} = 0.0053$ and $\alpha = 0.84$.

From Figures 3 and 4 it is quite clear that the molecular weight of the celluloses under study is more accurately determined when the universal calibration principle is used instead of the polystyrene standard calibration: with the direct calibration there is an important discrepancy between viscometry and SEC results, especially for the highest MW, while with the universal calibration the two methods give almost the same values.

A Case Study Based on Lignin Depolymerization

SEC is very useful to follow not only the depolymerization resulting from a thermo-mechano-chemical process but also to give evidence of the heterogeneity of a complex substance such as lignin.

Lignin Production

Glycol lignin was prepared from air-dried debarked aspen wood meal by solvolysis delignification using ethylene glycol at high temperature, followed by dilute acidification of the spent black liquor from the process to precipitate the lignin.

Kraft Indulin is from Westvaco Company, Charleston, NC.

The fractionation scheme is represented in Figure 5.

HPSEC Results

High performance size exclusion chromatography shows that the fractionating sequence with solvents gives fractions of increasing molecular weight. The MWD patterns show that the solvents used are more selective for fractionating glycol lignin than kraft lignin. Fr-3 from kraft lignin has a narrow molecular weight distribution pattern but accounts for only 5% of the starting material.

HPSEC analysis was completed with elemental analysis, FTIR spectroscopy and ^{13}C-NMR in order to compare glycol and kraft lignins. The results demonstrate the heterogeneity of both softwood and hardwood lignins in terms of functional groups, chemical structure, and molecular weight distribution. Fractionation by successive solvent extraction revealed that glycol lignin contains a high proportion of low MW material rich in guaiacyl units and aromatic hydroxyl groups and a medium MW fraction rich in syringyl units. Finally, we may conclude that in light of the data presented lignins are heterogeneous whether isolated on a semitechnical or technical basis by kraft or solvolytic procedures.

References

1. Garnier, G., Bouchard, J., Overend, R.P., Chornet, E., Vidal, P.F., "High Yield Production of Microcrystalline Cellulose by a Thermo-Mechanical-Solvolytic Treatment", Can. J. Chem. Eng., 1990, **68**, 113-8.
2. Vidal, P.F., Basora, N., Overend, R.P., Chornet, E., "A Pseudouniversal Calibration Procedure for the Molecular Weight Distribution of Celluloses", J. Appl. Polym. Sci., in press, to appear 1991.
3. Thring, R., Bouchard, J., Chornet, E., Vidal, P.F., Overend, R.P. "Evidence for the Heterogeneity of Glyco Lignin", Ind. Eng. Chem. Res., in press, to appear 1991.
4. Thring, R., Chornet, E., Heitz, M., Overend, R.P. "Production and Hydrolytic Depolymerization of Ethylene Glycol Lignin", in *Lignin: Properties and Materials*, W.G. Glasser, K.V. Sarkanen eds., ACS Symp. Ser., 1989a, pp 228-44.

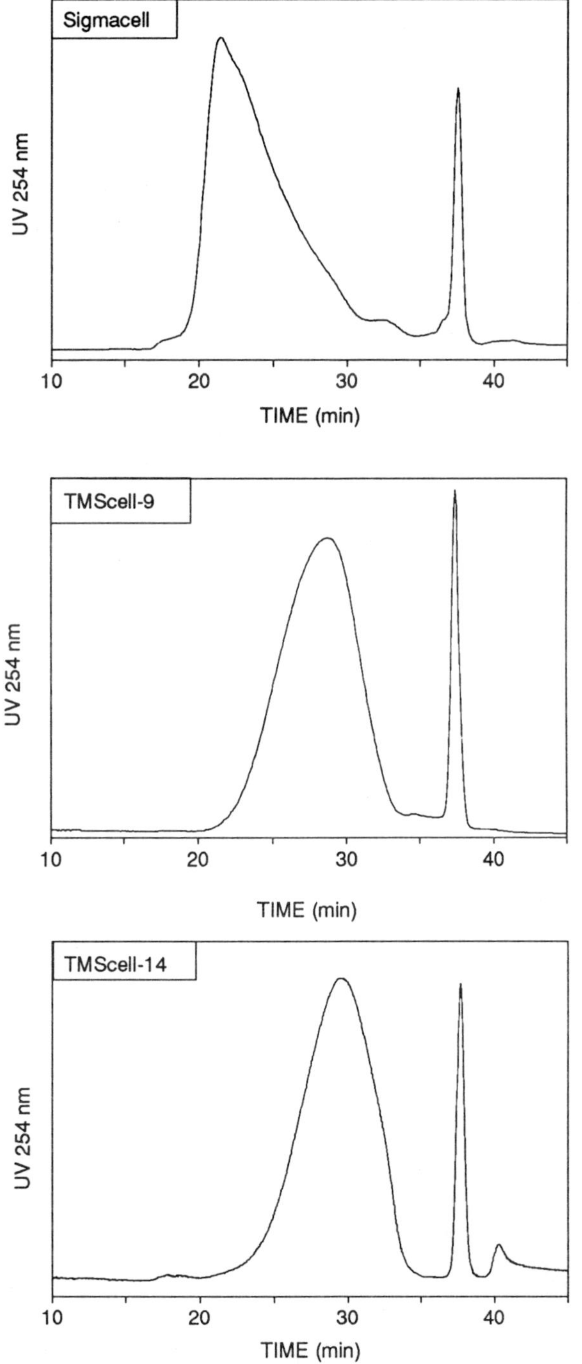

Figure 1: HPSEC chromatograms of a) Sigmacell; b) TMScell-9 ($DP_{v,vis,1,1.9}$ = 133); and c) TMScell-14 ($DP_{v,vis,1,1.9}$ = 90).

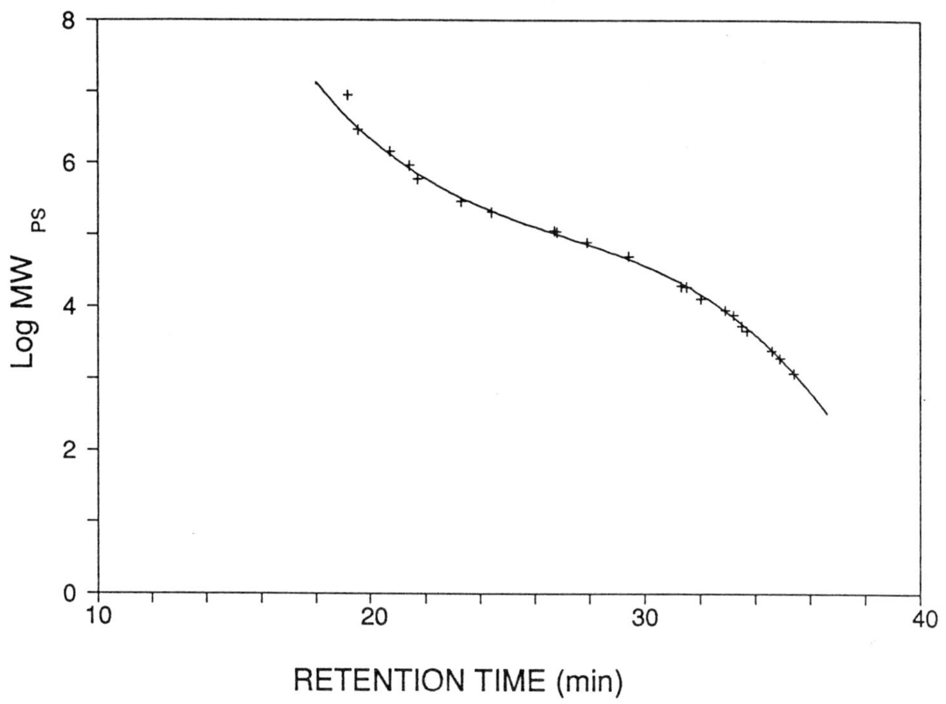

Figure 2: Calibration curve of the HPSEC system obtained using narrow distribution polystyrene standards.

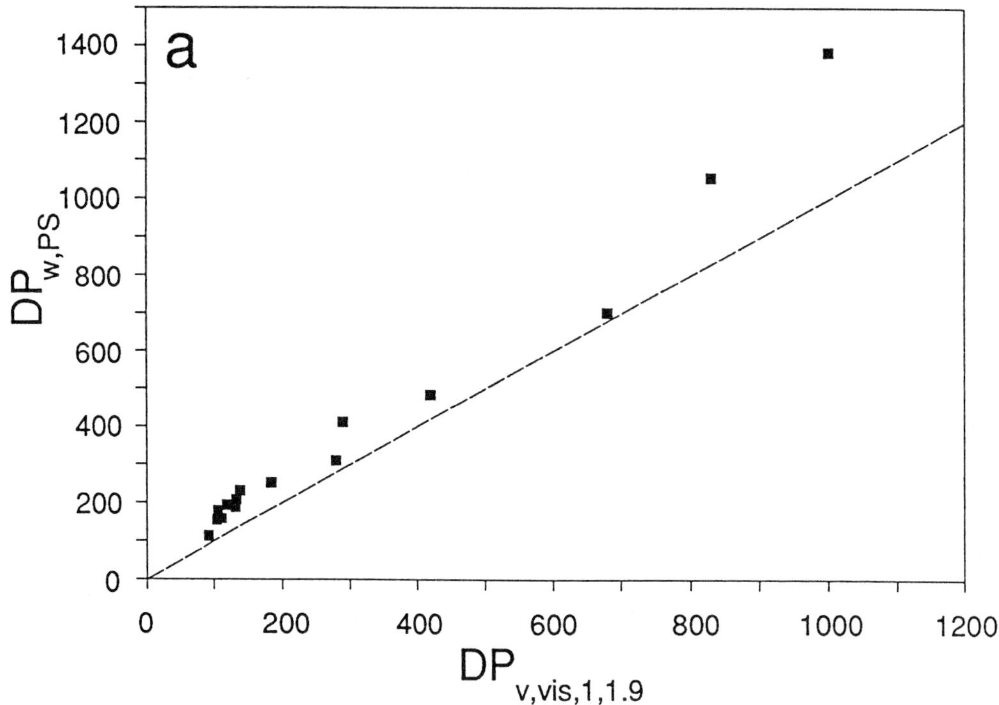

Figure 3: Weight-average DP obtained from the polystyrene calibration curve ($DP_{w,PS}$) in relation with the corresponding DP obtained by viscometry using: $DP_{v,vis,1,1.9} = 1,9 . [\eta]$.
The dashed line represents the theorical relationship: $DP_{v,SEC,1} = DP_{v,vis,1,1.9}$.

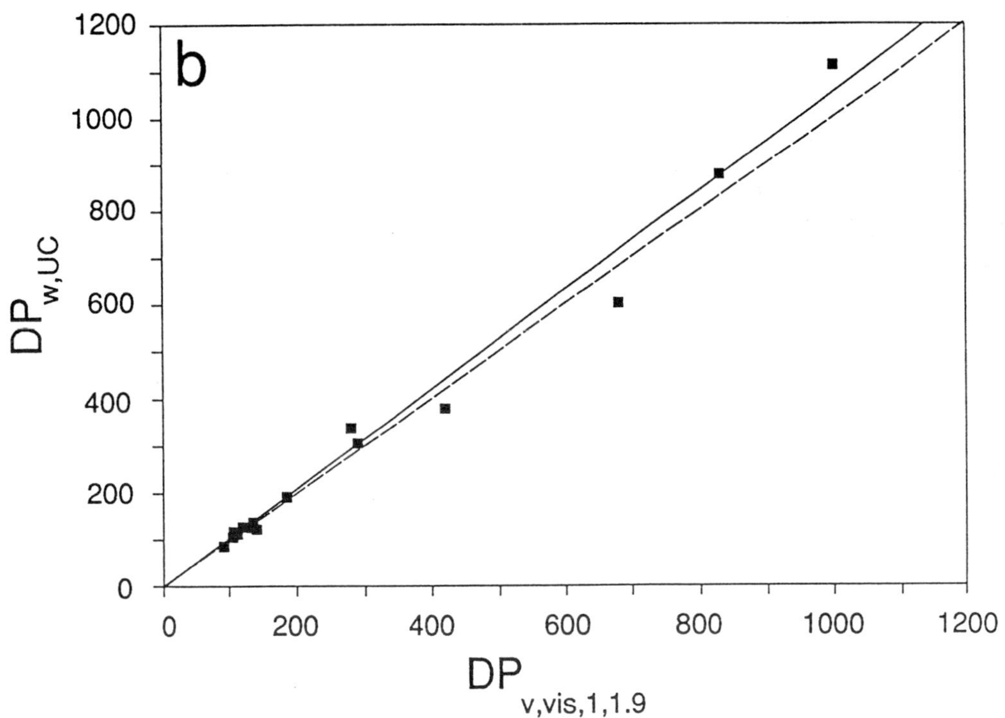

Figure 4: Weight-average DP obtained from the semi universal calibration curve ($DP_{w,UC}$) in relation with the corresponding DP obtained by viscometry using $DP_{v,vis,1,1.9} = 1,9 . [\eta]$
The dashed line represents the theorical relationship: $DP_{v,SEC,1} = DP_{v,vis,1,1.9}$.
The solid line represents the least squares linear regression.

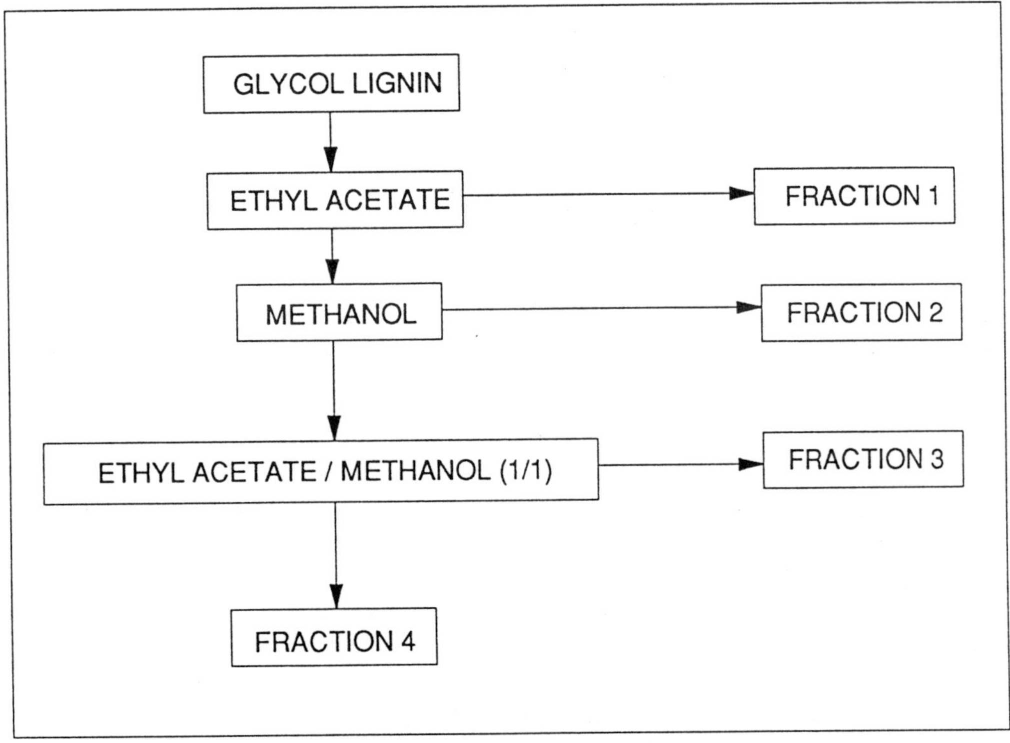

Figure 5: Scheme for the fractionation of glycol and kraft lignins.

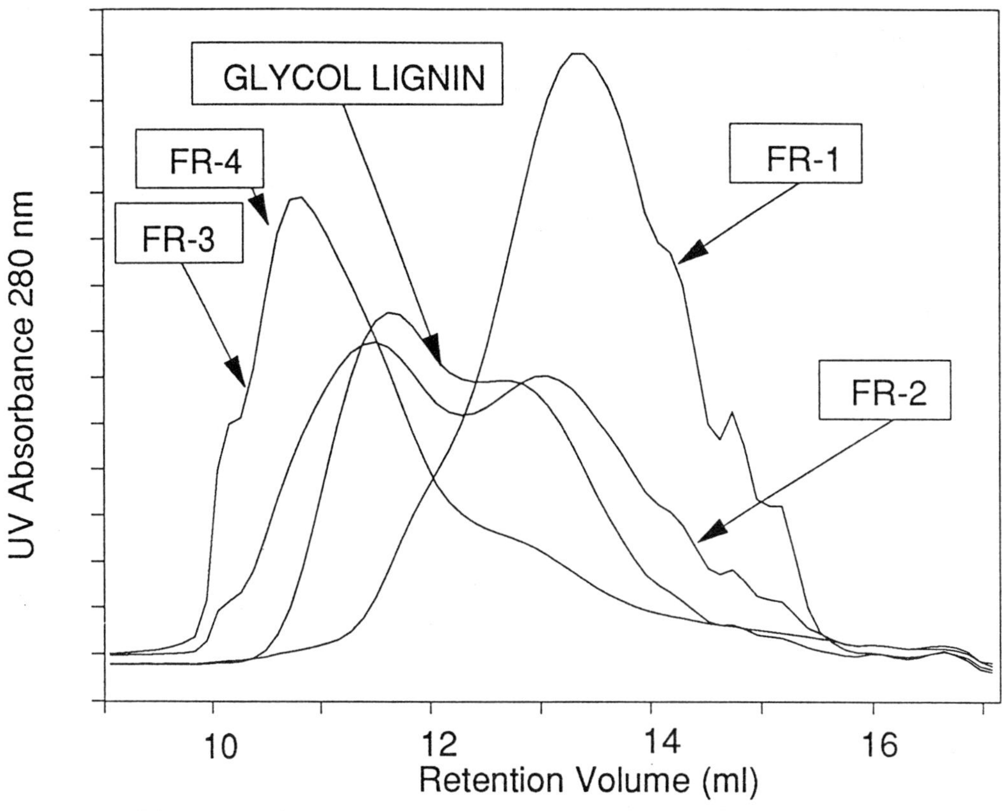

Figure 6: HPSEC chromatograms of acetylated glycol lignin and fractions.

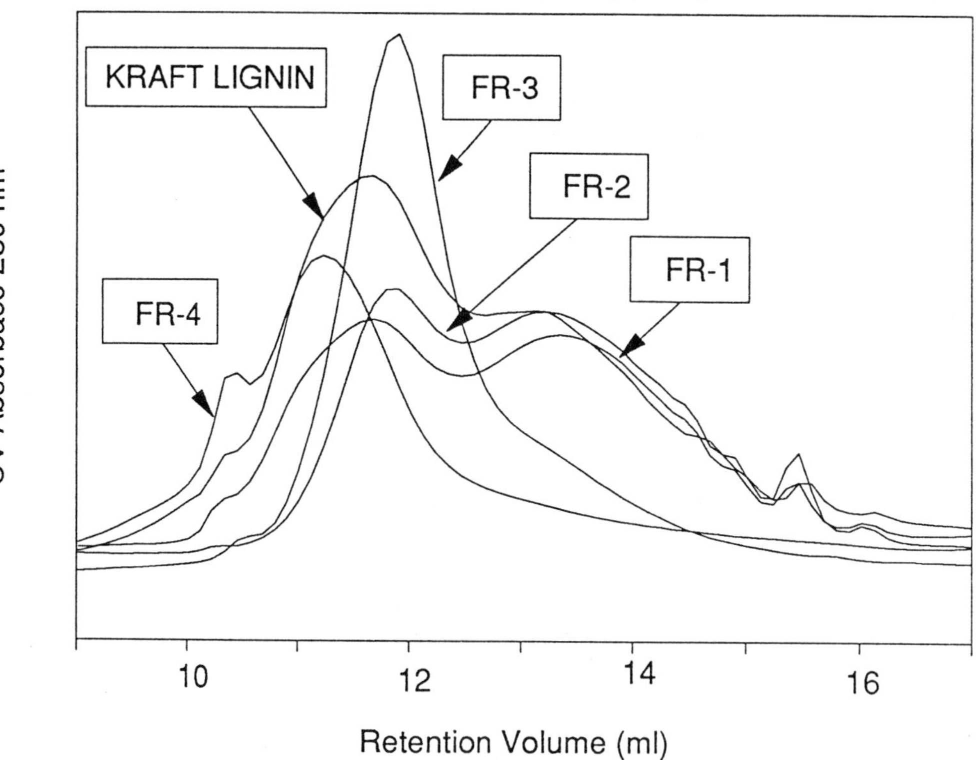

Figure 7: HPSEC chromatograms of acetylated kraft lignin and fractions.

Analytical Methodology for Aqueous/Steam Treatments

J. Bouchard[1], E. Chornet[1] and R.P. Overend[2]
[1]Dép. de Génie Chimique, Université de Sherbrooke, Québec
[2]Division of Biological Sciences, NRC, Canada
(Supported by Bioenergy Development Program, EMR, Canada - Contract no. 23283-8-6177/01-SZ)

Introduction

Aqueous/steam treatments for lignocellulosics can be considered as thermomechanical processes since heat, shear forces and hydrolysis (with or without added acids or bases) are combined to destructure the lignocellulosic matrices in order to meet different objectives: production of low cost sugars for use in biotechnological routes to fuels and chemicals; increase the accessibility of the cellulosic carbohydrates to cellulases; use of low cost residual lignocellulosics as feed for ruminants. Two categories of processes are distinguishable:

- treatments in which a large vapour phase is in contact with a wet solid. The latter is expelled out of the reaction vessel through intermittent decompression. These are the so-called steam explosion processes and detailed descriptions of these processes are well known **(1-4)**.
- treatments in which biomass is suspended as a slurry in water. Steam is usually employed to directly heat the mixture. Continuous systems as well as percolation reactors have also been well described **(5-8)**.

The fundamental principles leading to autohydrolysis of the hemicellulose in hardwoods by steam explosion processes are well established **(9, 10)** and kinetic models describing the solubilization profiles of the hemicellulose have been proposed and experimentally proven **(11,12)**.

Less well understood are the structural and functional modifications of the lignin and cellulose as a function of the severity of the aqueous/steam treatment.

A more recent objective of steam and aqueous pretreatments has been the fractionation of lignocellulosics to obtain each of the polymeric components in maximum yield and purity. We felt that a study of how the constitutive polymeric fractions change as a function of the treatment severity for a generic aqueous/steam process was a worthwhile goal to pursue in order to better understand the principles of fractionation as compared to the steam-explosion processes as well as the possible uses of the polymers themselves after treatment.

Our objectives for the present contract with the Bioenergy Development Program are to generalize the analytical methodology proposed for hardwood **(13)** to softwood treated by aqueous/steam fractionation process with and without use of acid as a catalyst. The analytical procedure is presented in Figure 1 and has been applied to the three series of experiments as a function of the severity of the treatment under conditions mentioned in the Figure 1. The generic steam/aqueous treatment used in our study has been described previously by Koeberle et al. **(7)** It is based on a methodology in which an aqueous suspension of finely divided lignocellulosics at medium solids consistency is rapidly heated by steam addition and passed through homogenizing valves **(14)**.

The present paper will be addressed to two specific points of our works. First, the problem we have encountered in quantification of cellulose and hemicelluloses in treated wood using the standard ASTM procedure and the method of correction we propose. Secondly, we will discuss about the autohydrolysis concept involved in the steam explosion process. A more complete presentation and discussion of our analytical methodology can be found in the relevant literature **(13,15,16)**.

Results and Discussion

The yield of soluble obtained for the three series of experiments are presented as a function of the severity in Figure 2. The logarithmic severity scale (log Ro) is a combination of the time and the temperature of the reaction in a single factor called Reaction Ordinate **(17)**.

While the yield of soluble reaches 32% of the initial hardwood and 90% of the hemicellulose initially present, the maximum obtained for softwood is 20%. These results are relevant to the structural barrier which is harder to destructurate in softwood. Adding sulfuric acid (0.06N) during the treatment increases greatly the yield of soluble products but when exceeding 30-35% of solubilization, cellulose is also converted to soluble products via hydrolysis and degradation of the solid residual cellulose also occurs as we have shown recently **(18)**. The objective of selective fractionation of wood is obviously not reached when cellulose is degraded.

Problem and correction of cellulose quantification in wood residues

The distribution of water-soluble products and the residual polymeric families are shown in Table 1 for the aqueous/steam fractionation of hardwood and softwood as a function of the severity. Cellulose, hemicellulose and lignin data have been determined by the ASTM procedures.

Table 1: Distribution of wood products obtained from aqueous/steam fractionation process as a function of the severity

Hardwood	Aqueous Fraction	Residue		
		Cellulose	Hemicell.	Lignin
H-1.3	1.8	42.0	32.1	24.2
H-2.6	5.0	40.1	31.5	23.8
H-3.2	11.2	41.6	23.6	22.2
H-3.6	17.4	42.5	15.7	21.6
H-3.9	26.1	**40.6**	**11.5**	20.5
H-4.3	31.9	**33.8**	**13.5**	20.7
Softwood				
S-1.4	6.0	41.8	29.3	25.6
S-3.2	12.1	42.8	15.0	29.2
S-3.8	16.4	41.3	12.2	30.9
S-4.1	18.4	**28.2**	**22.1**	30.7
S-4.4	20.4	**32.1**	**16.5**	30.8

The low solubilization yields of hemicelluloses coupled with the cellulose solubilization during the treatments at high severity suggest that the ASTM analytical procedures were not adequate since we felt they overestimate the hemicelluloses content in wood residues. We then decided to use a new approach based on the hypothesis that aqueous/steam treatment of wood can induce structural reorganisation of the cellulose network via hydrogen bonding modifications. Consequently, a part of the cellulose becomes soluble in sodium hydroxide solution used for the α-cellulose quantification. In this case, the solubilized cellulose will be "counted" as hemicellulose and give erroneous results.

As proposed by Theander and Westerlund **(19)** for dietary fiber, the correction procedure implies a reconstitution of the hemicellulose and cellulose fraction from the monosaccharides analyses:

- quantification of xylose and arabinose from furfural produced by extensive hydrolysis;
- correction of the monosaccharides concentration obtained by GLC quantification of their silyl derivatives following hydrolysis of the wood residues and using the correction factor calculated from the pentosan analyses;
- distribution of the galactose, mannose and glucose concentration as a function of their cellulosic or hemicellulosic origins;
- calculation of the hemicellulose and cellulose concentration in treated residues using the above data.

Figure 3 shows the difference in fractionation scheme between soluble and residual fraction as obtained by standard and corrected analytical procedures for the hardwood treated at the highest severity [4.3].

As compared with the initial wood, our corrected procedure clearly demonstrates that the fractionation objective has been reached. The solid residue is composed of cellulose and lignin while the hemicelluloses have been solubilized. This study demonstrates that the reconstitution of the cellulosic and hemicellulosic fractions from the monosaccharides distribution should be preferred to the standard procedure for cellulose determination in the case of aqueous/steam treated wood.

The aqueous/steam fractionation process and the autohydrolysis concept

The autohydrolysis concept as proposed by Wayman **(9,10)** in order to explain the hemicellulose removal from wood fiber by steam explosion processes is represented in Figure 4.

Formation of hydrogen ions is explained by dissociation of the wood acids and also by formation of acetic acid from almost complete deacetylation of the xylose present in the hemicelluloses. The hydrogen ions formed are able to hydrolyse the hemicelluloses present in the wood fibers. When the steam exploded wood fibers are washed with water, the oligomers and monomers obtained from the previous hydrolysis go into solution in the ratio indicated in Figure 4. Following this concept, it is rather improbable to consider that the solubilization of the hemicelluloses leads to large soluble polymer fragments.

In our experiments on aqueous/steam fractionation of wood, no monomeric carbohydrate has been detected in the soluble fractions by HPLC analyses. This has led us to the hypothesis that the autohydrolysis concept is perhaps not the only concept that can explain the mechanism of wood fractionation. An extensive study on the quantification of the acetyl groups linked to the hemicelluloses in the soluble and residual fractions has been carried out. The results are presented in Figure 5 as the % of deacetylation occuring as a function of the severity of the aqueous/steam treatment.

In the case of hardwoods, no deacetylation occurs before a severity of 3.8 and even at the highest severity (90% of solubilization of hemicelluloses) more than 75% of the acetyl groups remain linked to the hemicelluloses. Haw et al. **(20)** have also observed by ^{13}C NMR low deacetylation of red oak treated by RASH process. The deacetylation is more pronounced in the case of softwoods where 60% of deacetylation occurs at the highest severity. However, the acetyl content of softwoods is lower than hardwoods: 1.7% of initial wood as compared to 3.7% for hardwoods. A very small increase can be observed when acid is added during the treatment.

These low yields of deacetylation imply that the hemicelluloses are solubilized as polymeric materials without important depolymerisation due to an acidic attack on the fiber. The confirmation of this hypothesis has been obtained from the determination of the molecular weight distributions of the soluble hardwood hemicelluloses by size exclusion chromatography which are presented in Figure 6 as a function of the treatment severity.

The presence of high molecular weight polymers can be observed for low severity treatments and it is evident that depolymerization occurs in the soluble phase as the severity increases. However, even at the highest severity, more than 95% of the hemicelluloses remain in their polymeric or oligomeric forms. At an equivalent severity, the steam explosion process generates a soluble fraction which has a monomeric content greater than 60%.

In conclusion: we have demonstrated that solubilization and depolymerization of the hemicelluloses are distinct mechanism and that depolymerization seems to be the major mechanism involved in steam explosion treatment while the aqueous/steam fractionation process gives priority to solubilization with minor depolymerization.

References

1. Marchessault, R.H., Malhotra, S.L., Jones, A.Y., Perovic, A. In: *Wood and agricultural residues.* Soltes, J. (ed.), Academic Press, New-York, (1983) pp. 401-413.
2. DeLong, E.A., Canadian Patent 1,096,374 (1981).
3. DeLong, E.A., Canadian Patent 1,141,376 (1983).
4. Brown, D.B., Canadian patent 1,147,105 (1983).
5. Bobleter, O., Binder, H., Concin, R., Burtscher, E., In: *Energy from biomass.* Palz, W., Chartier, P., Hall, D.O., (Eds.), Elsevier Appl. Sci. Publ., (1981) pp. 554-562.
6. Bonn, G., Concin, R., Bobleter, O., *Wood Sci. Technol.* **17** (1983) 195-202.
7. Koeberle, P., Meloche, F., Chauvette, G., Heitz, M., Rubio, M., Torres, H., Ménard, H., Chornet, E., Jaulin, L., Overend, R.P., In: *Fifth Canadian Bioenergy R&D Seminar.* Hasnain, S., (Ed.), Elsevier Appl. Sci. Publ., (1985) pp. 263-266.
8. Young, R.A., Davis, J.L., In: *Fundamentals of Thermochemical Biomass Conversion.* Overend, R.P., Milne, T.A., Mudge, L.K., (Eds.), Elsevier Applied Science Publishers, (1985) pp. 121-142.
9. Wayman, M., Alcohol from cellulosics: the autohydrolysis-extraction process. NTIS document no conf B010302, U.S. Dept of Commerce, Springfield, VA, U.S.A., 1980.
10. Wayman, M., In: *International symposium on ethanol from biomass.* Royal Soc. Can. Proc. (1982) pp. 628-644.
11. Barnet, D., Autohydrolyse rapide du bois de peuplier (*Populus Tremula*). Thèse de 3ième cycle, Université de Grenoble, 1984.
12. Conner, A.H., Wood and Fiber Science **16** (1984) 268-277
13. Bouchard, J., Chornet, E., Overend, R.P., "Analytical methodology development for the characterization of products obtained by fractionation of biomass." Report of contract file # 24ST 23216-6-6168. Renewable Energy Division, EMR Canada, Ottawa, (1989) 128 pp.
14. Chornet, E., Vanasse, C., Lemonnier, J.P., Overend, R.P., In: *Research in Thermochemical Biomass Conversion.* Bridgwater, A., Kuester J.L., (eds.), Elsevier Appl. Sci. Publ. (1988) pp. 766-778.
15. Bouchard, J., Nguyen, T.S., Chornet, E., Overend, R.P., *Biomass* **23** (1990) 243-261
16. Bouchard, J., Nguyen, T.S., Chornet, E., Overend, R.P., *Biomass* (in press)
17. Overend, R.P., Chornet, E., *Phil. Trans. R. Soc. Lond.* **A321** (1987) 523-536.
18. Bouchard, J., Abatzoglou, N., Chornet, E., Overend, R.P., *Wood Sci. Technol.* **23** (1989) 343-355.
19. Theander, O., Westerlund, E.A., *J. Agric. Food Chem.* **34** (1986) 330-336.
20. Haw, J.F., Maciel, G.E., Biermann, C.J., *Holzforschung* **38** (1984) 327-331.

Discussion:

P. Sears (CANMET): The severity factor you have been using can you tell me how you came up with the form - the integral of temperature function with respect to time? How did you choose the actual form of that and the numbers that occur within it?

J. Bouchard (Speaker): The severity number is not essential in this study because the reaction time is fixed at two minutes. So, it is always the temperature of reaction which will make the severity index vary. But when you want to compare your data with other data obtained by steam explosion, with five minutes of residence time at 180 degrees, and another group working with ten minutes residence time, this is a way to unify the data on the paper Professor Chornet has published and you will see all the data are fixed on the same line. The conversion obtained is exactly the same for all different processes.

P. Sears: In other words you juggled the numbers until you got the results the same from different processes.

How did you choose the actual numbers? There's a 14.75 in there, it is E-100.

Speaker: Yes. This number, 14.75, originated from the work of Bash and Freid in the thermal mechanical pulping where they measure this number for the pulp and paper industry. It works with that. Initially we use exactly the same number, the same relationship. Now, there is a lot of work in Sherbrooke done in heavy oil as compared to wood on different processes, and in order to calculate the relevant number for the relevant polymeric products.

I am not an expert in the kinetics under all this discussion, but they found that they could determine the number if you have a polymeric structure as in the heavy oil, as in the wood, as in the peat, you will have a specific number which will fit in the severity scale.

P. Sears: Have you done any specific experiments where you chose different temperature/time relationships which gave the same R value and they gave you the same results?

Speaker: Yes. And we did it for work done here in Canada as compared to work done in Europe and in the United States. We take the published data and fit it in on the same scale.

P. Sears: But not work on the same equipment where you have taken two different --

Speaker: Yes.

P. Sears: You've done it in your equipment as well?

Speaker: Yes.

J. Bouchard: So, you could work with the residence time or the reaction temperature.

D. Boocock (University of Toronto): Do I understand you correctly that you are saying that the severity factor for the steam explosion process is higher than your steam/aqueous process, and that is why you get depolymerization in one case and not in your case?

Speaker: No. This severity index used in this study is very consistent with the operating condition of a steam explosion process, like the state technologies; it is very comparable.

D. Boocock: I guess my next question is: Do you have an explanation why you get depolymerization in their case and not in your process?

J. Bouchard: Yes. We think that to have a strong depolymerization in order to reach the monomers, you need some acid and the acid is coming from the de-acidulation which gives acetic acid, which gives enough hydrogen ion in order to do the hydrolysis.

We don't know why we don't have de-acidulation when the wood is suspended in water instead of when just receiving the steam. We don't know why we don't have de-acidulation. Probably it is because the water plays a productive role because all is well salivated than when you just add steam. When you have steam, you have a gas solid reaction; when you use wood suspension, you have a liquid solid reaction. So it could be different, but we have to do further investigation in order to propose a mechanism for that.

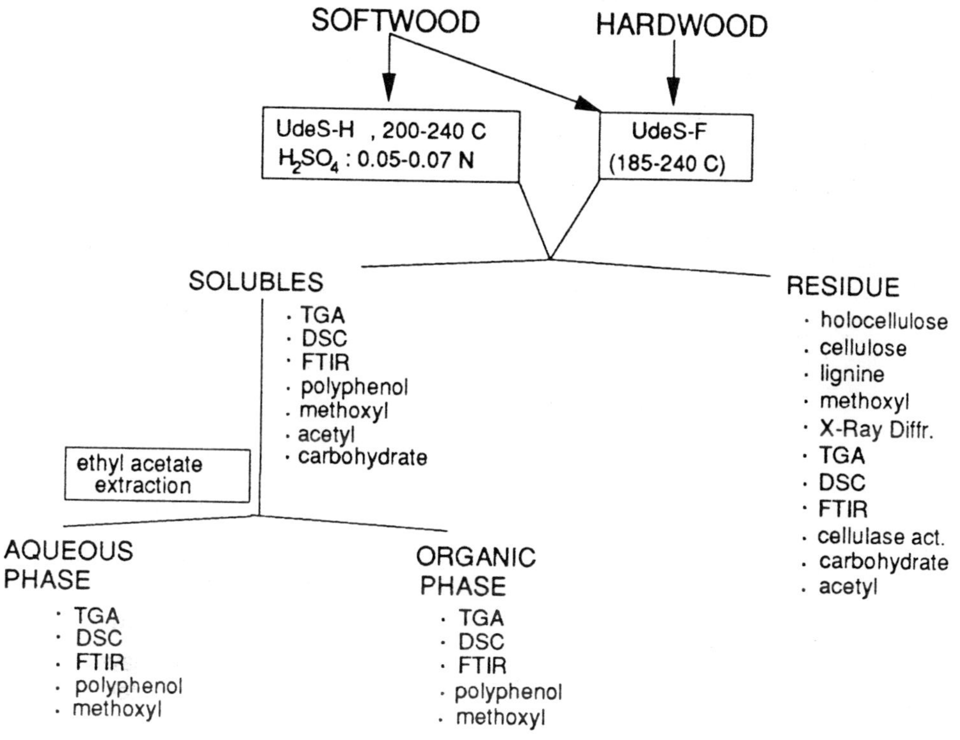

Figure 1: Aqueous/steam fractionation experiments and the analytical procedure used for characterization of the different fractions

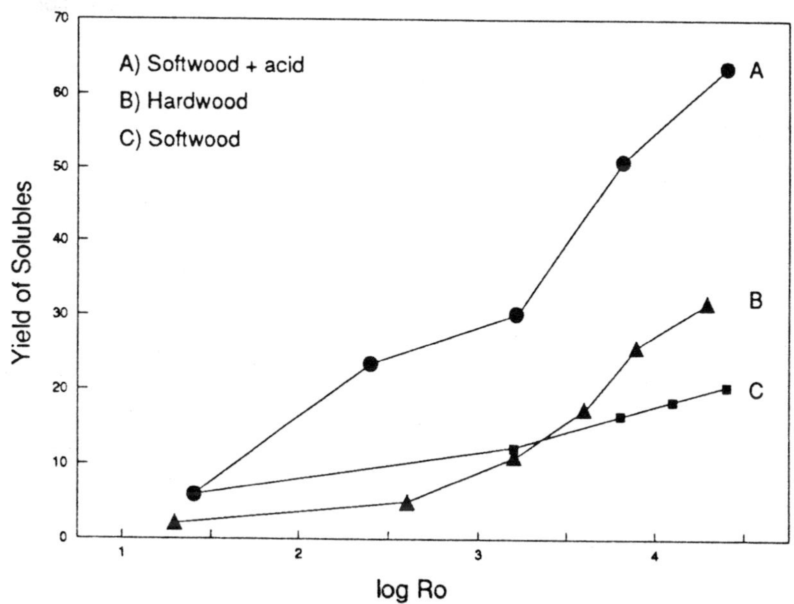

Figure 2: Yields of soluble as a function of the severity for harwood, softwood (with and without acid) treated by aqueous/steam fractionation process

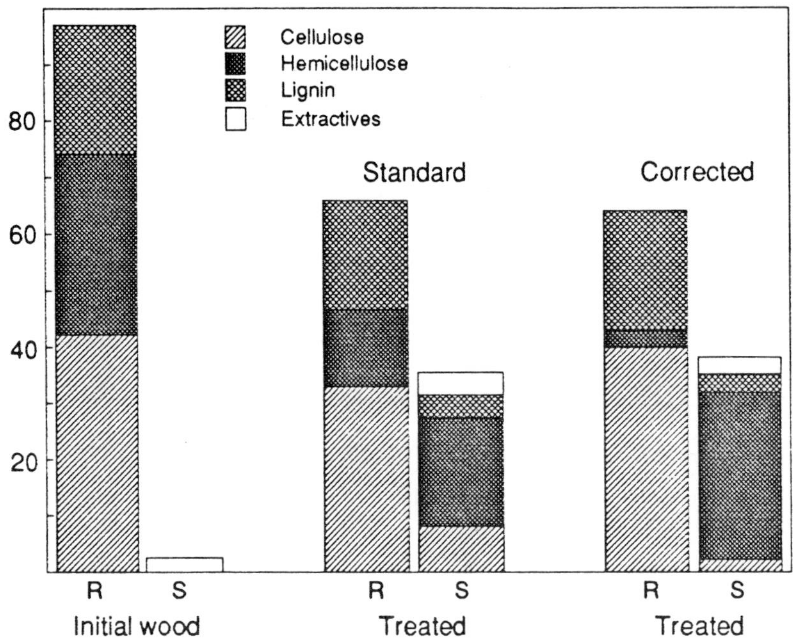

Figure 3: Distribution of products for initial and treated wood as quantified by the standard and corrected procedures

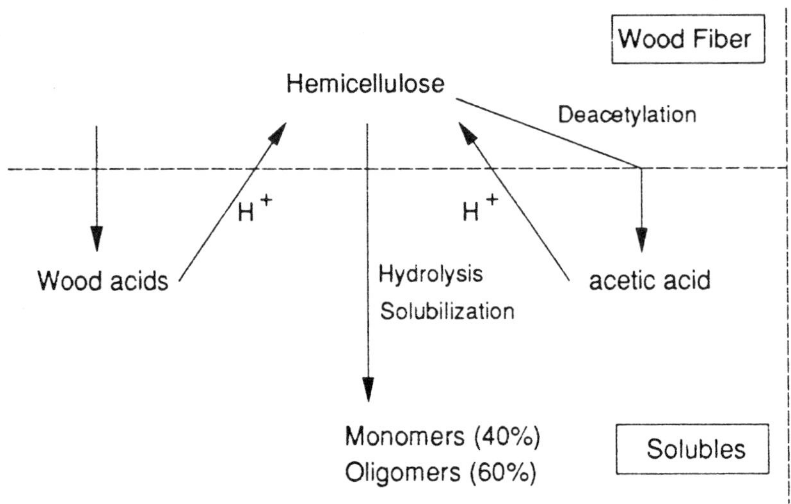

Figure 4: The autohydrolysis concept

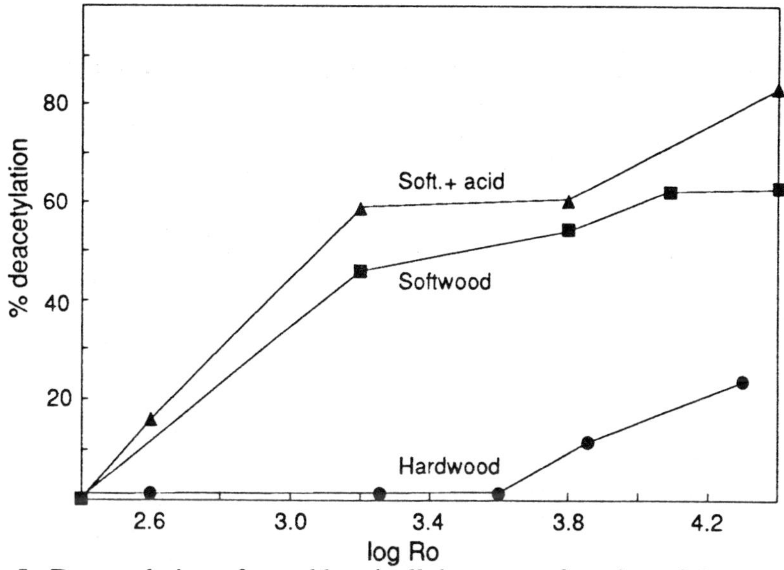

Figure 5: Deacetylation of wood hemicelluloses as a function of the treatment severity

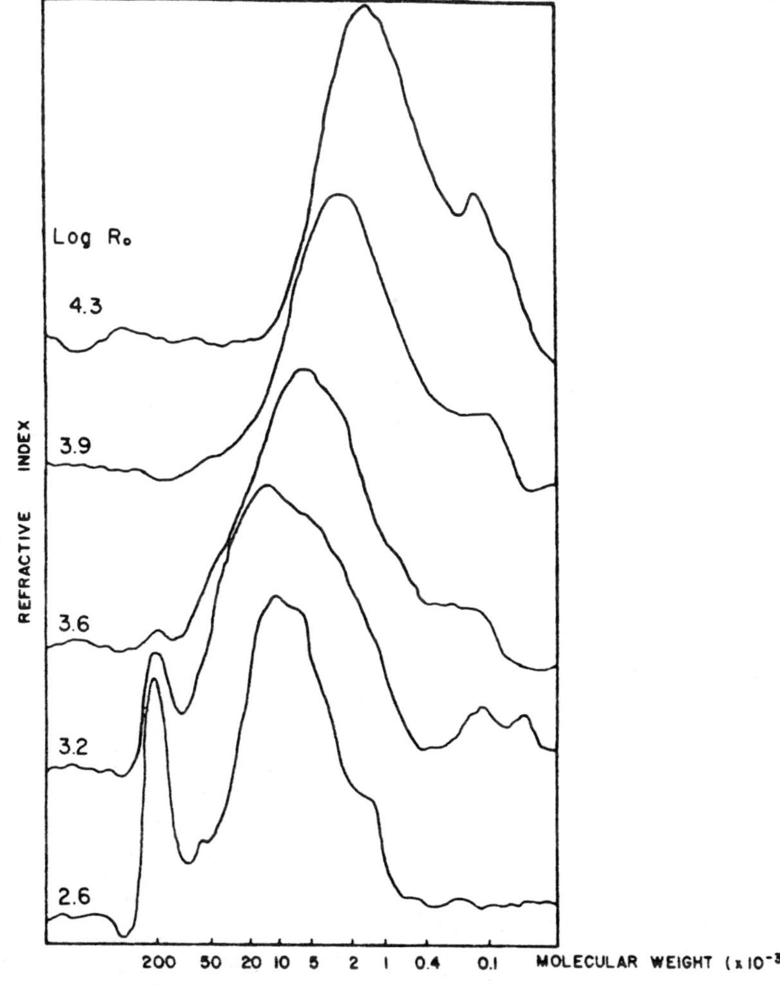

Figure 6: Molecular weight distributions of the soluble hardwood hemicelluloses obtained by size exclusion chromatography

Characterization of Vacuum Pyrolysis Oils of Diverse Origins

H. Pakdel and C. Roy
Dept. of Chemical Engineering, Université Laval
Sainte-Foy (Québec) Canada G1K 7P4

Introduction

Bioenergy research and development has been one of the promising and challenging fields over a decade in Canada. During this period, the contribution of bioenergy to Canada primary energy requirement has risen considerably from under 3% to a current level of approximately 7% **(1)**. This energy is generated from a wide range of materials, such as wood wastes, agricultural wastes, municipal solid wastes and refinery wastes. However, most of those wastes are discarded in dumps and landfill sites without potential application and have a great impact on the environmental pollution.

Major reliance on fossil fuel and the public awareness of the potential impacts of environmental pollution have initiated both the regulatory agencies and the private sector to the current environmental pollution as well as economical crisis considerably. For example, the pulp and paper industry has taken steps to recover energy from their daily produced wastes such as bark and lignin. Oil refineries are willing to treat and recycle oil residues produced by crude oil distillation.

Research is seriously needed to further develop those technologies for the next round of market opportunities which will continue to ensure a future mix of environmentally appropriate and economically competitive energy sources. The potential merit of vacuum pyrolysis in the conversion of various materials such as biomass and municipal solid wastes to recyclable solid and liquid fuels are discussed in this paper in terms of a source of energy and an efficient approach to control the environmental pollution.

Programs are currently underway in our laboratories for the development of analytical procedures and the preparative separation of chemicals from biomass. Chromatographic (GC, LC and HPLC) and spectroscopic techniques, (NMR, IR and GC/MS), were utilized. A preliminary large scale fractionation of pyrolysis oil into chemical types was made **(2)**. However, economical recoveries are still under investigation.

Wood Pyrolysis

Oils from Wood

Vacuum pyrolysis liquid products from wood are essentially composed of hydrocarbons, phenols, aldehydes, ketones, alcohols, furans, polyfunctional phenols, carboxylic acids, sugars, and water which were characterized in detail by GC and GC/MS. The oil, however, contains an additional fraction of about 40% by weight, which is either highly polar or non volatile, or less stable than the remaining portion, and could not be characterized easily.

Details on the pyrolysis system set-up and conditions can be found elsewhere **(2,3,4)**. Distillation, solvent extraction, liquid chromatography including LC, HPLC and GPC, gas chromatography equipped with capillary column, flame ionization and mass selective detectors (FID and MSD) were employed for the product characterization. The pyrolysis product yields are shown in Table 1.

Low molecular weight carboxylic acids, formic and acetic acids in particular, are one of the major groups of compounds in wood pyrolysis oils. Their method of analysis was developed recently **(3)**. Table 2 shows the carboxylic acid distribution in both oil and aqueous phases at different reactor pressures and condensation temperatures. Large scale extraction of carboxylic acids and their potential economical application as deicer, for example, has been studied recently **(6)**. Their recovery will certainly improve the pyrolysis feasibility.

Wood extractives are considered as a class of wood components and usually represent between 5 to 10% by weight of the dry biomass. They include volatile oils, terpenes, fatty acids, sterols, tannins, flavonoids, waxes, resins, coloring matters and gums. The role of various wood components in terms of pyrolysis oil, carboxylic acid and sugar yields were studied and summarized in the following section.

The Role of Wood Components

The role of extractives and wood cellulosic structure was investigated in terms of pyrolysis oil, carboxylic acids and sugar yields and is demonstrated in Figures 1 to 4. Figure 1 illustrates the "extractives" and "lignin" effect on the oil yield, at different pyrolysis temperatures. As the extractives are removed, the pyrolysis oil yield increased whereas lignin removal decreased the oil yield considerably. In contrast, extractives tend to increase formic acid and acetic acid yield (Figure 2 and 3). Lignin increases formic acid and decreases acetic acid yield considerably (Figures 2,3). Cellulose is the main source of formic acid in biomass. However, its low yield from Avicel, a crystallin cellulose, is attributed to the difference in crystallinity.

Levoglucosan, a dehydrated glucose, is another major pyrolysis oil constituent. Avicel is the most favorable feedstock for levoglucosan production. Lignin in the absence of extractives increased the levoglucosan yield by 10 times (Figure 4).

Chemical Type Separation

Class separation of wood and wood components pyrolysis oils was achieved by sequential elution solid chromatography on silica gel. Details on the experimental procedures will be found elsewhere **(7)**. Petroleum ether and dichloromethane enabled the separation of hydrocarbon and phenolic compounds, respectively. Phenolic fractions particularly, have industrial application whether as single pure compounds or in bulk. A softwood lignin pyrolysis oil yielded approximately 70% phenolic fraction which could be utilized directly in resin formulation **(8)**. Figure 5 illustrates the chemical type distribution of a softwood lignin and an aspen hardwood pyrolysis oil. Petroleum ether, dichloromethane, diethylether, water and 10% formic acid in methanol separated hydrocarbons, phenols, polyphenols/aldehydes/ketones, sugars/high polar compounds and acid/high polar compounds (the most polar fraction), respectively. This method of separation could further be considered as a standard method to evaluate different pyrolysis oils in terms of various classes of compounds and yields.

Lignin Pyrolysis

A lignin sample prepared by steam explosion treatment of aspen wood, kindly provided by Iogen Corporation, Ottawa, Canada, was pyrolysed under vacuum at 460°C in a batch reactor. The pyrolysis yields are shown in Table 1. The pyrolysis oil was fractionated on silica gel into various fractions. The total phenols, aldehydes, and ketones was about 42% of the oil. Phenolic yield from hardwood was found lower than from softwood but it contained instead a higher proportion of high polar compounds which could not be fully characterized by GC/MS due to their low or nil GC/MS responses.

Bark Pyrolysis

The bark of forest trees is a large renewable resource that is seldom used except as a fuel. Compared with wood, bark is even more complex in chemical composition. Bark itself is composed of various types of carbohydrates, extractives, lignin, wax and suberins, terpenes, monomeric polyphenols and their glycosides and polyflavonoids. Each of those chemical types has special application, e.g., pharmaceutical **(9)**. Polyphenols for example can be used to substitute phenol in synthesis of modified phenol-formaldehyde resins. The manufacture of fine chemicals from bark has been suffering from difficulties in obtaining economic separation of complex mixtures on an industrial scale.

Bark originating from a softwood sample was pyrolyzed in a batch reactor **(4)**. 51.7% of the pyrolysis oil phase was positively characterized and quantified by means of liquid chromatography and GC/MS (see Table 3). Due to its high polarity or unstability, 48.3% of the rest is unidentified at this stage. Low molecular weight carboxylic acids, phenols and sugars were found in significant proportions, and could be separated. Their economical recoveries require further investigation. Low molecular weight phenols in bark pyrolysis oil have been shown to have even higher polymerization tendencies with formaldehydes compared with the high molecular weight phenols or polyphenols in the original bark **(9)**.

Petroleum Residues Pyrolysis

Thousands of barrels of waste petroleum residues, classified as hazardous materials, are generated every day in North America and are accumulated on refinery sites. In comparison with conventional crude oils, petroleum residues and heavy oils contain a large proportion of asphaltenes, resins, heavy metals, sulfur and nitrogen compounds which are obstacles to upgrading processes **(10)**.

Asphaltenes have a tendency to form a high yield of coke and deactivate the catalyst. Previous studies have concentrated on catalytic upgrading of heavy oils and residues. Pyrolysis under reduced pressure and low temperature conditions, as an alternative upgrading approach, was developed recently. Petroleum waste samples with composition ranging from 36 to 82% water, 2 to 25% ash and 16 to 58% heavy oil were pyrolyzed at 425°C in a batch reactor. Details of the experimental set-up can be found elsewhere **(11)**. Light and heavy oils were recovered separately as the pyrolysis products. The initial oil contained about 30.9% oil, 55.1% water, 9.5% ash and 4.5% asphaltenes. The asphaltene concentration represented 12.7% of the water and ash-free feedstock. The pyrolysis oil contained only 0.5% water with no asphaltenes. The water was separated by decantation after the pyrolysis experiment. The maltene content of both feedstock and pyrolysis oil were fractionated into three fractions, aliphatic/alicyclic (F1), monoaromatics (F2) and polyaromatics/resins (F3). Their distributions are shown in Table 4.

Table 4 shows that the aliphatic/alicyclic and mono aromatic hydrocarbon content have increased whereas the polycyclic and resins have decreased after the pyrolysis process. The pyrolysis oil has also been analysed by Ultramar refinery and the results showed that it can by recycled as a refinery feedstock. Further analysis showed 1.3% sulfur in the initial oil which was reduced to 0.07% in the light pyrolysis oil, 0.1% in the heavy pyrolysis oil and 1.01% in the solid residue.

The results presented in this study revealed a significant reduction and changes in the nature of the asphaltenes, resins and other polar and heavy compounds after pyrolysis which were reflected in the physicochemical properties of the pyrolysis products. As a consequence, the high molecular weight saturated hydrocarbons will be more accessible for cracking in the absence of asphaltenes and heavy resins during refining.

The aqueous phase and solid residue recovered after pyrolysis were analysed in detail. The aqueous phase contained about 230 ppm soluble total organic matter (400 ppm total organic carbon, TOC) consisting of aliphatic and aromatic hydrocarbons, oxygenated compounds, e.g., mono and diphenols and nitrogenous compounds, e.g., pyridine which could be removed in the refinery water treatment plant. The solid residue was leached following the Quebec Ministry of Environment solid waste leaching procedure and the leachate was analysed. The soluble organic matter concentration was 8 ppm. Priority pollutants including polycyclic aromatic hydrocarbons and phenols were, however, absent in the leachate. The metals in the leachate were detected at a concentration below the regulated standards.

Conclusion

A major goal of this research was to increase our knowledge of the constituents of pyrolysis oils, utilising a combination of several techniques for separation and analysis. Particular attention has been paid in the present work to the potential merits of thermochemical conversion of various wastes including wood, lignin, bark and petroleum residue to reusable materials at moderate temperature and under vacuum conditions.

Vacuum pyrolysis was shown to be a potentially interesting technique to upgrade petroleum residues, particularly those with very high water and ash content. It is considered also a treatment process which could be applied for hydrocarbon-contaminated soils and activated sludges. The pyrolysis oil has a higher maltene content than the inital oil. Similarly, its asphaltene, resin and sulfur content was decreased.

The analytical results revealed that wood extractives have a negative effect on the pyrolysis oil yield and a positive effect on the carboxylic acid yield. The low molecular weight carboxylic acids were found to be the most abundant compounds in wood vacuum pyrolysis oils. Cellulose crystallinity plays a significant role in promoting sugar yield, levoglucosan in particular.

Wood, lignin and bark pyrolysis oils contain typically 40 to 60% by weight of a fraction mainly composed of formic and acetic acids, sugars and phenols. The remaining oil was difficult to characterize due to either its high molecular weight or high polarity or thermal lability. High-performance liquid chromatography (HPLC), however, with its ability to separate mixtures of high molecular weight, polar and thermally labile compounds is under investigation.

A large scale pyrolysis oil fractionation technique based on the liquid chromatography approach was tested and economical recovery of single pure compounds or in bulk is being assessed.

Acknowledgements

This study has been financially supported by Energy, Mines and Resources Canada/CANMET (Bioenergy Development Program), Energie, Ressources Québec and the Natural Science and Engineering Research Council.

References

1. Howe, B.,Introductory address. In: Proc. 7th Can. Bioenergy R&D Seminar. National Research Council of Canada Publication (1989) pp. 1-2.

2. Roy, C., H. Pakdel and D. Brouillard. The Role of Extractives during Vacuum Pyrolysis of Wood. J. Appl. Polym. Sci. (1990), 41, 337-348.

3. Pakdel, H. and C. Roy. Production and Characterization of Carboxylic Acids from Wood. Par 1: Low Molecular Weight Carboxylic Acids. Biomass, (1987) 13, 155-171.

4. Halchini, M. Caractérisation chimique de l'huile dérivée de la pyrolyse sous vide des écorces de bois. M.Sc. Thesis, Université Laval, Ste-Foy, Québec, Canada. (1990)

5. Perdrieux, S. and A. Lemoine. Les Bois National. 2 juillet (1983) 23-26.

6. Oehr, K.H. Production of Chemicals from Pyrolysis Oils. Thermochemical R&D Contractors Meeting. October 23-24, Ottawa (1990).

7. Pakdel, H. and C. Roy. Chemical Characterization of Wood Pyrolysis Oils Obtained in a Vacuum Pyrolysis Multiple-Hearth Reactor. In Pyrolysis Oils from Biomass, Producing, *Analysing and Upgrading.* Ed J. Soltes and Thomas A. Milne, eds. ACS Symposium Series 376, Washington, D.C. (1988) pp. 203-219.

8. Diebold, J. and A. Power. Engineering Aspects of the Vortex Pyrolysis Reactor to Produce Primary Pyrolysis Oil Vapors for Use in Resins and adhesives. In *Research in Thermochemical Biomass*

Conversion. A.V. Bridgwater and J.L.Kuester, eds. Elsevier Applied Science, Publisher, New York (1988) pp. 609-628.

9. Hemingway, R.W. Bark: Its Chemistry and Prospects for Chemical Utilization. *In Organic Chemicals from Biomass.* I.S. Glodstein, ed. 1978. CRC Press Inc. Boca Raton, Florida (1981) pp. 189-248.

10. Speight, J.G. Upgrading Heavy Oils and Residua: The Nature of the Problem. In *Catalysis on the Energy Scene.* S. Kaliaguine and A. Mahay, eds. Elsevier, Amsterdam (1984) pp. 515-527.

11. Pakdel, H., J.P. Blin and C. Roy. Upgrading of Petroleum Residues by Vacuum Pyrolysis. Fuel Science and Technology International. In press. (1990)

Discussion:

C. Alleyne (B.C. Research): In a couple of the slides that zipped past, you showed some GCMS identifications of compounds. I wonder what you do about confirming some of these identifications, because with GCMS it is very difficult to get specific isomeric identifications.

H. Pakdel (Speaker): Well, of course, with the standard compounds which we have, we have nearly 40 phenolic compounds, 40 oxygenated phenolics and phenol type compounds available in our laboratory. With the co-injections, of course, we have confirmed most of the compounds.

S. Czernik (Université de Sherbrooke): I would like to ask you several questions. The first one: what was the temperature of the pyrolysis of polyethylene you quoted from the distribution of the product?

Speaker: Between 400 and 450.

S. Czernik: All these products, were they evaporated from the reactor?

Speaker: Exactly.

S. Czernik: The next question is: what was your analytical strategy of the pyrolysis oil you got? You quoted several analytical methods used. But first, you have got some products. So, did you separate the product first for some fraction, or did you inject the product to ....

Speaker: No. This goes back to the last two or three years. We have been trying to develop methods of analysis, fractionation. We do not inject a whole lot into the GC because that is hopeless.

If you take a big column, a 100 metre column, of course we can do an analysis, but it is very costly in phenolics mainly - they destroy the column very quickly. So, we fractionate first, then we look at the composition.

S. Czernik: Of course, lignin and volatile, they would not pass through the chromatographic columns.

Speaker: Yes, exactly. Well that is, as I have mentioned, 50, 40, 30% of the material when we fractionate. They are not volatile or they are unstable. So, we cover them in separate fractions. That is why we are now switching into HPLC to characterize this type of fraction.

S. Czernik: You gave some data about the distribution of the pyrolysis product. It was some percentage of formic acid, some percentage of acetic acid, etc. So, what was the matter of quantification? Was it one method or were there several?

Speaker: No, we developed our acid measurement, which is based on the divitization into benzal esters. We divitized our oil before fractionation - the total. We divitize it, then we separate the simple purification and we inject in GC, which is all the acids there. Nothing goes into the system.

So, in order to analyze acids, we benzalate and we extract and we put it in GC and we quantify it. In GC you have only three peaks: formic, acetic and propionic, if there is another acid. You see three peaks and you quantify that.

S. Czernik: Then that technique comes from various methods of the determination of the quantities of the product I presume - some from gas chromatography, some from HPLC.

Speaker: Yes, mainly now for acids. It comes from divitization with GC - for gas it comes from GCMS - yes, GC and HPLC. The molecular weight distribution comes from GPC. A few years ago we used GPC to establish the molecular weight change distribution.

S. Czernik: Okay. The last question: you reported the yield of levoglucosan out of avicel cellulose is something like 13% or so. It is a fairly low yield from the vacuum pyrolysis, isn't it?

Speaker: Well, I just wanted to show how we can modify a sample of certain material in order to get whatever we want.

I don't argue about the levoglucosan content, I just wanted to say how the levoglucosan can be increased from wood to extractive-free wood, which extractive materials are maybe two or three per cent. We separate this, extract it and we increase ten times of levoglucosan content, whether the content of levoglucosan is at 10, 20, 50 - I do not comment on that one. That only shows if it was 3% or 5% in the original, initial wood, with the extractive-free, you can increase it to 50, ten times.

C. Roy (University of Laval): Just to complete a question which was asked by Stefan about the metal development, we are publishing as a chapter in a book, co-edited by

Drs. Lesage and Jackson from the Burlington National Water Research Institute.

On the request of Marcel Decarie Inc., there is a need for metal development for analysis of groundwater, contaminated groundwater and contaminated soils. So, we have had the contract over the last two years of about $300,000 for metal development and this is a rather lengthy chapter that actually Hooshang was writing, and it is a very, very valuable piece of material as an approach for an analysis of not only groundwater but oil and contaminated soii with oil and so on.

That, I guess, is going to be available in a matter of a few weeks or a couple of months. Anyone who is interested can contact either of us or write to Drs. Lesage or Jackson in Burlington. They are the editors for this book.

Table 1: Pyrolysis Products and Yields for Various Lignocellulosic Materials

Substance	Yield (% d.a.f.)			
	Oil	Water	Solid	Gas
Hardwood	60.7	14.1	14.4	10.8
Extractive-free wood	68.0	10.0	16.5	5.5
Holocellulose	42.9	19.3	20.0	n.a.
Avicel	83.1	9.7	4.3	2.9
Indulin AT	34.5	13.9	42.6	9.00
Aspen lignin	42.7	9.3	38.00	10.0
Lignosulfonate	8.8	28.6	41.2	21.4
Bark	24.1	23.6	33.3	19.0

d.a.f.: dry-ash-free basis
n.a.: not available

Table 2: Vacuum Pyrolysis of Wood: Summary of Operation Conditions and Yields (wt.%)

Experiment no.	Condensation temperature range (°C)	Pressure (mmHg)	Oil Yield %			Water yield %**			Acid yield (C_1 toC_7)		
			Oil Phase	Aqueous Phase	TOTAL	Oil Phase	Aqueous Phase	TOTAL	Oil Phase	Aqueous Phase	Total
C023-PDE-RL	50-55	12	23.31	26.63	49.94	0.99	20.08	21.07	1.39	7.50	8.89
C024-PDE-RL	30-35	30	27.54	18.82	46.36	1.01	20.36	21.37	1.99	6.05	8.04
C026-PDE-RL	15-25	10	33.63	15.74	49.37	2.21	18.12	20.33	3.20	5.78	8.98

* Moisture and ash-free wood basis
** As-received wood basis

Table 3: Bark Pyrolysis Oil Composition

Compounds	Yield (%)
Hydrocarbons	4.7
Carboxylic acids (LMW)	5.7
Carboxylic acids (HMW)	0.4
Sugars	8.3
Methyl esters	1.8
Phenolic compounds	11.4
Other oxygenated compounds	1.0
Water	18.4
Total identified	51.7

LMW = Low molecular weight
HMW = High molecular weight

Table 4
Chemical Composition of Maltenes (wt. %)

Sample	Aliphatic/Alicyclic Hydrocarbons (F1)	Monoaromatic Hydrocarbons (F2)	Polyaromatic/ Resins (F3)
Initial oil	41	32	27
Pyrolysis oil	46	36	18

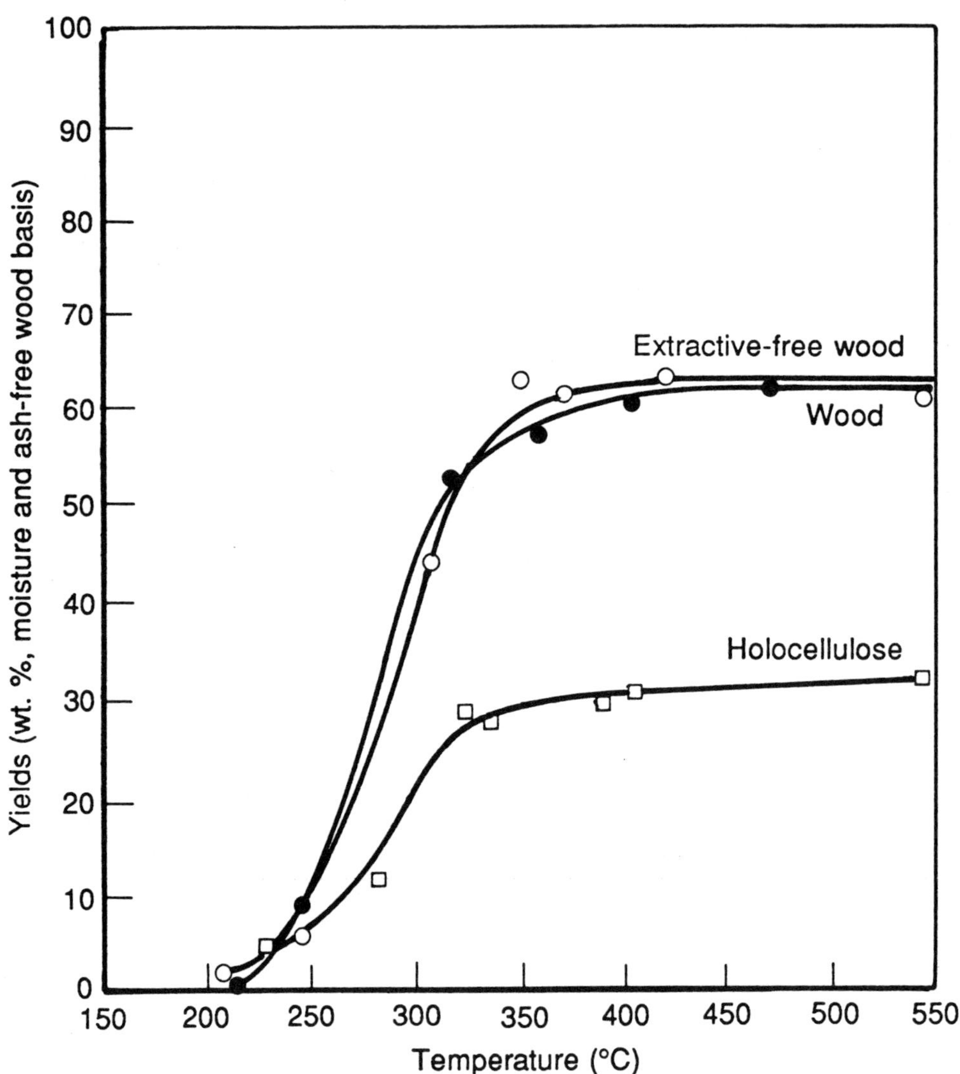

Figure 1 Oil Yield from Wood, Holocellulose, and Extractive-Free Wood Samples. Holocellulose was a Lignin-Free Wood Sample.

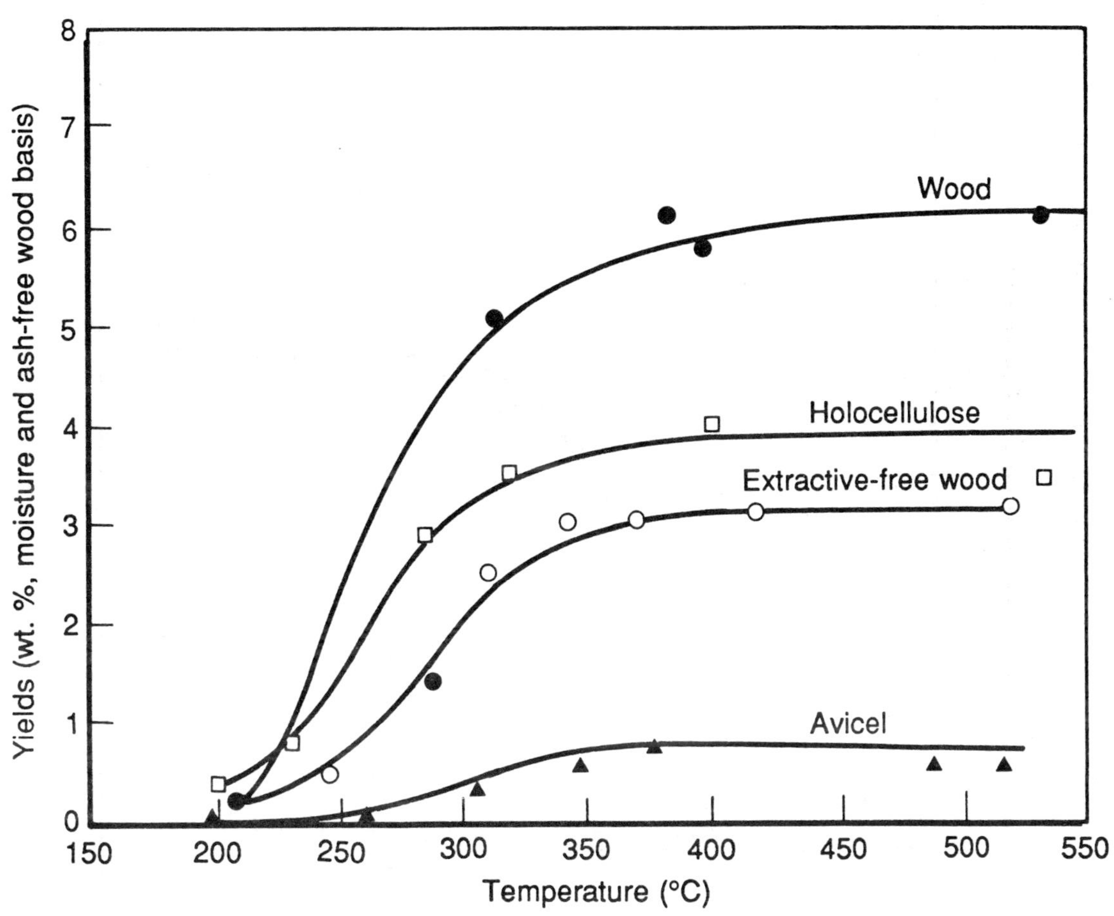

Figure 2

Yield of Formic Acid from Wood, Holocellulose, Extractive-Free Wood, and Avicel Samples. Holocellulose was a Lignin-Free Wood Sample.

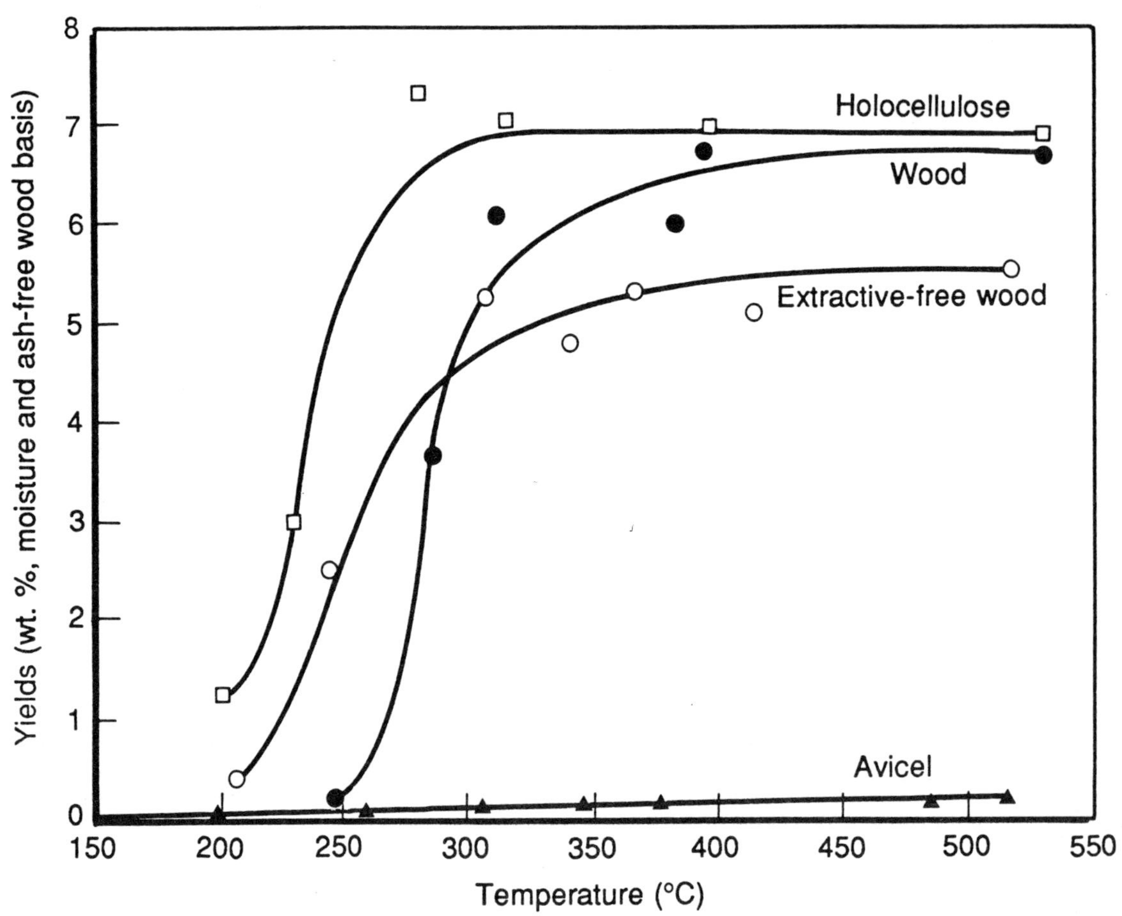

Figure 3

Yield of Acetic Acid from Wood, Holocellulose, Extractive-Free Wood, and Avicel Samples Holocellulose was a Lignin-Free Wood Sample.

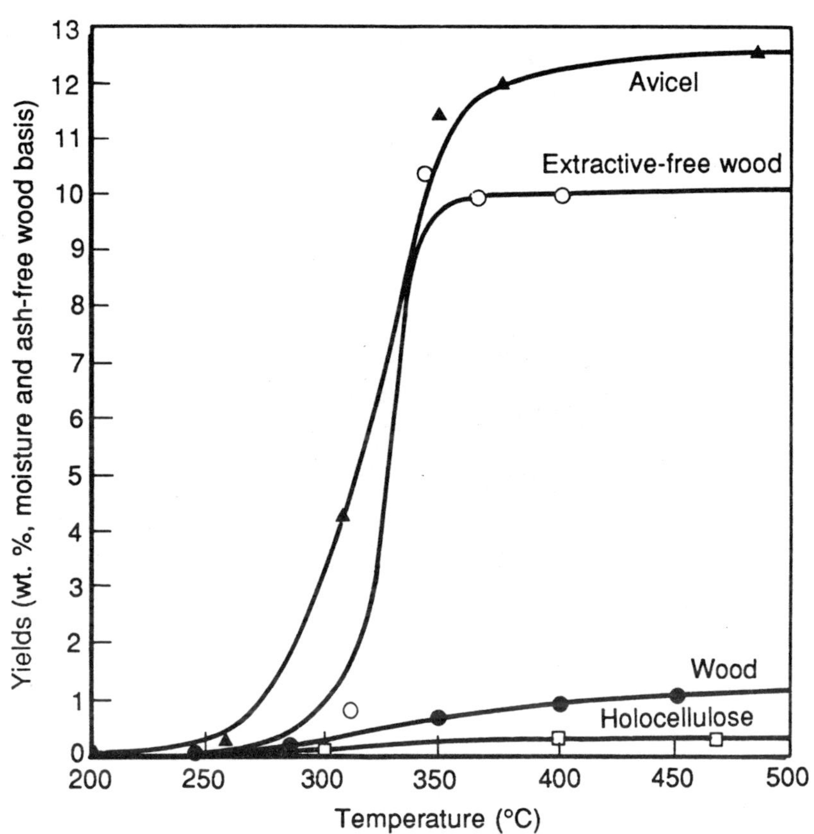

Figure 4

Yield of Levoglucosan from Wood, Holocellulose, Extractive-Free Wood, and Avicel Samples Holocellulose was a Lignin-Free Wood Sample.

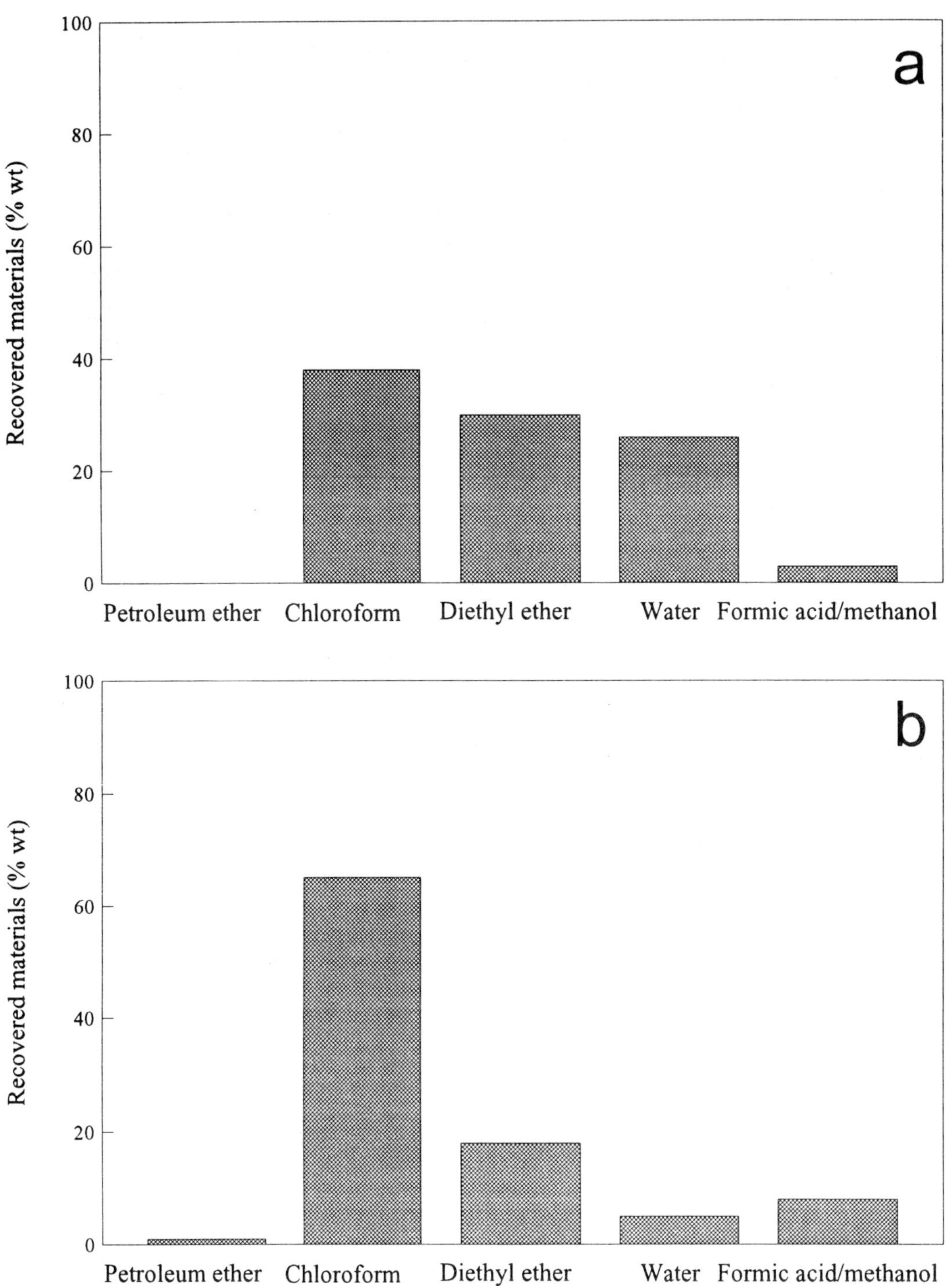

Figure 5

Sequential Elution Solvent Chromatographic Analysis of : (a) Wood and (b) Lignin Pyrolysis Oils.

Catalytic Upgrading of Pyrolytic Oils Over HZSM-5

R.K. Sharma and N.N. Bakhshi
Catalysis and Chemical Reaction Engineering Laboratory
Dept. Chemical Engineering, University of Saskatchewan, Saskatoon, S7N OWO

The upgrading of pyrolytic oils was studied over HZSM-5 catalyst. A fixed bed microreactor, operated at atmospheric pressure and in the temperature range 370-410°C was used. The oils were co-fed with tetralin. The upgraded product contained a large percentage of aromatic hydrocarbons as liquid. With high pressure liquefaction oil, the amount of pitch decreased by 70-80 wt.%. The phenolic content also decreased whereas the amount of aromatic hydrocarbons increased to a maximum. The gas product was usually less than 5 wt.%. At high temperatures, over 30 wt.% of the bio-oil was deposited as coke. With a low pressure pyrolysis oil, the coking reduced to less than 18 wt.% at 405°C. When a fluidized bed reactor was used, both the coke formation and the aromatics from the high pressure oil decreased. A reaction scheme was postulated based on the results.

On a étudié dans quelle mesure il était possible d'améliorer les huiles pyrolytiques à l'aide du catalyseur HZSM-5. On a utilisé un micro-réacteur à lit fixe qui fonctionnait à la pression atmosphérique et à une température comprise entre 370 °C et 410 °C. Les huiles étaient introduites dans le réacteur avec de la Tétraline. Le produit amélioré contenait une proportion élevée d'hydrocarbures aromatiques sous forme liquide. Dans le cas de l'huile obtenue par liquéfaction sous pression élevée, la teneur en brai était réduite de 70 - 80 % en poids. La teneur en composés phénoliques était également réduite, mais la quantité d'hydrocarbures aromatiques devenait maximale. Les produits gazeux constituaient généralement moins de 5 % en poids de la totalité. À des températures élevées, plus de 30 % en poids de la bio-huile était déposée sous forme de coke. Dans le cas de l'huile obtenue par pyrolyse sous faible pression, la quantité de coke produite tombait à moins de 18 % en poids à 405 °C. Lorsqu'on utilisait un réacteur à lit fluidisé, on obtenait moins de coke et de produits aromatiques à partir des huiles obtenues par pyrolyse sous pression élevée. On a proposé un mécanisme réactionnel en se basant sur les résultats obtenus.

Introduction

Pyrolytic oils are produced from biomass by catalytic liquefaction at high pressures (Eager et al., 1982) or by pyrolysis at high temperatures and low pressures (Scott et al., 1987). These bio-oils are highly corrosive, viscous and have a large amount of oxygen. Since the oils are also extremely unstable at high temperatures, their upgrading to useful products is difficult.

In the literature, two routes have been followed in order to produce hydrocarbons from the bio-oils. Baker and Elliot (1988) and Soltes (1987) used classical Co-Mo and Ni-Mo hydrotreating catalysts and H_2 or CO as reducing gases for upgrading. In the temperature range of 200-400°C, gasoline yield of 70-80 wt.% was achieved. Chantal et al. (1984) and Mathews et al. (1984), among others, used HZSM-5 at atmospheric pressure and 350-400°C to convert the low boiling fraction of the bio-oil to aromatic hydrocarbons.

In this study, HZSM-5 was used to upgrade the bio-oils. The oils were obtained both from the high pressure liquefaction and the low pressure-high temperature wood pyrolysis methods. Due to excessive catalyst coking, tetralin was used as a co-processing solvent with the bio-oils. Upgrading was carried out at atmospheric pressure and in the temperature range 370-410°C. A few experiments were also carried out using a fluidized bed reactor. The gas and liquid products were analyzed and the catalyst was examined for coke and tar deposition. Based on these results, a reaction scheme was postulated. The aspect of tetralin recovery was not considered.

Experimental

The high pressure bio-oil was prepared by the catalytic liquefaction of wood according to the method developed by Eager et al. (1982). The yield of the bio-oil was about 30 wt.% of the dried wood powder. The low pressure pyrolysis oil was obtained from the University of Waterloo where it was produced by the fast pyrolysis of wood at 510°C. The catalyst HZSM-5 was prepared by the method specified by Chen et al. (1973).

The experimental data were obtained at atmospheric pressure in a fixed bed microreactor, 520 mm long and 7.5 mm in diameter. The details of the experimental equipment, run procedure and analytical techniques have been described elsewhere (Sharma and Bakhshi, 1989). The fluidized bed reactor was 11 mm in diameter and the oil was fed from the top directly onto the fluidized catalyst bed. Argon was used as the fluidizing gas. The feed consisted of a mixture of bio-oil and tetralin. Since tetralin was found to be essentially non-reactive below 405°C (Sharma and Bakhshi, 1989), the analyses were reported on tetralin-free basis. The reaction products were separated into liquid and gas fractions by passing through an ice cooled trap. The liquid product was distilled for 30 mins. under a vacuum of 172 Pa at 200°C to recover the distillate product. The distillate product and the gas fraction were analyzed by gas chromatography. The spent catalyst was washed with hexane and then regenerated in air at 600°C. The difference in the weights before and after washing was termed as 'tar' and that before and after regeneration as 'coke'. The 'tar' and the 'residue fraction' (i.e. the fraction left over after the recovery of the distillate from the liquid product) were combined together and were designated as 'unconverted oil'. The detailed product collection scheme is shown in Figure 1.

Results and Discussion

High Pressure Liquefaction Oil

The oil was distilled at 150, 200 and 250°C under a vacuum of 172 Pa (corresponding to 410, 450 and 520°C at atmospheric pressure, respectively). A maximum of 60 wt.% of bio-oil was obtained as distillate at 200°C. The amount of residue in the oil was called 'pitch'. The analysis of the distillate is shown on Table 1. It is seen from the table that the phenols and polynuclear aromatic hydrocarbons constitute the bulk of the composition with a concentration of 20 wt.% each. It was also found that about 25 wt.% of the bio-oil was low boiling (below 220°C at atmospheric pressure).

The overall material balances in the upgrading of high pressure liquefaction oil are presented in Table 2. Over 60 wt.% of the pitch was converted at 370°C and the conversion increased to 82 wt.% at 410°C. The amount of coke formed also increased from 11 wt.% at 370°C to 35 wt.% at 410°C. These coke levels are comparable to those found by Chantal et al. (1984) but are much higher than those observed by Chen et al. (1987) in a fluidized bed reactor at low oil concentration in helium. The detailed composition of the distillate is given in Appendix A and the composition is summarized in Table 3. The aromatic hydrocarbons in the product increased from 24 wt.% at 370°C to 36 wt.% at 390°C before decreasing to 22 wt.% at 410°C. The concentration of the phenols in the distillate decreased from 31 wt.% to 12 wt.% as the temperature increased from 370 to 410°C. The phenols appear to be the primary products of upgrading. This has also been suggested by Soltes et al. (1987) for the pyrolytic oil upgrading over hydrotreating catalysts. It is quite likely that phenols also directly lead to coke formation. The conversion to gas product was usually less than 5 wt.% and mostly consisted of C_2 to C_5 paraffins and CO_2 (Table 4).

The product distributions indicated that these bio-oils may be excellent sources of useful chemicals such as phenols in addition to being suitable as feedstocks for the production of fuels. Although, tetralin was non-reactive under the operating conditions used in this study, there was some indication that a small degree of dehydrogenation to naphthalene occurred in the presence of bio-oil. The amount dehydrogenated was not taken into account since it was small as compared to the amount of tetralin fed. These results also showed that excessive catalyst coking occurred at the reactor temperature of 410°C. This suggests a two stage reactor where the bio-oil is first stabilized at a lower temperature (350-380°C) followed by a second stage upgrading at a higher temperature (380-410°C).

Low Pressure Pyrolytic Oil

The composition of low pressure pyrolytic oil, obtained from the University of Waterloo, is presented in Table 5. The oil contains about 18 wt.% water and about 28 wt.% of water insoluble lignin. The amount of oil which remained unidentified was 31 wt.%.

Only two experiments were carried out for the upgrading of the low pressure pyrolysis oil using tetralin as solvent. The overall material balances in these experiments are presented in Table 6. The product contains 18-21 wt.% water. The amount of coke is less than 20 wt.% at 405°C which is lower than the coke formed with the high pressure oil. This may be due the large amount of water present in the low pressure oil. The amount of gas product is 2-4 wt.%. The detailed composition of the distillate is given in Appendix B. A large number of components were identified but further work is needed for their confirmation. The composition of gas product is shown in Table 7 which indicates that C_3 and C_4 hydrocarbons are the main components with a combined concentration of over 70 wt.%

Fluidized Bed Reactor

The overall material balances from the upgrading of the high pressure oil in the fluidized bed reactor are shown in Table 8. The feed consisted of a mixture of bio-oil and tetralin. The amount of coke at 390°C was 21 wt.% which increased to 27 wt.% at 415°C. These values are much lower than those in the fixed bed reactor indicating that the pitch conversion and coke formation were dependent on the mode of processing. The amount of unreacted oil decreased from 24 wt.% to 15 wt% as the temperature increased from 390 to 415°C. An increase in the amount of unreacted oil and a decrease in coking in this reactor relative to the fixed bed reactor may be due to a lower residence time in fluidized bed reactor (0.2 s compared to over 500 s in the fixed bed). The amount of distillate was 50-54 wt.%. The composition of distillate (Table 9) indicates that the concentration of phenols was about 46 wt.% which is much higher than in the fixed bed reactor. The concentration of aromatic hydrocarbons was 22-26 wt.%. It appears that there is an insufficient secondary cracking and deoxygenation of the oil molecules in the fluidized bed reactor. The composition of gas product (Table 10) shows that it contains 30-40 wt.% olefins. The concentration of ethylene increased from 15 wt.% at 390°C to 33 wt.% at 415°C. The product also contained nearly 50 wt.% concentration of C_6-C_8 hydrocarbons at 390°C which decreased to 22 wt.% at 415°C.

The fluidized bed reactor has the advantage that the catalyst may be regenerated continuously in order to maintain a high bio-oil conversion. On the other hand, the amount of bio-oil lost as coke may be minimized using a two-stage fixed bed reactor.

Reaction Scheme

Based on the results in this study, a reaction scheme was postulated and is shown in Figure 2. According to this scheme, the non-volatiles in the pitch are first cracked and partially deoxygenated to volatile components which are precursors to the coke and gas formation. The coke forming reactions appeared to be much more temperature sensitive compared to the conversion of non-volatiles. As suggested by Baker and Elliot (1988) for upgrading at high pressures, a two step process in which the pitch is first converted at lower temperatures (350-380°C) followed by the cracking of the high boiling components at higher temperatures (380-410°C) seems more appropriate.

Conclusions

Significant reduction of the pitch in the high pressure bio-oil was achieved in the fixed bed reactor using HZSM-5. The concentration of phenols and aromatic hydrocarbons was highest at 377 and 390°C, respectively. The gas product consisted of C_2-C_5 paraffins and increased with temperature. For the low pressure pyrolytic oil, the coking was lower as compared to the high pressure oil apparently due to the large amount of water in the low pressure oil. The amount of bio-oil lost as coke may be minimized by using a two-stage fixed bed reactor.

In the fluidized bed reactor, the coking with the high pressure oil decreased but the amount of unreacted oil also decreased. High bio-oil conversions may be achieved in the fluidized bed reactor by continuous regeneration of the catalyst.

References

1. Baker, E.G., Elliot, D.C., Catalytic upgrading of biomass oils. in *Research in Thermochem. Biomas Conv.*, Elsevier Appl. Sci. Pub., New York, 883-(1988).
2. Chantal, P., Kaliaguine, S., Grandmaison, J.L., Mahay, A., Production of hydrocarbons from aspen poplar pyrolytic oils over HZSM-5. *Appl. Catal.* 10(3), 317-32 (1984).
3. Chen, N.Y., Maile, J.N., Regan, W.J., Preparation of zeolites. *U.S. Patent 411 2056* (1973).
4. Chen, N.Y., Walsh, D.E., Koenig, L.R., Fluidized bed upgrading of wood pyrolysis liquids. *ACS, Div. Fuel Chem.* 32(2), 264-75 (1987).
5. Eager, R.L., Mathews, J.F., Pepper, J.M., Liquefaction of aspen poplar wood. *Can. J. Chem. Eng.* 60, 289-94 (1982).
6. Mathews, J.F., Tepylo, M.G., Eager, R.L., Pepper, J.M., Upgrading of aspen poplar wood oil over HZSM-5. *Can. J. Chem. Eng.* 63(4), 686-89 (1985).
7. Scott, D.S., Piskorz, J., Grinshpun, A., Graham, R.G., The effect of temperature on liquid product composition from the fast pyrolysis of cellulose. *ACS, Div. Fuel Chem.* 32(2), 1-11 (1987).
8. Sharma, R.K., Bakhshi, N.N., Upgrading of biomass-derived pyrolytic oils over HZSM-5 catalyst. *Report to Bioenergy Development Program, Energy, Mines and Resources, Canada,* July 1989, pp. 78.
9. Soltes, E.J., Lin, S.C.K., Sheu, Y.H.E., Catalytic specificities in high pressure hydro-processing of tars. *ACS, Div. Fuel Chem.* 32(2) 229-39 (1987).

Discussion:

J. Piskorz (University of Waterloo): With this massive coke deposit on your catalyst don't you upset one of the initial bursts of activity of your catalyst and later there is nothing going on?

N. Bakhshi (Speaker): Yes, usually this is what happens. When you say initial activity, initially it is active, but after awhile the activity does drop and the way we measure that is by the gas from it.

J. Piskorz: Did you investigate your product distribution with the time of your reaction, because I think the longer the residence time you will have a totally different spectre.

Speaker: Yes, I think your question is well taken. You see, that kind of activity, that you are talking about, happens if you have a so-called regular type of catalyst, maybe nickel, alumina, cobalt, for example, typical catalysts.

This is a zeolite and as in the case of zeolite, this activity which you are seeing is on account of the acid - the special kind of pores which give you what you call the configuration of the fusion. So you are not going to see the same kind of bursts of activity initially and then it keeps on changing.

Here you are going to see a sort of steady decline. We are doing work, to a certain extent, where for example, in this case our time was about 50 minutes or so and now what we are doing is looking at just what we are talking about - what is the effect of time. For example, we will stop our run at 20 minutes and then see what our product distribution is. We are doing that, but I don't think it will make much of a difference --- if a different catalyst.

E. Chornet (Université de Sherbrooke): One of our students has completed his PhD thesis on lignin conversion, and one topic that he addressed was the conversion of lignin in the presence of tetralin, which I think is relevant to what you say; that is why I make this intervention.

He was able to show that just putting lignin with tetralin, and using conditions very similar to the ones you have shown, with no catalyst added, his results indicate 70 to 75 percent of the lignin becomes monomeric phenolic materials and the rest are gums and some coke.

So my suggestion is that perhaps the catalyst is not that needed when you use tetralin as a vehicle. I know that this is a controversial subject, but I want to refer these results to you, just that you know that perhaps, in the presence of tetralin, we can break the COC bonds in a singular manner. My first comment.

My second comment is that what we call phenols are not only phenol, they are mitoxylated (phonetic) phenols, catacols (phonetic), and a bunch of materials. They are not a homogenous material.

J. Diebold (SERI): I was wondering what the method of addition of the oils to the reactor was? Were you adding the oils as a liquid or were you evaporating the oils before introduction to the reactor?

Speaker: The reactor is about 20 to 25 centimeters long, and the catalyst sites in the middle. So you have about 10 to 12 centimeters, sort of a preheating zone if you wish to call it. The oil itself, the feed itself and tetralin plus the oil that we were adding as liquid. So what we are hoping is that by the time it gets about 370-380 degrees then some cracking and vaporization takes place. Before that, it is my feeling that it is not vaporized.

J. Diebold: So you were trying to evaporate the oils prior to their entering the catalyst zone. Do you have any ideas to relative amounts of coke in the preheater and the relative amount of coke in the catalyst zone?

Speaker: Well, we didn't see anything on the reactor walls. Most of the coke simply sits right on the catalyst itself.

J. Diebold: The temperatures that you were measuring, were those in the catalyst bed - and several zones, and were they different places in the bed, or was it a single temperature involvement?

Speaker: Well, three zones, one was at the top, one was in the centre, and one was on the bottom and this is the average of these three temperatures, plus/minus two or three degrees, that variation.

Table 1. Composition of high pressure liquefaction oil, wt.%

Acids	1.9
Alcohols	4.4
Aromatic hydrocarbons	19.1
Ethers	1.9
Furans	5.6
Aliphatic hydrocarbons	2.7
Naphthenes	4.1
Phenols	19.7
Non-volatile pitch	40.0
Unknown	0.6
Total	100.0

Table 2. Overall material balance in the upgrading of high pressure liquefaction oil, wt.%

WHSV, h^{-1}	2.5	2.5	2.5
Temperature, ° C	370	390	410
Bio-oil/tetralin, wt.	1.2	1.2	1.2
Run time, min	50	60	65
Distillate	72.4	60.4	50.0
Aqueous	1.4	1.5	1.7
Gas	1.9	3.0	3.3
Unconverted oil	12.6	13.0	7.2
Coke	11.4	22.1	37.7
Unaccounted	0.3	-	0.1
Total	100.0	100.0	100.0

Table 3. Composition of the distillate product from high pressure liquefaction oil

WHSV, h^{-1}	2.5	2.5	2.5
Temperature, °C	370	390	410
Product distribution, wt.%			
Aliphatic hydrocarbons	3.7	2.8	4.6
Aromatic hydrocarbons	32.5	57.4	44.5
Phenols	43.5	23.3	23.9
Naphthenes	2.0	2.4	1.4
Alcohols	6.8	4.1	12.3
Furans	8.6	3.1	10.6
Acids	2.0	0.6	1.9
Ethers	0.9	6.3	0.8
Total	100.0	100.0	100.0

Table 4. Composition of gas product from the upgrading of high pressure liquefaction oil, wt.%

Temperature, ° C	370	390	410
WHSV, h^{-1}	2.5	2.5	2.5
Methane	1.6	-	-
C_2	71.5	-	-
C_3	-	6.5	21.4
Isobutane	0.8	1.4	14.4
Butene	-	-	-
Butane	-	32.2	21.5
C_5	0.6	54.2	35.6
$CO + CO_2$	25.5	5.7	7.1
Total	100.0	100.0	100.0

Table 5. Composition of low pressure pyrolytic oil, wt. % (Piskorz et al., 1988)

Cellobiosan	1.2
Glucose	0.2
Fructose	0.5
Glyoxal	0.9
Levoglucosan	1.5
Hydroxyacetaldehyde	9.3
Formaldehyde	2.9
Acetic acid	5.5
Ethylene glycol	0.6
Water	18.4
Water insoluble lignin	28.0
Unknown	31.0
Total	100.0

Table 6. Overall material balance in the upgrading of low pressure pyrolytic oil, wt.%

WHSV, h^{-1}	2.5	5.8
Temperature, $^{\circ}$ C	405	405
Bio-oil/tetralin, wt.	2.2	2.2
Run time, min	45	52
Distillate	42.5	59.5
Aqueous	21.2	18.5
Gas	1.9	4.3
Unconverted oil	16.4	6.6
Coke	17.6	11.1
Unaccounted	0.4	-
Total	100.0	100.4

Table 7. Composition of gas product from the upgrading of low pressure pyrolytic oil

Temperature, °C	405	405
WHSV, h^{-1}	5.8	2.5
Methane	-	6.5
Ethane + ethylene	0.1	8.1
Propylene	44.0	23.8
Propane	32.5	26.4
Isobutane	0.7	5.8
1-Butene	20.4	3.4
Butane	-	5.5
2-Butene	-	7.3
C_5	0.3	4.9
$CO + CO_2$	2.0	8.3
Total	100.0	100.0

Table 8. Overall material balance in the upgrading of high pressure liquefaction oil in fluidized bed reactor, wt.%

WHSV, h^{-1}	2.5	2.5
Temperature, ° C	390	415
Bio-oil/tetralin, wt.	3.3	3.3
Run time, min	40	47
Distillate	49.1	54.1
Aqueous	-	-
Gas	6.1	4.1
Unconverted oil	24.0	14.1
Coke	20.8	27.1
Unaccounted	-	-
Total	100.0	100.8

Table 9. Composition of the distillate from the upgrading of high pressure liquefaction oil in fluidized bed reactor

WHSV, h^{-1}	2.5	2.5
Temperature, °C	390	415
Product distribution, wt.%		
Aliphatic hydrocarbons	3.6	3.3
Aromatic hydrocarbons	22.1	26.0
Phenols	45.6	46.4
Naphthenes	0.7	0.6
Alcohols	11.3	4.3
Furans	13.2	12.1
Acids	2.8	5.7
Ethers	0.7	1.6
TOTAL	100.0	100.0

Table 10. Composition of gas product from the upgrading of high pressure liquefaction oil in fluidized bed reactor, wt.%

Temperature, °C	390	415	
WHSV, $^{-1}$	2.5	2.5	
Methane	-	-	
Ethylene	14.8	33.7	
Ethane	-	-	
Propylene	6.5	5.5	
Propane	1.8	8.6	
Butene	4.6	7.7	
Butane	-	-	
Pentene	4.6	-	
Pentane	15.0	15.7	(total C5)
C_6	5.1	6.4	
C_7	32.9	16.5	
C_8	11.8	-	
$CO + CO_2$	1.9	5.9	
Total	100.0	100.0	

Appendix A

Composition of the distillate from the upgrading of high pressure liquefaction oil

Temperature, °C	370	390	410
WHSV, h^{-1}	2.5	2.5	2.5
Hexane	0.4	0.6	0.6
Benzene	0.3	0.4	0.6
Heptane	2.0	1.7	0.6
Methyl cyclohexane	1.2	1.3	--
Toluene	9.5	44.6	14.3
Octane	0.8	0.5	0.5
Ethyl cyclohexane	1.0	1.2	1.5
Ethyl benzene	2.0	1.1	2.0
m,p Xylene	6.9	4.0	6.5
o- Xylene	1.5	0.6	0.8
Propionic acid	2.1	0.7	2.0
Cyclohexanol	3.3	1.5	3.1
Furfural	6.2	2.7	8.1
1- Decene	0.6	--	3.0
Methyl furfural	1.4	0.7	2.7
Cyclopentanol	0.8	0.3	1.1
Phenol	3.9	2.2	2.8
2-Hydroxy 3-methyl cy. pentanol	3.1	2.6	8.4
o-Cresol	1.3	0.8	0.6
Guaiacol	2.8	1.5	1.4
m,p- Cresol	2.7	4.8	2.3
2,4 Dimethyl phenol	1.5	0.9	1.6
2,6 Dimethyl phenol	--	0.5	2.3
Dimethyl guaiacol	5.8	0.4	1.2
C3 Alkyl phenol	0.9	0.6	0.7
4- Ethyl guaiacol	4.9	1.1	5.5
Methyl naphthalene	2.1	0.8	0.8
Eugenol	4.6	1.5	3.7
Iso-eugenol	14.8	8.4	0.4
2-Methyl 4 propyl phenol	1.1	1.1	0.8
Methyl Eugenol	1.2	0.9	0.7
Flourene	5.3	4.6	10.4
Methyl dibenzofuran	0.8	--	--
Methoxy flourene	1.0	6.6	0.7
Phenantharene	2.8	2.5	2.3
Anthracene	1.2	0.5	0.7
Methyl phenantharene	1.5	0.9	3.9
Dimethyl phenantharene	0.7	0.7	1.3
Total	104.0	105.8	99.9

1-INDENE, 2,3-DIHYDRO-1,3-DIMETHYL-	0.22	0.3
NAPHTHALENE, 1,2-DIHYDRO-6-METHYL-	0.34	0.31
BENZENE, 1,4-DIMETHOXY-2-METHYL-	0.11	0.2
1-INDEN-1-ONE, 2,3-DIHYDRO-	0.45	0.2
1-INDENE, 1-ETHYLIDENE-	0.68	0.72
NAPHTHALENE, 1,2,3,4 - TETRAHYDRO-1,1,6-TRIMETHYL-	1.82	1.64
PHENOL, 2,6-DIMETHOXY-	0.57	0.92
PHENOL, 2-METHOXY-5-(1-PROPENYL)-	0.45	0.2
BENZENEMETHANOL, .ALPHA.-ETHYL-4-METHOXY-	0.79	0.3
1-NEPHTHALENONE, 3,4-DIHYDRO-	0.79	0.31
BENZALDEHYDE, 4-HYDROXY-3-METHOXY-	2.73	1.95
NAPHTHALENE, 1-ETHYL-	0.91	3.19
NAPHTHALENE, 2-ETHYL-	1.71	1.54
BENZENE, 1,2,3-TRIMETHOXY-	1.93	0.41
PHENOL, 2-METHOXY-4-(1-PROPENYL)-, ACETATE	5.81	0.2
BENZENE, 1-METHYL-2-(PROPYLTHIO)-	1.14	17.92
NAPHTHALENE, 1,4,6-TRIMETHYL-	5.7	0.3
BENZOIC ACID, 3,4-DIMETHOXY-	0.57	0.82
2-PROPANONE, 1-(4-HYDROXY-3-METHOXYPHENYL)	0.34	1.44
PHENOL, 2,6-DIMETHOXY-4-(2-PROPENYL)-	0.91	2.57
BENZOIC ACID, 2-NITRO-	0.68	0.82
PHENOL, 2,6-DIMETHOXY-4-(2-PROPENYL)-	2.73	0.20
BENZALDEHYDE, 4-HYDROXY-3,5-DIMETHOXY-	0.22	0.51
PHENOL, 2,6-DIMETHOXY-4-(2-PROPENYL)-	0.11	0.2
PHENOL, 1,4-DIMETHOXY-2,3,5,6-TETRAMETHYL-	0.11	0.82
1-BUTANONE, 1-(2,4,6 TRIHYDROXY -3-METHYLPHENYL)-	0.22	1.54
ETHANONE, 1-(4-HYDROXY-3,5-DIMETHOXYPHENYL)-	0.11	0.41
BENZOIC ACID, 2-NITRO-	0.45	0.82
3-INDAZOL-3-ONE, 1,2-DIHYDRO-2-PHENYL-	0.34	1.23
1-CYCLOPROPANAPHTHALENE- OCTAHYDRO-1	0.11	0.41
SPIRO[CYCLOBUTANE-1,1(2-PHENANTHRENE)-2-ONE	0.11	0.41
1-PHENANTHRENECARBOXALDEHYDE, OCTAHYDRO-	0.11	0.3
METHANONE, [4-(1,1-DIMETHYLETHYL) PHENYL](4-HYDROXYPHENYL)-	0.11	0.31
PHENOL, 2-(METHYLSULFONYL)-	0.34	0.72
3-INDAZOL-3-ONE, 1,2-DIHYDRO-2- (4-METHOXYPHENYL)-5-METHYL-	0.11	0.20
TOTAL	99.37	99.48

Appendix B

Composition of distillate from pyrolysis oil upgrading

Temperature, °C	405	405
WHSV, h^{-1}	5.8	2.5
4-HEXENOIC ACID, 3-METHYL-2,6-DIOXO-	1.93	3.60
ETHANETHIOIC ACID, 5-PROPYL ESTER	0.45	0.82
CARBONIC ACID, ETHYL 2-MERCAPTOETHYL ESTER	0.68	1.23
PENTANOIC ACID, 2-METHYL-	2.28	1.54
THIOSULFURIC ACID, 5-(2-AMINOETHYL) ESTER	0.34	0.51
1,3,5, CYCLOHEPTATRIENE	2.5	0.61
6-HEPTEN-1-OL, 3-METHYL-	1.59	3.09
2-FURANCARBOXALDEHYDE	0.57	2.98
2-FURAN, 2,5-DIMETHYL-	1.14	5.15
2,4-PYRIMIDENEDIONE, DIHYDRO- 5-HYDROXY-5-METHYL-	3.19	0.72
BENZENE, 1,2-DIMETHYL-	0.57	1.13
BENZENE, 1,3-DIMETHYL-	1.02	0.51
2-CYCLOPENTEN-1-ONE, 3-METHYL-	0.57	2.98
HEXANOIC ACID	2.73	2.36
2-FURANONE, DIHYDRO-5-METHYL-5-PHENYL-	2.96	0.51
BENZENE, 1,3,5-TRIMETHYL-	0.45	2.67
BENZENEACETIC ACID, 2-CARBOXY-	1.48	1.13
2-CYCLOPENTEN-1-ONE, 2-HYDROXY-3-METHYL-	1.02	1.23
PHENOL, 2-METHYL-	0.91	1.64
BENZENE, 1-ETHENYL-2-METHYL-	2.62	0.82
BENZENEPROPANENITRILE, .BETA.-HYDROXY-	0.57	1.54
PHENOL, 2-METHOXY-	1.25	0.92
BENZENE, 1-ETHENYL-4-ETHYL-	8.32	0.51
SPIRO[4.5]DECANE	3.07	0.61
PHENOL, 3,4-DIMETHYL-	1.14	1.95
1,4,8-METHANO CYCLOPENTAZULENE, -OCTAHYDRO-	0.91	1.33
NAPHTHALENE, 1,2,3,4-TETRAHYDRO-	-	-
1,3-BENZENEDIOL, 4-ETHYL-	0.91	0.82
AZULENE-	18.12	10.09
BENZENE, 1-ETHYL-4-(1-METHYLETHYL)-	0.79	0.61
OXIRANECARBOXYLIC ACID, 3-ETHYL- 3-PHENYL-, ETHYL ESTER	0.45	0.41
PHENOL, 2-ETHYL-4-METHYL-	0.22	0.41
BENZENE, 1-ETHYL-2,4,5-TRIMETHYL-	0.11	0.2
1,2-BENZENEDIOL, 3-METHYL-	0.57	0.51

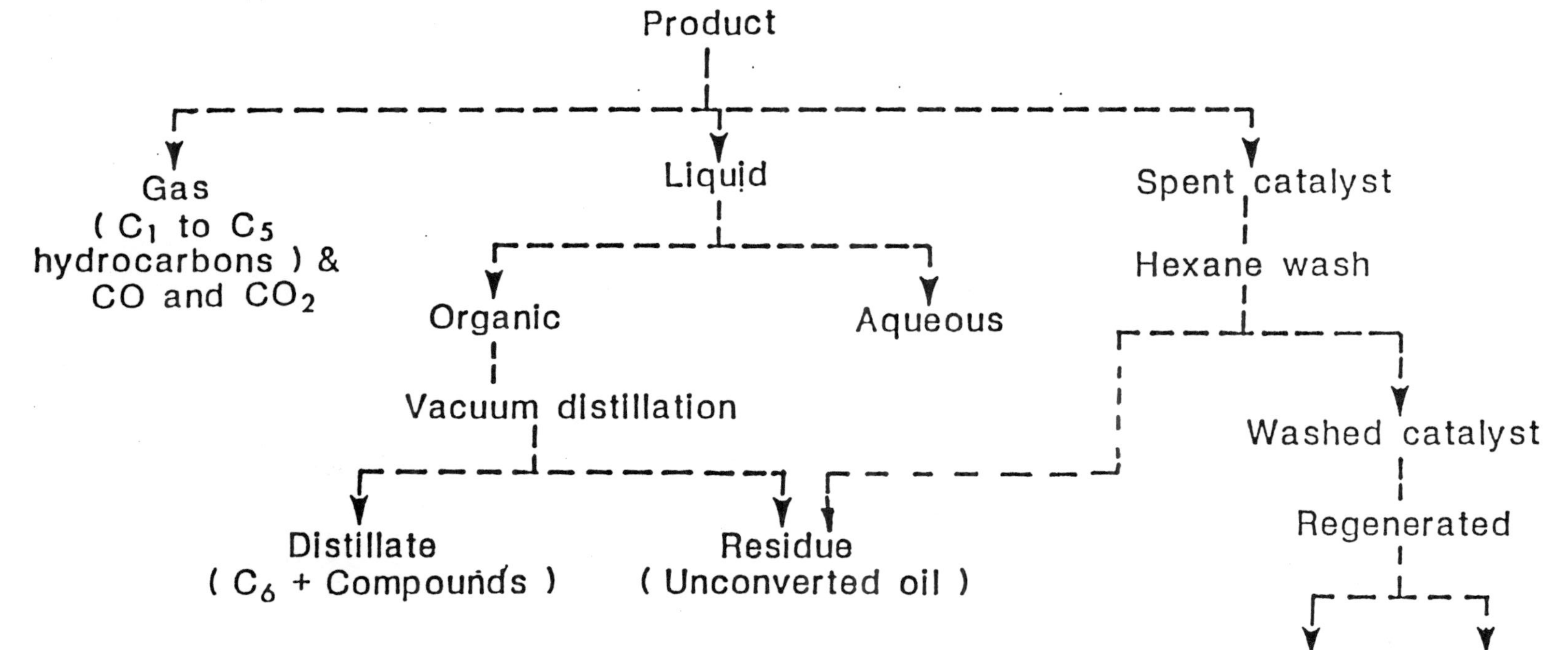

Figure 1. Product collection and analysis system.

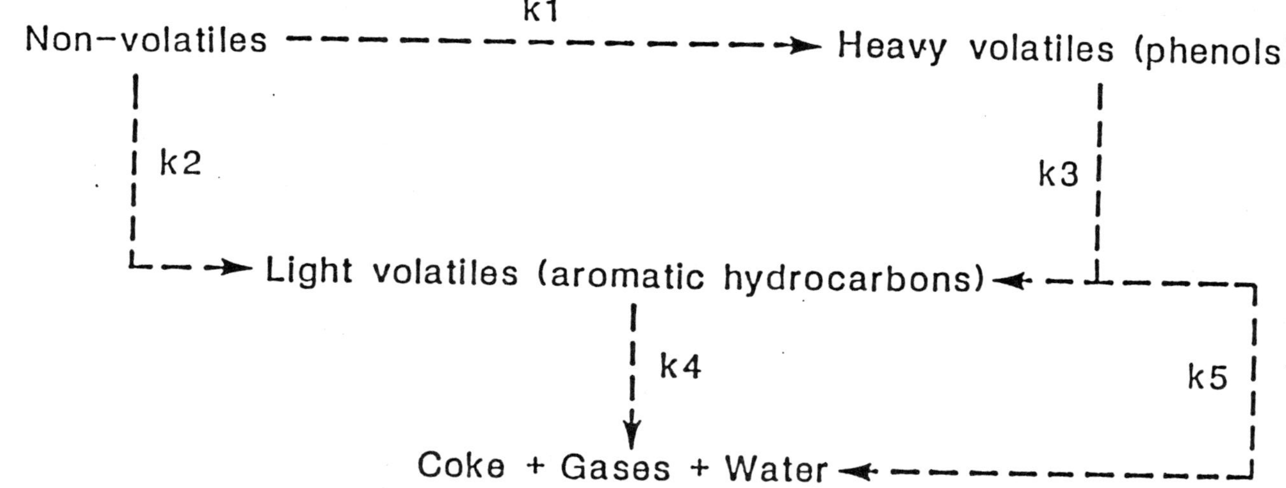

Figure 2. Postulated reaction scheme in pyrolytic oil upgrading.

Potential of Fast Pyrolysis for the Production of Chemicals

D.S. Scott, D. Radlein, J. Piskorz and P. Majerski
Dept. of Chemical Engineering, University of Waterloo

Introduction

The original objective of the fast pyrolysis process developed at the University of Waterloo (the WFPP - the Waterloo Fast Pyrolysis Process) was to obtain a maximum conversion of biomass to a liquid product which could serve as an alternative fuel oil. When this was achieved in our atmospheric pressure fluidized bed process (1980-84) with liquid yields from biomass among the highest reported for a pyrolysis process, attention was turned to determining the composition of these liquids.

Our analysis of the pyrolysis liquids revealed some surprising facts. Although the decomposition products of slow biomass pyrolysis had been identified by earlier workers, very little of this work was relevant to fast pyrolysis. The liquids from the WFPP were found to contain large amounts of a relatively few chemicals, with a number of minor components. These findings suggested that the liquids could be a profitable source of some of these chemicals, and indeed could be considered as valuable chemical feedstocks. In addition, it was clear that the reaction mechanisms of fast pyrolysis were not clearly understood.

As we continued investigation of the WFPP, it was found that some degree of control over reaction pathways and therefore product composition could be achieved not only by varying process parameters, but also by the use of pretreatment of the biomass, by the use of catalysts and by the use of additives. These findings have greatly increased the potential of the WFPP as a source of chemical feedstocks for the recovery of a number of valuable chemicals. Processes for extracting some of these compounds from the crude pyrolysis liquid have already been developed, and a number of others are the subject of current research.

Chemicals from Wood

An average composition of a WFPP pyrolysis oil prepared from poplar wood is shown in Table 1. The high proportions of hydroxyacetaldehyde, acetic acid, formic acid, acetol and glyoxal are of interest. A considerable sugar content is also present as anhydrohexoses, hexoses and anhydrodisaccharides.

The effect of pretreating the wood by removal of hemi celluloses with a mild acid hydrolysis is shown in Table 2. The major effect is a very large increase in yields of sugars with a corresponding decrease in the yield of low molecular weight carbonyls. These results led to some speculations on the mechanisms of fast pyrolysis as expressed in the Waterloo reaction model (Figure 1). The recovery of anhydrosugars from pyrolysis oils is a difficult task but progress is being made. Alternatively, these sugars can be readily hydrolyzed, and after removal of the acidic and phenolic fractions of the liquid, can be quantitatively fermented to ethanol.

If the sand normally used in the fluidized bed is replaced by a suitable catalyst, with either a hydrogen atmosphere or an inert atmosphere at normal pressure, then either a synthesis gas, a high (95%) methane gas, or a gas rich in C_4 + hydrocarbons can be obtained (Table 3).

An aromatic fraction derived primarily from biomass lignin can be readily separated from the pyrolysis liquid (pyrolytic lignin). This material is reactive and of low molecular weight for a lignin. It can readily replace up to 50% of the phenol in phenol-formaldehyde wood adhesives with good results. In addition, at normal hydrotreating conditions, this lignin can be converted in 64% yield to hydrocarbon liquids in the gasoline/diesel oil range (Table 4).

New Developments

A pretreatment of poplar wood which did not remove the hemicellulose fraction gave previously unobserved peaks in the chromatogram. On hydrolysis this liquid product yielded significant amounts of xylose.

Figures 2 and 3 present the structure of possible anhydroxyloses which might be derived from hardwood hemicelluloses, and Figure 4 presents mass spectra evidence to show that anhydroxyloses are very likely being formed. When standards can be synthesized, a positive confirmation will be made. To the best of our knowledge, this is the first report of significant yields of anhydropentoses from fast pyrolysis.

Pyrolysis of wood with different catalysts and conditions can lead to somewhat different modes of decomposition, particularly of the lignin fraction. Of particular interest is the significant amounts of phenolpropenyl compounds which can be produced. Since these are the "monomers" of wood lignins, and are highly reactive multifunctional aromatic compounds, their presence in pyrolysis liquids may present another new opportunity for chemicals recovery (Figure 5).

If feedstocks other than biomass are used with the WFPP, then additional chemical products are possible. For example, pyrolysis tests in our laboratories with waste polyethylene feed using an activated carbon catalyst yielded over 60% of volatile hydrocarbon liquids containing an appreciable amount of aromatic compounds (Table 5).

Economics

The cost of production (including profit) of pyrolysis liquids from waste wood has been estimated to be in the range of 10¢ to 15¢/per kg. Recovery of only four of the major chemical components would give a value added of 8 to 10 times the feedstock cost. Sugars for fermentation could be produced by pyrolysis, according to our preliminary estimates, for a cost no greater than enzymatic hydrolysis, and at a lower capital cost. Obvious opportunities exist to modify the pyrolysis process to allow production of many other specialty chemicals of high value.

Acknowledgements

The authors would like to acknowledge the assistance of Dr. Stefan Czernik with aspects of both experimental and analytical work. The financial support of the Alternative Energy Division of CANMET is also gratefully acknowledged. The authors would also like to thank Prof. Robert Helleur of Memorial University, Newfoundland, for his contribution in carrying out mass spectra analyses for us.

TABLE 1.

OXYCHEMICALS - POPLAR WOOD

% OF ORGANIC LIQUID		
OLIGOSACCHARIDES	1.06	
CELLOBIOSAN	1.98	
GLUCOSE	0.61	
OTHER HEXOSES	1.99	
GLYOXAL	3.31	
1,6 ANHYD. GLUCOFURANOSE	3.69	
LEVOGLUCOSAN	4.62	
HYDROXYACETALDEHYDE	15.24	
ETHYLENE GLYCOL	1.60	
ACETOL	2.13	
FORMIC ACID	4.70	} { CaMg ACETATE & FORMATE
ACETIC ACID	8.25	
METHYLGLYOXAL	0.99	
FORMALDEHYDE	1.76	
LIGNIN DERIVED AROMATICS	24.62	→ PYROLYTIC LIGNIN
	76.55	

TABLE 2.

PYROLYSIS OF RAW & TREATED POPLAR WOOD

	PP59	A-2	A-4	A-3	A-1	STAKE
	Pilot Plant	Bench Unit	Hot H_2O ext.	Very Mild Hyd.	Mild Hyd.	Steam & Cold Acid
Temperature °C	504	497	504	503	501	512
Yields of Tar Components % mf. Feed						
Oligiosaccharides		0.70	2.58	3.80	1.19	0.16
Cellobiosan		1.30	3.18	10.08	5.68	5.70
Glucose		0.40	1.00	1.67	1.89	1.21
Fructose (?)		1.31	2.35	4.00	3.89	1.19
Glyoxal		2.18	3.68	4.10	0.11	2.04
1,6 Anh. Glucofuranose		2.43	4.12	3.08	4.50	2.88
Levoglucosan	<1	3.04	5.17	15.70	30.42	23.37
Hydroxyacetaldehyde	8.86	10.03	12.61	5.35	0.37	1.45
Formic Acid		3.09	3.42	2.54	1.42	0.22
Acetic Acid	4.33	5.43	5.20	1.46	0.17	1.17
Ethylene Glycol		1.05	0.78	0.43	-	0.26
Acetol	2.93	1.40	1.20	0.06	0.06	0.36
Methylglyoxal		0.65	1.28	0.41	0.38	0.49
Formaldehyde		1.16	1.78	0.72	0.80	1.16
Aromatics (lignin)	by G.C.	16.20	-	18.00	19.00	15.44
Totals		51.50	-	71.40	69.90	57.10
% of Pyrolysis Oil		78.30	-	91.20	87.80	77.90
Sugars, % of cellulose fed (glucose eg.)		20.40	43.90	61.80	83.40	61.20

TABLE 3.

THERMOCATALYTIC CONVERSION OF WOOD TO HYDROCARBONS IN WFPP PYROLYSIS

SINGLE STEP PROCESS, H_2 ATMOSPHERE

Yields	WT. % mf Feed	Yields	WT. % mf Feed
Organic Liquid *	11.7	CO	11.5
Water	32.2	CO_2	4.5
Gas	44.2	CH_4	4.1
Char	16.4	C_2H_4	2.4
	104.5	C_2H_6	2.0
		C_3H_6	4.5
		C_3H_8	0.7
		C_4+	14.5

* Aromatic hydrocarbons plus oxychemicals

% C Fed Converted to C_2 - C_7 43.1%

% C in Fed Converted to Hydrocarbons (all types) 56%

TABLE 4.

CATALYTIC UPGRADING OF LIGNIN FROM B.C. HOG FUEL

FEED 18% H_2O, C=73.22 mf, H=5.95 mf

Amount Fed 102.1 grams

Yields, grams

Water	35.0	
Light Organics	53.7	C=88.48%; H=11.06%
Heavy Residue	1.3	
CO	Tr	
CO_2	2.67	
CH_4	2.63	
C_2H_4	Tr	
C_2H_6	0.80	
C_3H_8	0.72	
C_4+	0.76	
Total Gas	7.67	% m.f. Feed Converted to Light Organics 64%
Total Recovery	95.7%	% C in Feed Converted to Light Organics 93%

TABLE 5.

FAST PYROLYSIS OF LLD POLYETHYLENE IN A CARBON - IRON BED

Run Number		P14 5%Fe	P16 5%Fe	P17 5%Fe	P20 3%Fe	P22 1%Fe
Temperature, C °		690	600	500	605	600
Feed Particle Size, mm		0.85-1.19	0.85-1.19	0.42-0.85	0.42-0.85	0.42-0.85
Yields, % feed						
Char		5.9	14.4	31.3	13.1	11.8
Liquid		3.7	10.3	11.3	15.7	9.3
Absorbed in Filter		40.7	57.1	35.3	18.6	16.5
Gas		44.6	15.4	4.1	41.8	43.9
Gases :	H_2	1.0	3.0	2.0	1.3	1.6
	CH_4	7.6	3.3	0.8	1.9	3.1
	C_2H_4	12.7	3.6	0.4	1.9	2.6
	C_2H_6	5.4	3.5	0.7	1.6	3.1
	C_3H_6	13.0	1.6	0.2	2.5	5.4
	C_3H_8	3.7	0.4	0.0	0.9	2.7
	C_4	1.0	0.0	ND	6.2	6.6
	C_5+	0.2	0.0	ND	25.5	18.8
Total		94.9	97.2	82.0	89.2	81.5

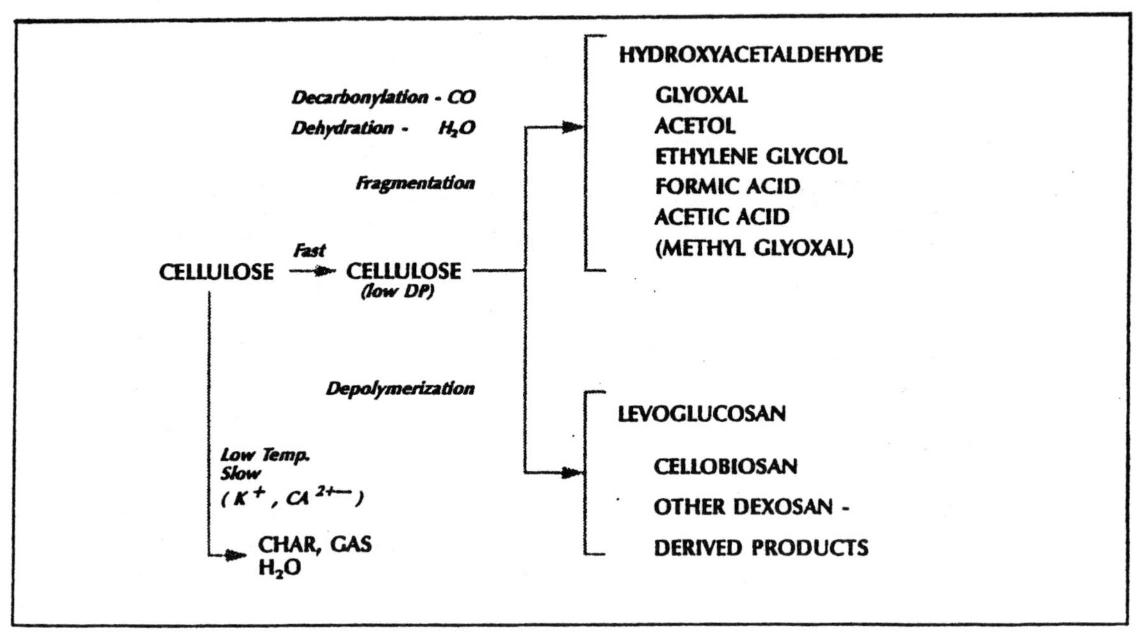

Figure 1. Proposed Decomposition Mechanism of Cellulose in Fast Pyrolysis

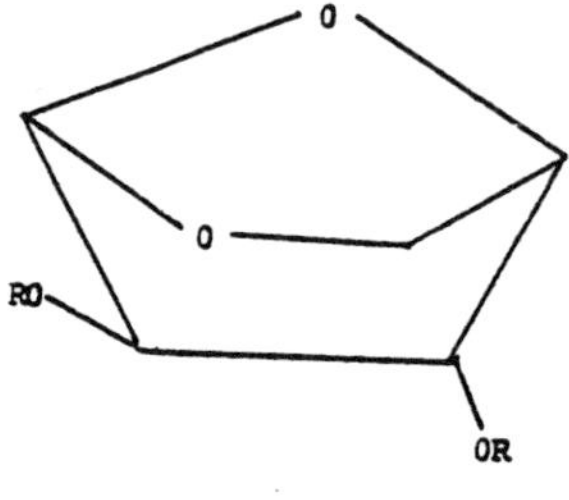

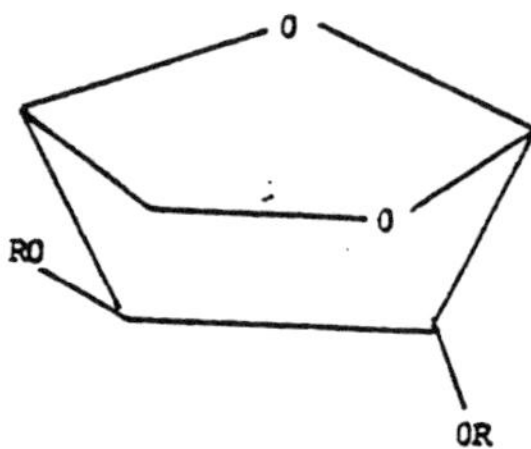

Figure 2. Isomeric Forms of 1.4-Anhydroxylofuranose

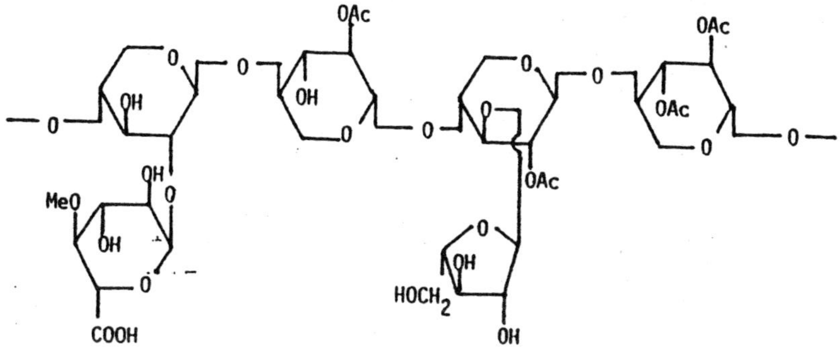

Figure 3. 0-Acetyl-(4-0-Methylglucurono) xylan

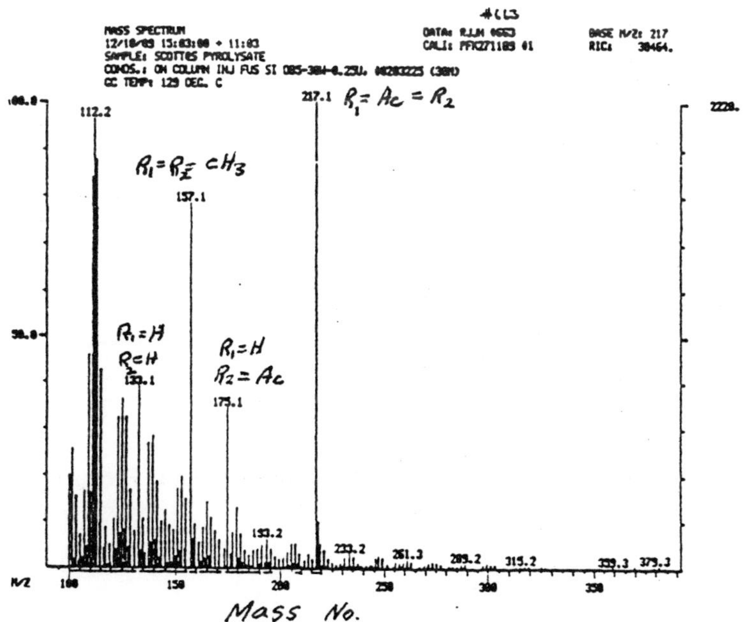

Figure 4. Chemical Ionization Spectrum of Anhydroxyloses

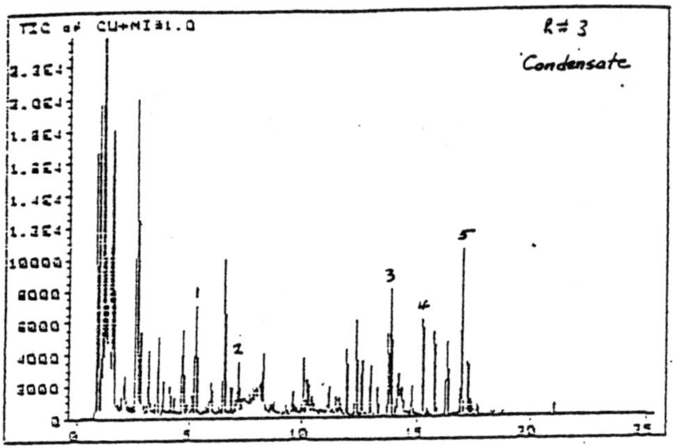

1. Cycle hexanone
2. 2-Cyclopenten-1-one, 2-hydroxy, 3-methyl
3. Phenol, 2-methoxy-4-(1-propenyl)
4. Ethanone, 1-(2-hydroxy-5-methiphenyl)
5. Phenol, 2, 6-dimethoxy-4-(2, propenyl)

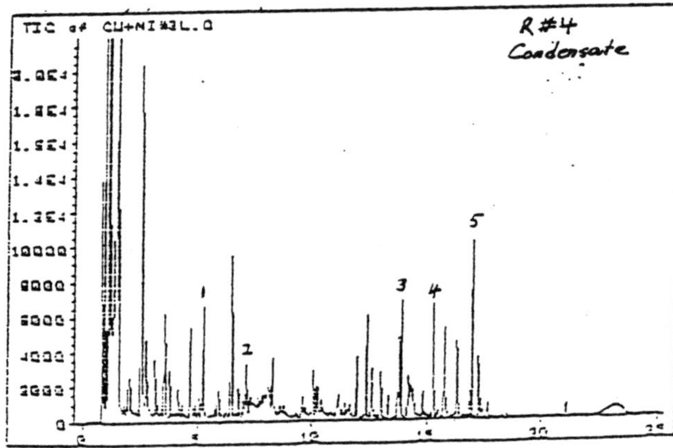

Hydropyrolysis of Poplar Wood
Modified Catalyst

Figure 5. GC-MS Chromatographs for Liquid Product (Runs 3 and 4)

Levoglucosan from Pyrolysis Oils: Isolation and Applications

C.J. Longley, J. Howard and A.E. Morrison
British Columbia Research Corporation, Vancouver, B.C.

A new method for isolating high yields of crystalline levoglucosan from pyrolysis oils has been developed. The method is superior to previously published isolation procedures, and is suitable for scale up for commercial production of levoglucosan. The non-levoglucosan components of the oil are separated as solids which are easy to handle. The levoglucosan is extracted into an inexpensive organic solvent which may subsequently be recovered and recycled in the process. One application for biomass-derived levoglucosan is in the production of biodegradable plastics. Derivatives of levoglucosan suitable for incorporation into polymers such as polystyrene have been prepared.

On a mis au point une nouvelle méthode permettant d'isoler, avec un rendement élevé, le lévoglucosane cristallin contenu dans des huiles de pyrolyse. Cette méthode est supérieure aux méthodes publiées précédemment, et on peut en modifier l'échelle pour l'appliquer à la production commerciale de lévoglucosane. Les constituants non lévoglucosaniques de l'huile sont séparés sous forme de solides que l'on peut facilement manipuler. Le lévoglucosane est extrait dans un solvant organique peu coûteux que l'on peut ensuite récupérer et réutiliser dans le procédé. La production de plastiques biodégradables constitue l'une des applications du lévoglucosane tiré de la biomasse. On a préparé des dérivés du lévoglucosane pouvant être incorporés dans des polymères comme le polystyrène.

Objectives

The objectives of the project were:

- to develop a practical method of separating levoglucosan from wood pyrolysis oils.
- to investigate end-uses for biomass-derived levoglucosan, in particular in plastics applications.

Introduction

Levoglucosan is the largest single component in acid-pretreated wood pyrolysis oils. It can represent 25-35% by weight of the oil. If a chemicals-from-biomass industry is to succeed, it is important that levoglucosan can be extracted from pyrolysis oil using an efficient process. It is also important that potential markets for levoglucosan be identified or developed.

Background

Until recently, the best available methods for extracting levoglucosan from pyrolysis oils were those patented by Weyerhaeuser (**1,2**). However, these methods require large volumes of expensive and toxic methyl iso-butyl ketone. In addition, they generate syrupy by-products which are difficult to handle and retain extraction solvent. These drawbacks, coupled with the high energy requirement of the processes, have limited their commercial utility.

Results and Discussion

Levoglucosan Extraction

The process developed at B.C. Research involves the following steps:

1) Removal of lignin and phenolic compounds from pyrolysis as solids.
2) Solvent extraction of levoglucosan.
3) Crystallisation.

The method generates solid by-products and residues, and employs an inexpensive organic solvent which could readily be recycled in the system. The process is more economical in terms of energy input than the Weyerhaeuser methods.

Excellent yields of levoglucosan have been obtained from oils provided by the University of Waterloo Flash Pyrolysis Process (WFPP):

Oil	% Wt. Levoglucosan in Oil	% Recovery Levoglucosan
WFPP #57 (Avicell)	22	77
WFPP #132 (Acid-pretreated Aspen)	31	65

A patent application concerning the isolation of levoglucosan from wood oil has been submitted to Canadian Patents and Development Ltd.

Polymer Applications

Levoglucosan is a sugar derivative and is therefore most likely to be biodegradable. A biodegradable plastic based on biomass-derived levoglucosan is an attractive concept. Allyl ether derivatives of levoglucosan which form a copolymer with styrene have been prepared. The **chemical** incorporation of levoglucosan into polystyrene is expected to give a biodegradable plastic in which extensive polymer chain degradation will occur. Once the properties of the polymer, including its biodegradability, have been tested, suitable products will be targeted, e.g., disposable polystyrene containers, agricultural mulch films.

Future

One aspect of levoglucosan which should be exploited in its applications is its inherent chirality. This could be valuable in the commercial synthesis of biologically active molecules, such as pharmaceuticals **(3)**.

The successful demonstration of levoglucosan as a chiral building block or in a chiral catalyst system could put it in high demand in a wide range of specialty chemical industries.

Conclusions

1) Levoglucosan is the largest single component in many wood pyrolysis oils. As such, it will play a key role in determining the economic feasibility of a chemical industry based on biomass.
2) Levoglucosan may be separated from pyrolysis oil using a process developed at B.C. Research. The process gives an excellent recovery of crystalline levoglucosan, and is believed to be commercially viable.
3) A priority is to develop applications for levoglucosan which will make it a valuable commodity in world markets.
4) Opportunities for levoglucosan as a biodegradable plastics component, and in the synthesis of chiral high value specialty chemicals are being examined.

Acknowledgement

We thank Energy, Mines and Resources, Renewable Energy Division for providing funding for this project. We are grateful to Mr. Ed Hogan for his administrative guidance and Dr. David Fung for his assistance as the Scientific Authority. The co-operation of Dr. Don Scott and Mr. Jan Piskorz of the University of Waterloo, in providing pyrolysis oils, is also acknowledged.

References

1. Canadian Patent 818,497, "Separating Levoglucosan and Carbohydrate Acids from Aqueous Mixtures Containing the Same By Solvent Extraction." To A.K. Esterer of Weyerhaeuser Co., Tacoma, (1969).
2. U.S. Patent 3,374,222, "Separating Levo-glucosan and Carbohydrate Acids from Aqueous Mixtures Containing the Same by Treatment with Metal Compounds." To Q.P. Peniston of Weyerhaeuser Co., Tacoma,(1965).
3. R.E. Merrill, "A Chemist's Tool Kit", CHEMTECH, Feb. 1981, 118.

Discussion:

S. Czernik (Université de Sherbrooke): Your presentation is entitled "Isolation of Levoglucosan" and the only thing I heard about the isolation was that you could recover 70% of the stuff and this methyl iso-butyl ketone is involved in the process. Would you elaborate some more about the methodology of this isolation of levoglucosan?

C. Longley (Speaker): Methyl iso-butyl ketones is not involved, actually that was in the previous inferior processes. At this stage I cannot really elaborate much more on the process because it is currently being patented, so I am not really in a position to divulge too much about it.

You could talk to David Fung or Ed Hogan - I don't know how much we should say at his stage, but I deliberately didn't want to give you all the details.

The important thing is that we can isolate 70%.

E. Chornet: Why is there a barrier for a start production of levoglucosan?

Speaker: It just doesn't give a very good yield.

Biomass Derived Alkaline Carboxylate Road Deicers

K.H. Oehr and G. Barrass
British Columbia Research Corporation, Vancouver, B.C.

Alkaline carboxylate road deicers such as calcium magnesium acetate are beginning to replace conventional chloride containing deicers since they are less damaging to concrete, metal surfaces and living systems. The purpose of this project was to develop economical methods for the recovery of these superior road deicers from aqueous biomass pyrolysis liquors. Calcium formate/acetate/propionate deicers were successfully prepared in high yield from aqueous pyrolysis liquors derived from wood. These deicers are expected to be superior to calcium magnesium acetate deicers due to their high formate content. Deicer recovery should become commercially viable if the recovery of high value industrial chemicals such as hydroxyacetaldehyde can be integrated with deicer recovery. Although technically straightforward, deicer production without byproduct chemicals recovery is uneconomical. Total pyrolysis liquor, derived from cellulose rich biomass waste is a good source of deicer and other chemicals.

Les produits de déglaçage des routes à base d'un carboxylate alcalin, comme l'acétate de calcium et de magnésium, sont en voie de remplacer les produits classiques à base de chlorure, car il sont moins dommageables pour le béton, les surfaces métalliques et les systèmes vivants. Le but de ce projet était de mettre au point des méthodes économiques permettant de récupérer ces produits de déglaçage supérieurs présents dans des liqueurs aqueuses obtenues au cours de la pyrolyse de biomasse. On a préparé avec succès et avec un rendement élevé des produits de déglaçage à base de formiate/acétate/propionate, à partir de liqueurs aqueuses obtenues par pyrolyse de bois. Ces produits seront supérieurs, prévoit-on, aux produits à base d'acétate de calcium et de magnésium, en raison de leur teneur élevée en formiate. La récupération des produits de déglaçage devrait être commercialement rentable si elle est intégrée à la récupération de produits chimiques possédant une grande valeur sur le plan industriel, comme l'hydroxyacétaldéhyde. Bien qu'il s'agisse d'un procédé techniquement simple, la fabrication de produits de déglaçage sans récupération des sous-produits n'est guère économique. La liqueur entière obtenue au cours de la pyrolyse de déchets de biomasse riches en cellulose constitue une bonne source de produits de déglaçage et d'autres produits chimiques.

Objective

The objective of this research program was to develop economical methods for the production of carboxylate deicers from aqueous biomass pyrolysis liquors.

Background

Conventional calcium and sodium chloride salt deicers corrode vehicles, bridges and other steel-reinforced concrete structures. They can also damage nearby vegetation and fresh water resources. An immediate market opportunity exists for inexpensive replacements for these deicers. Alkali metal carboxylate deicers such as calcium formate and calcium-magnesium acetate have shown good deicing properties without the metal corrosion problem **(1-4)**. The relative efficiency of deicers is shown below:

Deicer Salt	Kg Deicer Needed to Melt 100 kg Ice at -10°C
Calcium formate	12
Calcium acetate	25
Calcium propionate	25
Sodium chloride	16
Magnesium acetate	25

Calcium formate has especially good deicing properties and has been found to be non-toxic to animals **(5)**.

Results and Discussion

Aqueous wood pyrolysis liquors were obtained from the University of Waterloo Flash Pyrolysis liquors had the following % composition (weight/weight):

Component	Laval	Waterloo
Acetate as acetic acid	1.6	2.7
Formate as formic acid	0.8	0.9
Propionate as propionic acid 0.1	0.1	

Attempts were made to concentrate acetic acid in the pyrolysis liquors and their distillates by ion exchange and activated carbon absorption. However, these methods were not satisfactory: acetic acid recovery was incomplete even at high adsorbent loadings. A novel, proprietary technique was developed for producing calcium carboxylate deicers from these liquors. Crude deicers prepared by this technique from both the Laval and Waterloo aqueous pyrolysis liquors had the following % composition (weight/weight):

Component	Laval	Waterloo
Calcium	20.0	22.4
Acetate	38.4	39.4
Formate	12.6	10.4
Propionate	2.9	1.6
Water	-	6.8
Other	-	19.4
(Water + Other)	26.1	-

Total	100.0	100.0

Percent recovery of pyrolysis liquor carboxylic acids as crude deicer calcium carboxylates was high, in spite of the low acid content of the aqueous pyrolysis liquors, as shown below:

Component	Laval	Waterloo
Acetate	90	98
Formate	61	80
Propionate	78	97

Refined deicers, prepared by cleaning the crude Laval and Waterloo deicers, had the following % composition (weight/weight):

Component	Laval	Waterloo
Calcium	25.7	23.8
Acetate	45.0	52.7
Formate	19.0	13.9
Propionate	2.9	2.0
Water	-	5.7
Other	-	1.9
(Water + Other)	7.4	-
Total	100.0	100.0

Percent recovery of pyrolysis liquor carboxylic acids as refined deicer calcium carboxylates was as shown below:

Component	Laval	Waterloo
Acetate	55	82
Formate	48	67
Propionate	42	77

Conclusions

1) Economic modelling revealed that deicer production from aqueous pyrolysis liquors derived from wood would only be feasible with byproduct chemical production, since the carboxylic acid content of the aqueous pyrolysis liquor was too low.

2) The deicer recovery technology should be given serious consideration as one component of a biomass "refinery" which would optimise production of a wide range of high-value chemicals, such as hydroxyacetaldehyde, derivable from the full selection of pyrolysis liquor components.

3) Commercial viability of deicer production will depend on the availability of cellulosic material in high volume near a given location, such as low grade waste paper, the economics of byproduct chemicals recovery, the water content of the pyrolysis liquor, and the degree to which the deicer must be purified for sale.

Recommendations

1) Integrate the recovery of high value industrial chemicals, such as hydroxyacetaldehyde, with deicer salt recovery from total pyrolysis liquor, of low water content and higher acid content, derived from cellulose-rich waste biomass.

2) Once economic modelling shows that such refining of biomass pyrolysis liquors is economical, e.g., >20% return on investment cost below $50,000,000, the calcium carboxylate deicers should be tested to determine their value and characteristics as compared to CMA and other deicers.

Acknowledgements

We thank Energy, Mines and Resources, Renewable Energy Division for providing funding for this project and Mr. Ed Hogan for guidance as the Scientific Authority and in the administration area. The co-operation of Dr. Don Scott, Dr. Desmond Radlein and Mr. Jan Piskorz of the University of Waterloo and Dr. Christian Roy of the University of Laval in providing pyrolysis liquor feedstocks is appreciated.

References

1. Dunn, S.A. and R.U. Schenk, Alternate Highway Deicing Chemicals. Final Report prepared for Federal Highway Administration, Offices of Research and Development, Department of Transportation, Washington, D.C. NTIS Report PB82-242629, 1979, 157 Pages.

2. Wyatt, J. and C. Fritzsche, The Snow Battle: Salt vs. Chemicals/American City and County. April, 1989. 4 Pages.

3. Wyatt, J. (Chevron Chemical Company). Status Report on the New Highway Deicer, Calcium Magnesium Acetate (CMA). July, 1988, 6 Pages.

4. Coghlan, A., 1990. A Salt Free Diet for Ailing Roads. New Scientist. February 17, 1990, 34.

5. Malorny, Von G., Acute and Chronic Toxicity of Formic Acid and Formates. Zeitschrift fur Ernaehrungswissenschaft 1969, 9(4):332-339.

Discussion:

C. Roy (Université Laval): One reason for the treatment of the water phase from the vacuum pyrolysis process is to clean it out because this is a weight stress that has to be treated and it would make sense if there are any chemicals like that acetate and formate which are removed. Now, you haven't spoken about the quality of the final residual water after extraction of chemicals. Would it be of quality good enough to be sent or just discarded?

K. Oehr (Speaker): There is no such thing as "just discard". As far as I am concerned, anything that you don't recover is a waste, unless you can burn it. There is no other option.

C. Roy: So what you say, is that you extract valuable chemicals but there is still another treatment, a further treatment to be applied?

Speaker: There would always be a final treatment on an aqueous liquor after recovery.

C. Roy: So, your process, in a magic box, will not handle 100% of the chemicals, there is still another.

Speaker: No, we are going to go for the maximum amount necessary to generate enough revenue on the sale of the products to pay for the treatment and still get a good return on investment.

C. Roy: Can you give us figures about the residual amount of organic chemicals still remaining in the water phase?

Speaker: Okay. I think it is possible, based on total liquor, to recover in excess of 40 to 45% of the chemicals, if you are starting from a cellulose feedstock. I also believe it is possible to take something in the neighbourhood of about 15 chemicals and end up with about three commercial products, the deicer would be one of them.

The other thing I would suggest is, if you are trying to make pyrolysis liquor for chemical recovery, you should consider drying your feedstock before you process it to try and decrease the water concentration of your aqueous liquor so that you can boost your concentration of your polar organics in the liquor.

I should also mention the refined deicer. We have a sample of the water low deicer that I showed here, and I will pass it around and you can have a look at it. It is a powdered material. I suggest you just open the jar and smell it and just notice the fact that you won't be able to smell the original liquor in that sample. So, it is actually quite a clean smell; it's got a bit of an acetic acid odour to it, but it is actually very similar to the Chevron product.

Upgrading of Biomass Oils to Cetane Enhancers

D. Soveran
Petroleum Research, Saskatchewan Research Council

Since the invention of the diesel engine there have been many attempts to utilize biomass oils (in particular vegetable oils) as diesel fuel extenders. After the energy crisis of 1973, research in this area increased dramatically. Initially most of the research focused on using raw vegetable oils in diesel engines. There were successes but several problems prevented the extended use of vegetable oil as a diesel fuel. Conversion of vegetable oils using relatively unconventional conversion technology was tried, again with some success. In 1986, researchers at the Saskatchewan Research Council recognized that an unconventional technology with a biomass feedstock has limited commercialization potential. A project was started with funding from the Bioenergy Development Program to establish if conventional refining technology was suitable for vegetable oil conversion.

A patented hydroconversion process was developed in the first project. Vegetable oil was successfully converted to diesel fuel material. The product also proved to have a high cetane value.

The process is modified conventional petroleum hydroconversion technology. It shares many common elements with typical petroleum refining processes. With the proper operating conditions, most hydrotreating units in Canadian refineries could convert biomass oil today.

Two refining processes were tested in the first project, fluid catalytic cracking (FCC) and hydroconversion. The selected emphasis was to be hydroconversion because a better quality product was expected from it. The experimental results proved the initial expectation. The high yield of diesel range material and its cetane value were unexpected benefits. The FCC process was also successful but it made lower value products.

Depuis l'invention du moteur diesel, nombreuses ont été les tentatives d'utiliser des huiles tirées de la biomasse (et plus particulièrement des huiles végétales) comme additifs des carburants diesel. On a observé une hausse dramatique des activités de recherche dans ce domaine après la crise de l'énergie de 1973. Au départ, la plupart des recherches portaient sur l'utilisation d'huiles végétales brutes dans les moteurs diesel. Certaines recherches ont été fructueuses, mais plusieurs problèmes empêchaient l'utilisation à grande échelle d'huile végétale comme carburant diesel. On a tenté de convertir des huiles végétales à l'aide de techniques relativement peu classiques, et là encore certains travaux ont été fructueux. En 1986, des chercheurs au *Saskatchewan Research Council* ont reconnu que le potentiel de commercialisation d'une technologie non classique utilisant la biomasse comme charge d'alimentation était limité. On a entrepris un projet financé dans le cadre du Programme de développement de la bioénergie, en vue d'établir si les techniques classiques de raffinage pouvaient être appliquées à la conversion des huiles végétales.

Dans le cadre du premier projet, on a mis au point un procédé d'hydroconversion breveté. L'huile végétale était convertie efficacement en un produit utilisable comme carburant diesel. Ce produit possédait aussi un indice de cétane élevé.

Le procédé est une technique modifiée d'hydroconversion du pétrole. Bon nombre de ses éléments sont les mêmes que ceux des procédés typiques de raffinage du pétrole. La plupart des unités d'hydrotraitement dont disposent les raffineries canadiennes pourraient, dans les bonnes conditions, convertir aujourd'hui l'huile tirée de la biomasse.

On a évalué deux procédés de raffinage au cours du premier projet, soit le craquage catalytique fluide et l'hydroconversion. On a choisi de se concentrer sur l'hydroconversion, car ce procédé donne, du moins le prévoyait-on, un produit de meilleure qualité. Les résultats expérimentaux obtenus ont confirmé nos prévisions. Le rendement élevé en produit utilisable comme carburant diesel et l'indice de cétane de ce produit constituaient des avantages inattendus. Le craquage catalytique fluide était aussi un procédé efficace, mais il donnait des produits de qualité moindre.

This process has been patented by EMR and licensed to Arbokem Inc. For further information contact:

A. Wong
Arbokem Inc.
358 East Kent Ave. S. (Suite 101)
Vancouver, B.C., V5X 4W6
Tel: (604) 321-9331
Fax: (604) 321-3043

La dépolymérisation d'une lignine de vapocraquage

M. Heitz, J. Lapointe, E. Chornet et R.P. Overend
Dép. de génie chimique, Sciences Appliquées
Université de Sherbrooke, Sherbrooke, P.Q. J1K 2R1

L'utilisation de la lignine comme source de produits chimiques est une voie prometteuse. Cependant il est impossible présentement de définir ce que sera son rôle dans un futur proche. En effet si en théorie le chimiste peut convertir toute cette entité en n'importe quel autre produit chimique désiré, dans la pratique le problème est plus complexe. D'une part, la composition pondérale et chimique de la lignine varie non seulement avec l'espèce botanique mais aussi avec la région où pousse l'arbre. D'autre part les estimations de rentabilité sont difficiles à effectuer dans la mesure où cette filière n'est quasiment pas exploitée et où différentes avenues sont possibles.

Dans une première étape, pour obtenir de la lignine (qu'il faut le préciser, n'est pas pure mais mélangée à des cendres, des sucres et des constituants secondaires) à partir des liqueurs noires il faut concentrer la suspension (opération relativement dispendieuse).

Dans une seconde étape, la transformation de cette substance ligneuse en phénols entre en compétition avec celle issue de l'industrie pétrolière ou du charbon. Dans la dégradation du polymère en fragments on se heurte à la non compétitivité de la réaction; en effet la ramification importante et l'hétérogénéité du polymère, la solidité des liaisons de constitution font que la plupart du temps un grand nombre de produits sont formés (d'où l'intérêt d'orienter la réaction de dépolymérisation par l'utilisation de catalyseurs). Un autre fait important qu'il faut souligner est que les lignines sont avant tout des sous produits de l'industrie papetière. Celle-ci libère des résidus ligneux dont la masse représente des millions de tonne par an. Mais ces résidus sont brûlés après concentration afin de fournir de l'énergie. Il est évident que les produits chimiques issus de la lignine (substance intéressante car renouvelable) pourront devenir compétitifs face à ceux obtenus à partir du pétrole quand ce produit se raréfiera.

Il est à noter d'autre part que si une technologie adéquate permettant l'obtention d'une lignine débarassée de l'holocellulose se développait au niveau industriel, cette technique réduirait les coûts de production des phénols et mettraient ces produits en compétition avec ceux issus de l'industrie pétrolière.

Ces technologies existent. En effet nous savons que les voies papetières traditionnelle sont aujourd'hui concurrencées par les voies thermomécaniques où la défibration nécessaire à la mise en pâte s'effectue à l'aide de forces mécaniques intenses. L'objectif principal étant de désagréger l'ultrastructure de la matière lignocellulosique pour pouvoir ensuite fractionner la biomasse en ses principales unités.

Il existe trois principaux traitements thermomécaniques: en phase vapeur, en phase aqueuse, et en phase organique. En opérant sous des conditions opératoires précises, il s'avère possible de solubiliser les hémicelluloses et la lignine, la cellulose restant comme résidu. La séparation subséquente des hémicellulose et de la lignine peut s'accomplir par les techniques du génie chimique.

Les objectifs de nos travaux ont été :

1. la production de nouvelles lignines de vapocraquage représentatives d'un procédé industriel,
2. la dépolymérisation alcaline de ces lignines en phénols.

Matières Premières

La matière première utilisée au cours de cette étude est le *Populus tremuloides* provenant des Cantons de l'Est. Le bois à été écorcé puis mis en copeaux (2.0 cm * 2.0 cm * 2.0 cm).

La composition du bois déterminée par les méthodes ASTM est la suivante:

- extrait à l'alcool toluène	3.6 %
- hémicelluloses	24.2 %
- cellulose	51.6 %
- lignine	20.6 %
- cendres	0.4 %

L'humidité des copeaux est comprise entre 45 et 55 %.

Production d'une lignine de vapocraquage

Le fractionnement du bois à été accompli par traitement thermomécanique en phase vapeur. Cette technologie de vapocraquage est un traitement à la chaleur de courte durée effectué en présence de vapeur d'eau suivi d'une décompression flash (décompression soudaine à travers une vanne).

L'appareil consiste en un digesteur conventionnel alimenté en continu (c'est-à-dire dont le chargement, le déchargement et le fonctionnement ont lieu simultanément) par un système coaxial formé d'une vis d'alimentation et d'un piston. Le temps de résidence de la biomasse dans le digesteur est contrôlé par le jeu de la vis d'alimentation et celui de la vis de transfert dans le digesteur ainsi que par la séquence d'ouverture et de fermeture de la vanne de sortie. Avec un canal d'alimentation de 25.4 cm, le système STAKE peut traiter environ 100 t de biomasse par jour. (figure 1)

Le prétraitement du bois par la vapeur est suivi de deux lavages à l'eau dans le but de solubiliser les hémicelluloses ($T=70^oC$, $t\approx30$ min)

Le résidu issu de ces lavages est ensuite extrait par une solution alcaline (NaOH)($T=100^oC$, $t\approx30$ min, $pH\approx13$) afin de solubiliser la lignine, laquelle est ensuite précipitée par une solution acide (H_2SO_4)($T=80^oC$, $t\approx10$ min, $pH\approx2$)puis récupérée.

Le résidu solide résultant est la cellulose.

Le rendement en lignine est de l'ordre de 80 %.

L'analyse centésimale de la lignine (C=64.0 %, H=6.5 %, N=0.15%) indique que cette lignine est relativement pure.

Le mode opératoire est présenté sur la figure 2.

Dépolymérisation d'une lignine de vapocraquage

La dépolymérisation de la lignine à été effectuée dans un autoclave "batch" de 500 ml muni d'un agitateur magnétique. On utilise 5 g de lignine, 98 ml d'eau et 2 g de soude. Le réacteur est pressurisé avec de l'azote à une pression de 300 psi et la vitesse de l'agitation est fixé à 500 rpm. Le réacteur est plongé dans un bain de sel préchauffé, et la température et le temps de la réaction sont enregistrés par micro ordinateur. Après le traitement, le réacteur est trempé dans un bain d'eau puis ramené à pression atmosphérique. Aucune analyse du gaz n'a été effectuée jusqu'à date car la quantité de gaz produite est négligeable (de l'ordre de 3 %). La séparation des produits est présentée sur les figures 3a et 3b. L'analyse par chromatographie en phase gazeuse des phénols préalablement acétylés, a été réalisé sur un chromatographe Hewlett-Packard équipé d'un détecteur à ionisation de flamme et d'une colonne capillaire de DB-1.

Taux de conversion de la lignine

Le taux de conversion et défini par le ratio :

$$\frac{[\text{masse de lignine originelle - masse de lignine résiduelle}]}{[\text{masse de lignine originelle}]}$$

qui est une fonction complexe dépendant du temps et de la température de la réaction c'est-à-dire du facteur Ro défini selon l'équation ci-dessous :

$$\log Ro = \int \exp\left(\frac{T-100}{14.75}\right) dt$$

t en min.
T en °C

La figure 4 montre l'évolution de la conversion de la lignine versus la sévérité du traitement. La conversion augmente linéairement avec la sévérité du traitement : elle atteint 60% à une sévérité telle que log (Ro/min) ≈ 8, c'est-à-dire $T=350^oC$ et $t\approx10$ min.

Le solide résiduel issu de la dépolymérisation a été caractérisé par analyse élémentaire. La figure 5 présente la variation du rapport C/O versus la sévérité du traitement. Nous observons deux domaines: un domaine à basse sévérité $2 \leq (\log Ro) \leq 5$ où le rapport C/O augmente graduellement et un domaine à haute sévérité $6 \leq (\log Ro) \leq 8$ où le rapport C/O croît fortement avec la sévérité, ce qui indique une augmentation du degré de charrification.

Obtention du guaiacol, du syringol et des catecholes

La figure 6 montre la variation de la fraction soluble à l'éther ainsi que la fraction des phénols identifiables par chromatographie versus le facteur de sévérité.

La quantité maximale de substances solubles à l'éther (20 %) a été obtenue pour une sévérité telle que log (Ro/min) ≈ 8.

Dans toutes les gammes de sévérité, on n'a pas réussi à identifier par chromatographie en phase gazeuse plus des 3/4 de la fraction soluble à l'éther.

a) Traitement à basse sévérité. (figure 7)

A basse sévérité, c'est-à-dire telle que $2.3 \leq \log Ro \leq 7$, les produits majoritaires sont le syringol et le guaiacol. Pour un log Ro ≈ 5, on obtient 5% de syringol et de guaiacol.

b) Traitement à haute sévérité. (figure 8)

A haute sévérité, c'est-à-dire $7 \leq \log Ro \leq 8.2$, les produits majoritaires sont les catechols (catéchol, 4-méthylcatéchol, 4-éthylcatéchol). Ils représentent 9% pour une sévérité telle que log Ro ≈ 8.

Influence de l'imprégnation de la lignine par la soude sur le rendement en phénols

On a noté qu'une imprégnation de la lignine par la soude pendant 3 jours permet d'augmenter les rendements en phénols identifiables de 30 % ainsi que les rendements en catéchols de 26 % par rapport à une imprégnation de 5 minutes. (tableau 1)

Influence de l'origine de la lignine sur le rendement en phénols

Nous avons aussi dépolymérisé une lignine provenant d'un bois résineux (épinette noire) obtenu par traitement thermomécanique en phase vapeur ainsi qu'une lignine d'origine papetière (Indulin AT) provenant d'un bois résineux (pin).

Aucune différence significative est observable entre les diverses lignines (*Populus tremuloides*, épinette noire et Indulin AT). Le comportement lors de la dépolymérisation de ces trois lignines s'avère similaire bien qu'une soit issue d'un bois angiosperme (*Populus tremuloides*) et que les deux autres proviennent de bois gymnospermes (épinette noire et pin). (tableau 2)

Influence de la concentration de soude sur le rendement en phénols

Seule la concentration en soude varie, toutes les autres conditions opératoires restent constantes (log Ro ≈ 7.8). Les résultats sont présentés dans le tableau 3. La fraction éthérée ainsi que les rendements en phénols identifiables augmentent jusqu'à une concentration de soude de 2 % puis diminuent avec l'accroissement de la concentration de soude. Le maximum de catechols est obtenu pour une concentration en soude voisine de 2 %. Le maximum de syringol et guaiacol est obtenu pour une concentration en soude voisine de 0.3 %.

Donc, pour une sévérité proche de 7.8, afin d'obtenir un bon rendement en catechols, il faut utiliser une concentration de soude proche de 2 %.

Si au contraire on veut favoriser la production de syringol et de guaiacol, il faut travailler à une concentration de soude voisine de 0.3 %.

Conclusion

Cette étude nous a montré que l'obtention de lignines de vapocraquage à grande échelle est possible. Le rendement est de l'ordre de 80 %. La pureté de ces lignines est parfaite.

La dépolymérisation de ces lignines a été effectuée à l'échelle laboratoire en présence de soude.

A basse sévérité, c'est-à-dire 2 ≤ log Ro ≤ 7 on produit majoritairement du guaiacol et du syringol. (On obtient 5 % de ces deux produits à une sévérité telle que log Ro ≤ 5 et pour une concentration voisine de 2 %).

A haute sévérité c'est-à-dire 7 ≤ log Ro ≤ 8.2 c'est le catéchol qui s'avère le produit prépondérant. On obtient 9 % de ce produit à une sévérité telle que log Ro ≈ 7.8 et pour une concentration voisine de 2 %.

Certes les rendements obtenus sont faibles mais cette étude de base sera suivie d'une recherche portant sur la dépolymérisation de lignines en présence de catalyseurs dans des réacteurs plus spécifiques ce qui devrait conduire à de meilleurs rendements.

Précisons que le catechol ou le guaiacol peuvent être les produits de base de toute une industrie et en particulier de celle de la vanilline et de ses dérivés.

Remerciements

- Le groupe de recherche sur les technologies et procédés de conversion du département de génie chimique de l'Université de Sherbrooke.
- Centre Québecois de vaolorisation de la biomasse, CQVB.
- Fonds de chercheurs et actions de recherche, FCAR
- Conseil de recherche en Sciences Naturelles et en Génie, CRSNG.

Discussion:

D. Boocock (University of Toronto): Penses-tu que le guaiacol est converti à catechol?

M. Heitz (Conférencier): Oui, le guaiacol peut être converti en catechol, dépendant de la température et des conditions opératoires. Il y a une transformation possible. On l'a vu en particulier à une température de 330 degrés. Dépendant de la concentration de soude, on obtient soit une grande quantité de catechol, soit une grande quantité de guaiacol et de syringol.

Influence de l'imprégnation de la lignine par la soude sur le rendement en phénols

	conversion (%)	phénols identifiables (%)	catéchols (%)
- lignine (P.T.) imprégnation 5 min.	16.8	10.0	6.4
- lignine (P.T.) imprégnation 3 jours	15.6	13.4	8.7

log (Ro/min) = 7.8

T = 330 °C
t = 10 min

concentration de soude = 1%

N.B. tous les pourcentages sont exprimés par rapport à 100 g de lignine originelle.

Tableau 1

Influence de l'origine de la lignine sur le rendement en phénols

	conversion (%)	phénols identifiables (%)	catéchols (%)
- lignine (Populus tremuloides)	16.8	10	6.4
- lignine (Epinette noire)	14.5	9.0	5.3
- lignine papetière (Indulin AT)	13.4	8.7	5.6

log (Ro/min) = 7.8

T=330 °C
t=10 min

concentration de soude = 1%

N.B. tous les pourcentages sont exprimés par rapport à 100 g de lignine originelle.

Tableau 2

Influence de la concentration de soude sur le rendement en phénols

concentration de soude (%)	fraction éthérée (%)	phénols identifiables (%)	catéchols (%)	syringol + guaiacol (%)
0.3	15.0	9.1	2.4	5.0
1.0	16.8	10.0	6.4	1.5
2.0	20.6	10.1	8.8	0.4
7.0	17.9	7.5	4.1	0.0

log (Ro/min)= 7.8

T = 330 °C
t = 10 min

N.B. tous les pourcentages sont exprimés par rapport à 100 g de lignine originelle.

Tableau 3

MACHINE STAKE II

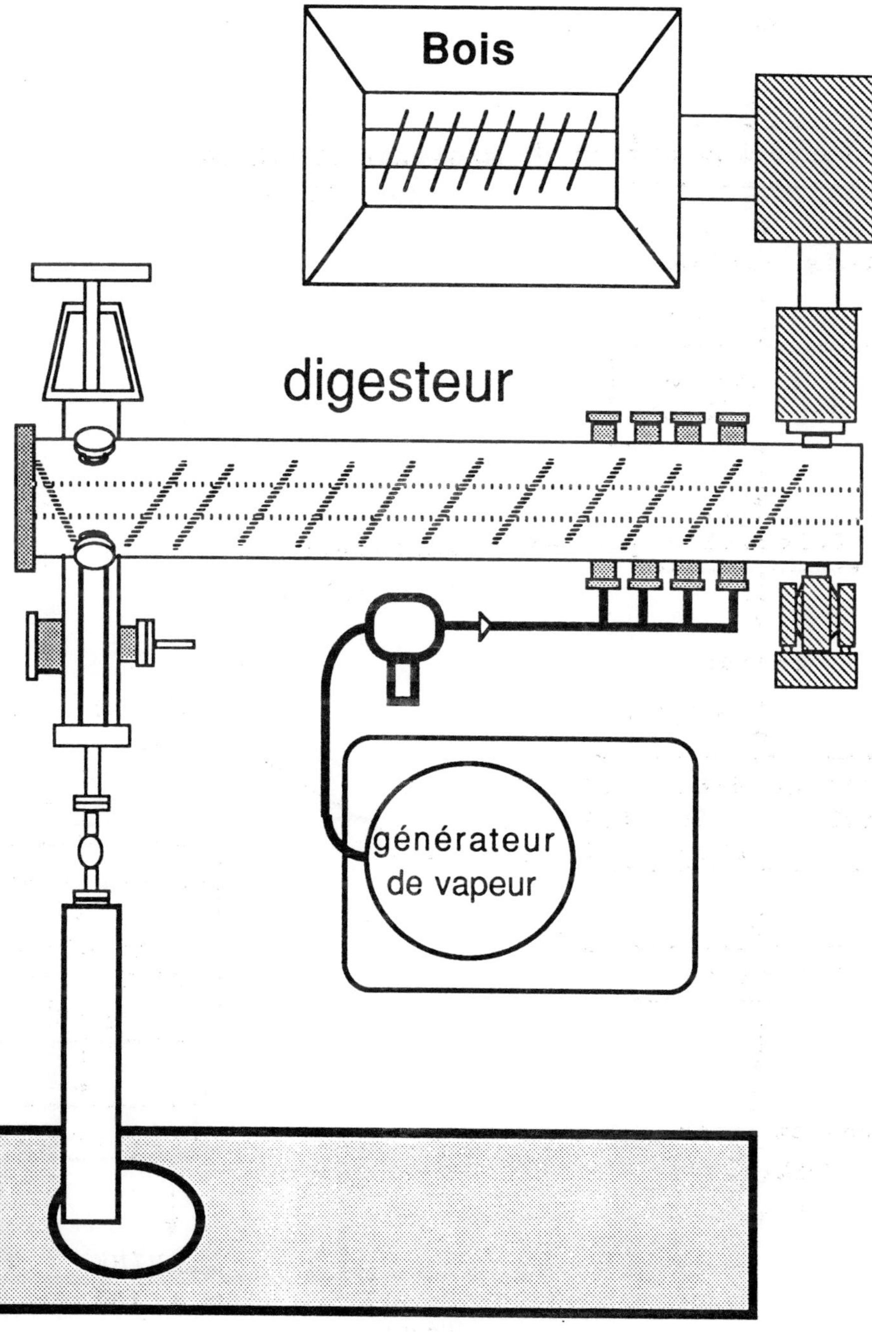

figure 1

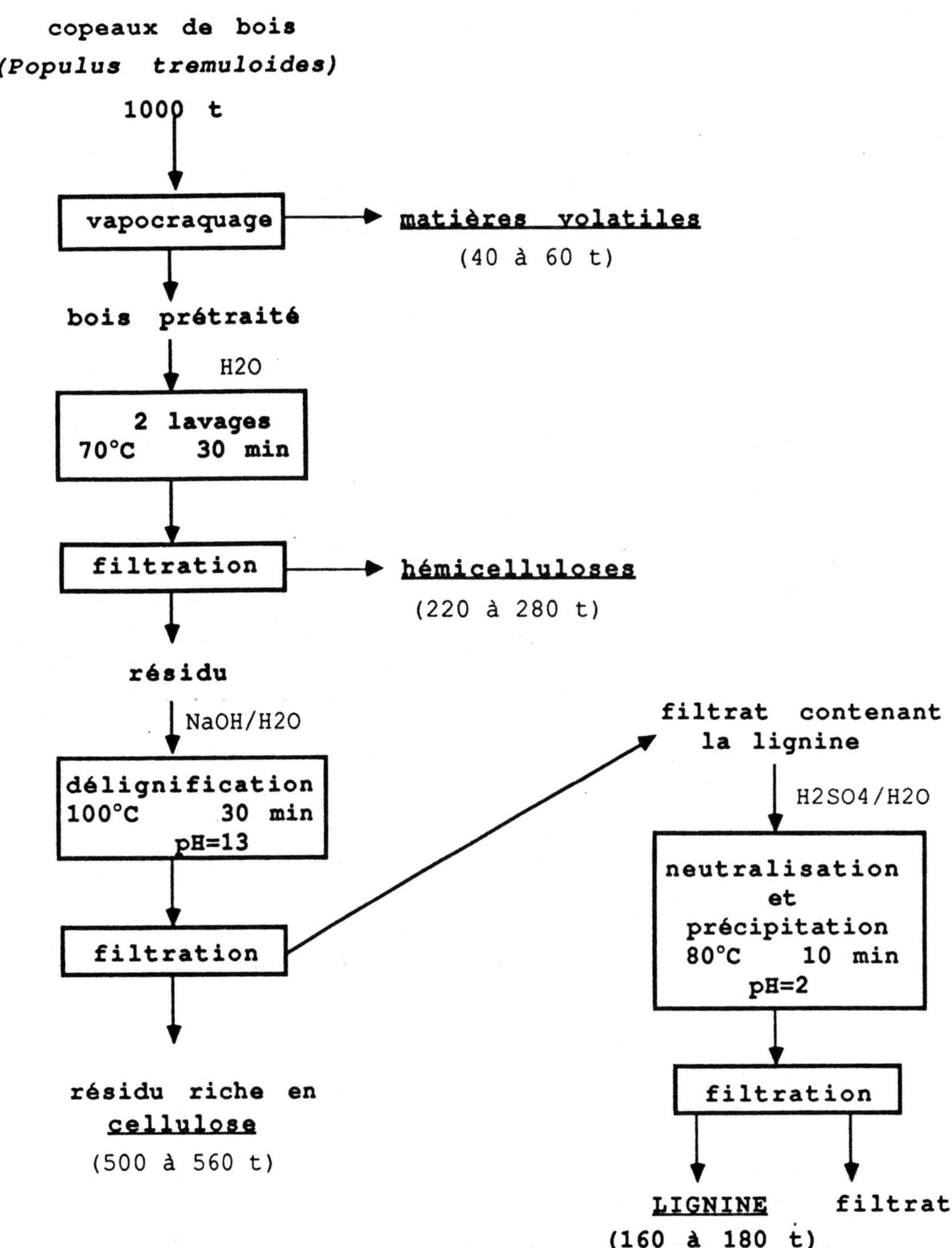
FRACTIONNEMENT PAR VAPOCRAQUAGE
copeaux de bois
(Populus tremuloides)
1000 t
vapocraquage
matières volatiles
(40 à 60 t)
bois prétraité
H2O
2 lavages
70°C 30 min
filtration
hémicelluloses
(220 à 280 t)
résidu
NaOH/H2O
délignification
100°C 30 min
pH=13
filtration
résidu riche en
cellulose
(500 à 560 t)
filtrat contenant
la lignine
H2SO4/H2O
neutralisation
et
précipitation
80°C 10 min
pH=2
filtration
LIGNINE
(160 à 180 t)
filtrat

Figure 2

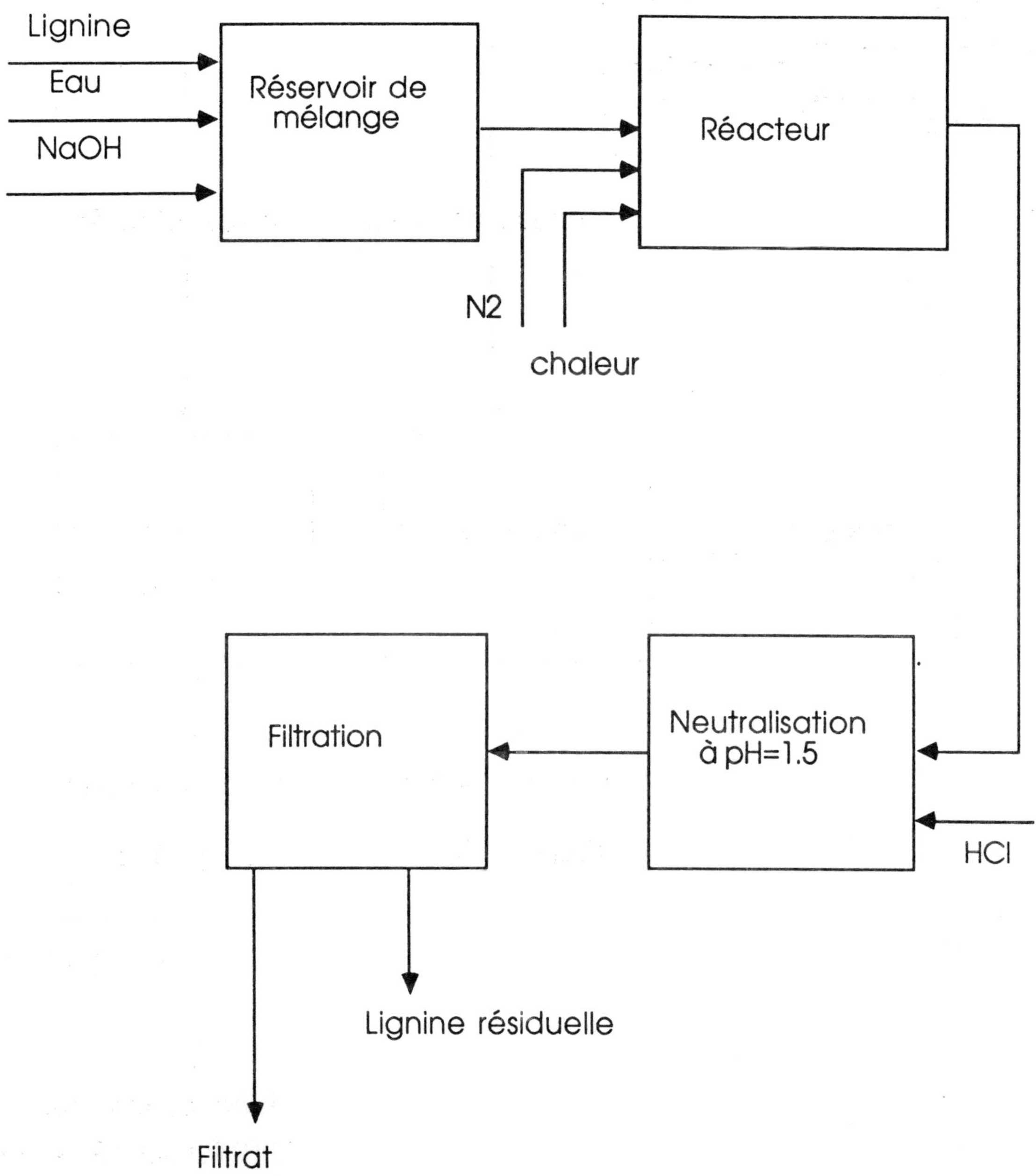
Lignine
Eau
NaOH
Réservoir de mélange
Réacteur
N2
chaleur
Filtration
Neutralisation à pH=1.5
HCl
Lignine résiduelle
Filtrat

Figure 3a

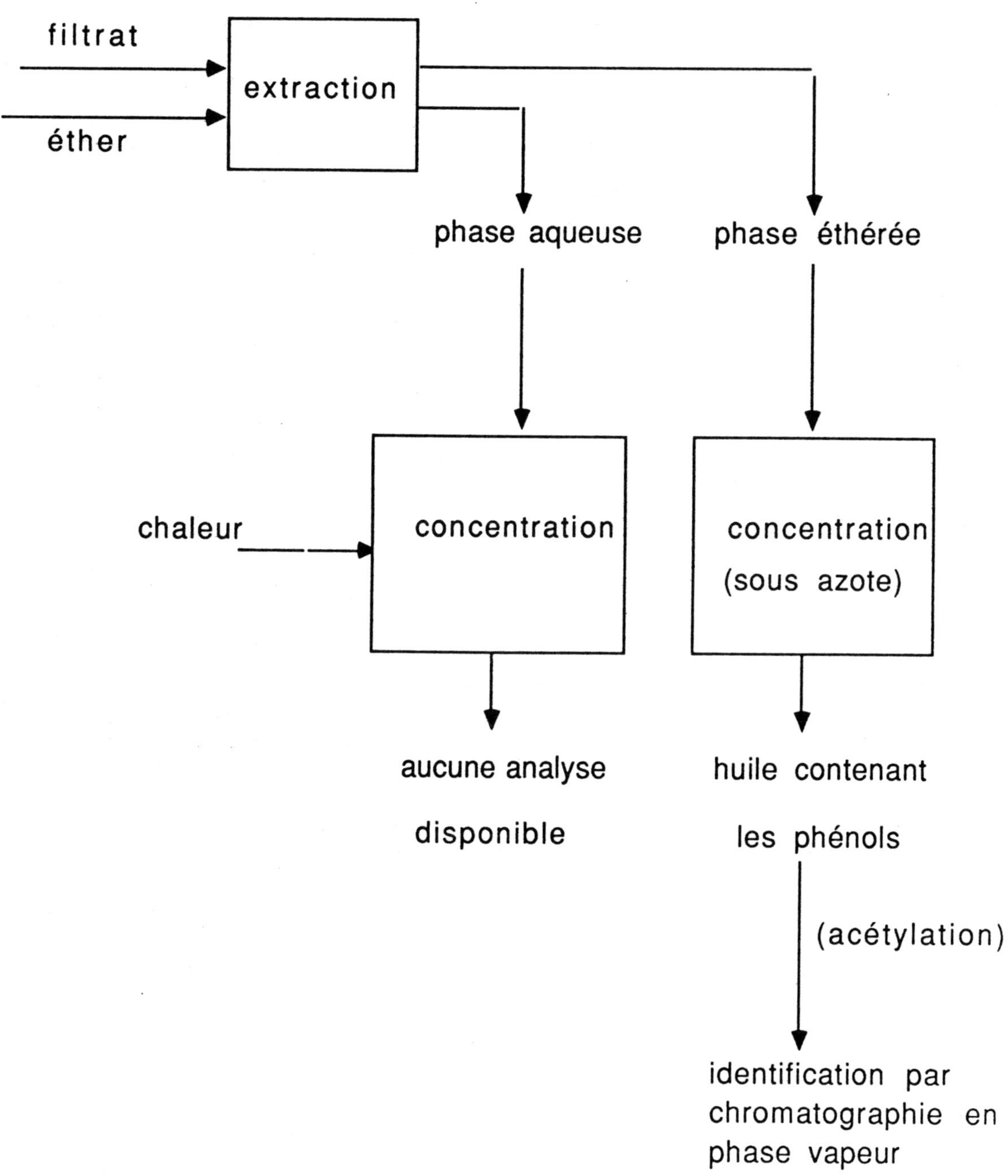
filtrat
extraction
éther
phase aqueuse
phase éthérée
chaleur
concentration
concentration
(sous azote)
aucune analyse
disponible
huile contenant
les phénols
(acétylation)
identification par
chromatographie en
phase vapeur

Figure 3b

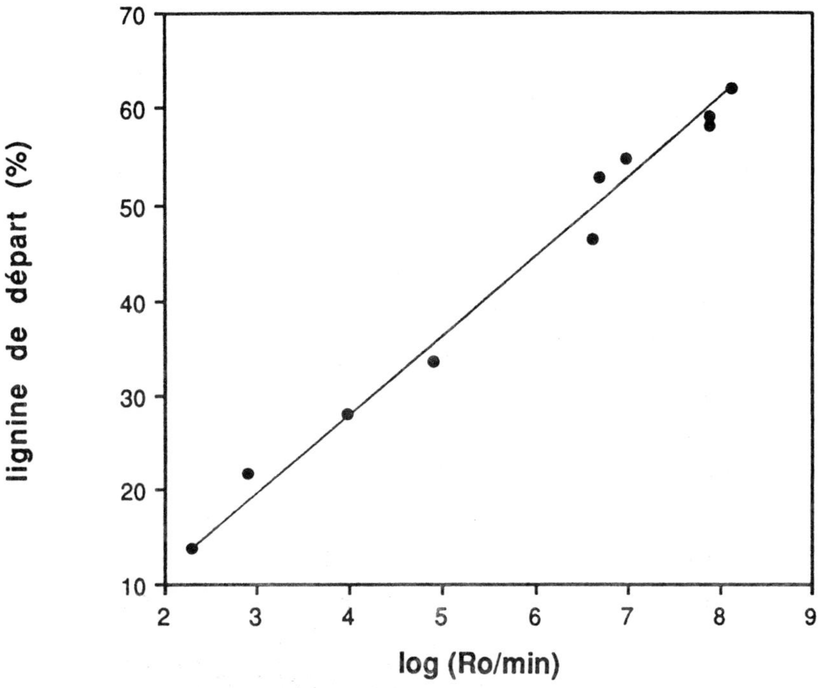

conversion versus sévérité (NaOH = 2%, t = 10 min)

Figure 4

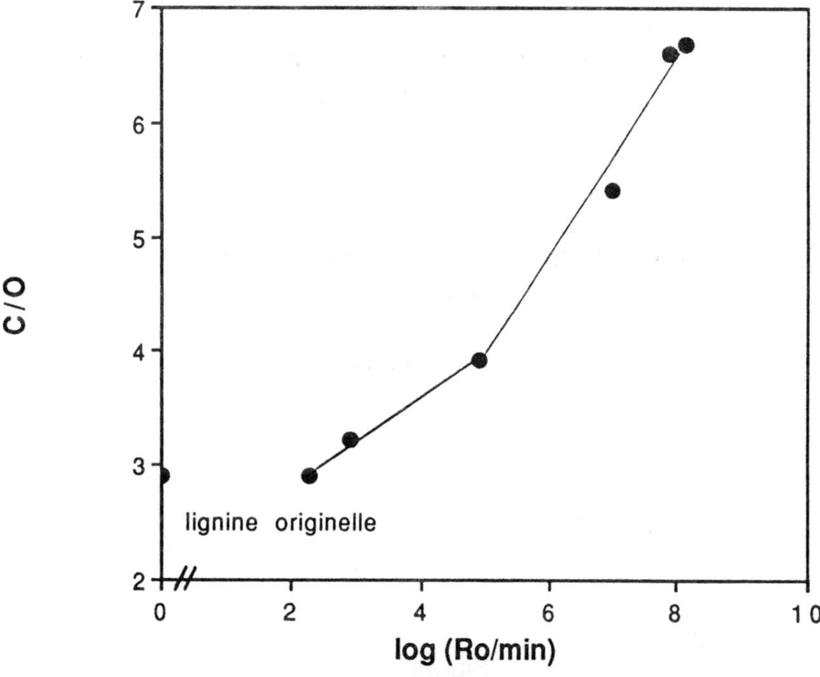

C/O versus sévérité (NaOH = 2%, t = 10 min)

Figure 5

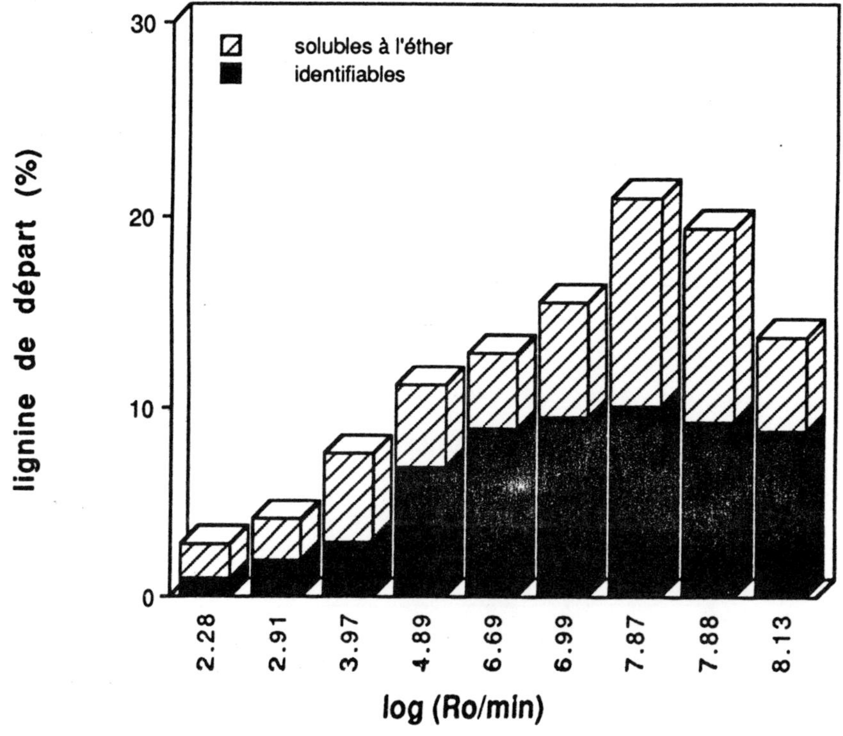

fraction éthérée et phénols identifiables versus sévérité

Figure 6

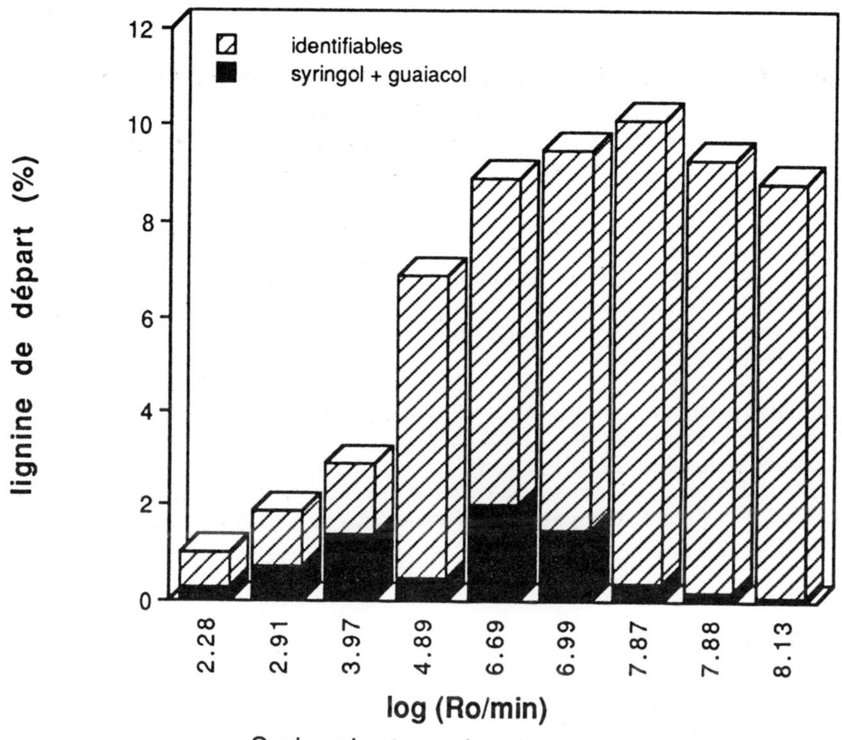

Syringol et guaiacol versus sévérité

Figure 7

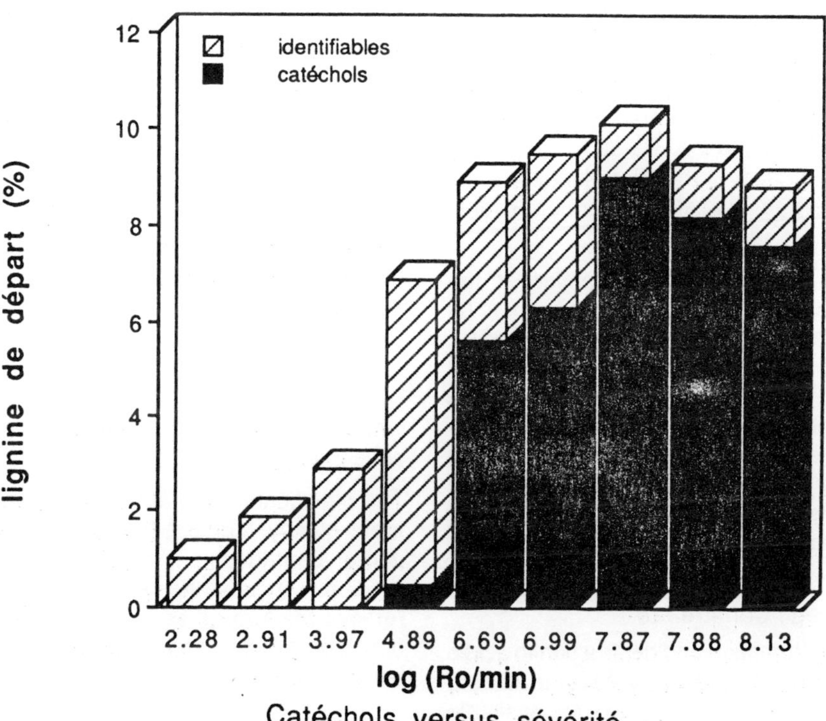

Catéchols versus sévérité

Figure 8

The Conversion of Pentoses into Furfural Via a Novel Jet Reactor

C. Dubois, N. Abatzoglou, R.P. Overend and E. Chornet
Département de génie chimique, Université de Sherbrooke
Sherbrooke, Qué, Canada, J1K 2R1

A novel approach for the acid catalyzed conversion of pentoses into furfural is described. The strategy followed focuses on the rapid removal of the furfural from the liquid reaction medium. A high interfacial area two-phase reactor has thus been designed and operated. The residence time of the dispersed liquid phase is adjusted so that the conversion of the pentoses into furfural at high temperatures (~ 250°C) is followed by the transfer of the furfural into the continuous gas-vapour phase thus minimizing the furfural degradation. Selectivities close to 100% furfural have been thus obtained at pentose conversions higher than 80%.

Cette présentation décrit une nouvelle méthode permettant de convertir le pentose en furfural à l'aide d'un catalyseur acide. La stratégie adoptée consiste à extraire rapidement le furfural du liquid en réaction. L'on a conçu pour ce faire un réacteur deux phases à haute interface. La durée de la phase liquide est ajustée pour que la conversion des pentoses en furfural à haute température (plus ou moins 250 °C) soit suivie par le transfert du furfural en phase gazeuse pour minimiser la dégradation du furfural. Des processus donnant plus de furfural ont ainsi été obtenus avec un taux de conversion de plus de 80 %.

Introduction

In previous works we have proposed that the liquefaction of lignocellulosics is best conducted following a fractional approach in which the hemicelluloses, lignin and cellulose are first separated (i.e., fractionated) and each of the fractions is subsequently converted to products of interest, be primary, intermediary or energy-related chemicals as well as fibers. Such a strategy is compatible with thermochemical and enzymatic approaches and opens the way for the biomass refinery of the future.

In a recent paper (Heitz et al., 1991) we have demonstrated how the fractionation strategy can be applied to a prototype hardwood, *Populus tremuloides*. Recoveries of 65-70% pentoses, 75-80% lignin and 100% cellulose are obtained, at the pilot plant scale, by simple aqueous/steam treatment at a convenient severity (220-225°C, 2 min.).

The possible use of the pentoses-rich fraction as the feedstock for the conversion to furfural had prompted us to study how this fraction could be treated to maximize furfural yields. Since the hemicellulose (pentose-rich) fraction is recovered as an aqueous phase following water extraction of the treated fiber, the conversion pentoses -> furfural had to be carried out as a homogeneous reaction in the liquid phase. This led us to the analysis of the kinetics and to the development of a novel reactor concept for the reaction.

Kinetic Constraints

The kinetics of furfural formation and destruction in acidic aqueous media have been studied by several authors through the past decade. Thorough reviews are available (Kwarteng, 1983; Sproull, 1986). The latter author had proposed the following model as representative of the reacting system:

$$\text{Pentosans} \xrightarrow{\text{fast}} \text{Pentoses} \xrightarrow{k_1} \text{Furfural} \xrightarrow{k_2} \text{Degradation products}$$

$$\text{Pentoses} \xrightarrow{k_4} \text{Condensation products} \qquad \text{Furfural} \xrightarrow{k_3} \text{Condensation products}$$

Our own work (Abatzoglou et al., 1990) has provided evidence that an alternate model also fits well the data:

$$\text{Pentosans} \xrightarrow{k_0} \text{Pentoses} \xrightarrow{k_1} \text{Furfural} \xrightarrow{k_2} \text{Degradation products}$$

$$\text{Furfural} \xrightarrow{k_3^1} \text{Condensation products}$$

First order reactions, based on the reacting species, are considered in all models. The specific rate constant assumes the following formalism:

$$k_i = k_{io}\, c_A^{ni} \exp(-E_i / RT)$$

where n_i is to be found experimentally, and C_A is the acid concentration.

The significant point about kinetic models is that all of them clearly suggest that once furfural is formed, its partial degradation is inevitable in the liquid reaction medium. An advantageous strategy would thus be to remove the furfural as soon as it is formed. By so doing one would block, indirectly, the degradation reactions which would then not take place.

Reactor Concept

Our strategy was to use a two-phase reaction system (phase 1 = aqueous medium; phase 2 = carrier gas +

vapour) and create a high interfacial area to facilitate rapid mass transfer of the formed furfural from the liquid phase to the gas + vapour phase.

The engineering aspects of the novel reactor are described in a thesis by Dubois (1991). It consists in adequately mixing an acidic aqueous solution containing the pentoses with a carrier gas (N_2 was used); heating rapidly the two-phase mixture to a prescribed reaction temperature; creating a very high interfacial area by means of a high shear device (the jet-reactor); letting the reaction proceed in the droplets (dispersed phase) while the formed furfural is rapidly transferred to the gaseous continuous phase where no degradation occurs; condensing the vapours present in the gaseous phase and recovering the furfural as an aqueous condensate ready to undergo distillation.

Experimental

Three series of experiments using pure xylose solutions up to 5% sugar concentration were carried out. One additional series of experiments was performed using hemicellulose liquours (pentose-rich) derived from aqueous/steam treatment of *Populus tremuloides* at 220°C for 2 min in a STAKE II digestor.

Quantitative analysis of the input and output products to the reactor was carried out using HPLC following established procedures in our laboratory.

The experimental conditions were as follows: (i) liquid flow rates = 0.2 ι/min; gas (N_2) flow rates = 3-8 ι/ STPmin; condensing steam was used to raise the temperature of the gas/liquid mixture within the range 200-262°C; a single fluid nozzle (proprietory design) was used as a high shear device to create a high interfacial area between liquid and gas by discharging the two-phase mixture into a 1 ι chamber (at reaction T) where the furfural is rapidly transferred from the liquid droplets to gas/vapour phase. Cooling the two-phase mixture is carried out by means of a tubular heat exchanger placed at the exit of the reservoir. The time at which the two-phase mixture is exposed at high temperatures is of a few seconds. This is effectively the time available for the homogeneous reactions to proceed. This reaction time can be varied by a proper choice of liquid and gas flow rates as well as the chamber geometry and nozzle design.

Results

The data obtained using pure xylose solutions (3% w/w) and H_2SO_4 (3% w/w) as catalyst are shown in the following table:

Reaction Temperature (°C)	Furfural yield (as % of potential)	Furfural selectivity (as % all products)
220	24.7	98
240	43.4	93
250	69.2	93
256	73.8	92
262	85.7	99

A few experiments have been conducted so far with real hydrolyzates in the 200-230°C temperature range. The composition (%wt) of the dissolved solids present in the hydrolyzate derived from *Populus tremuloides* (see experimental section for the pretreatment conditions) is as follows: xylose, 80.0%; mannose, 9.5%; glucose, 6.8%; furfural, 3.6%; arabinose, traces. The results have shown furfural yields (expressed as % of potential) of 35% at 224°C and 51% at 230°C. Further optimization is still needed with the real hydrolyzates to determine the most favourable combination of pretreatment and pentoses cyclization conditions.

The furfural yields obtained are clearly above any reported values. The achievement of high selectivities even at the high conversions attained with pure xylose is a direct consequence of the inhibition of furfural degradation reactions in our two-phase flow reactor.

Conclusions

We have demonstrated that using an improved reactor configuration, selectivities above 90% can be obtained in the conversion of pentoses (xylose) to furfural even at > 80% conversion levels.

The reactor configuration chosen is a two-phase flow system in which a large interfacial area is obtained by means of a high shear device.

The preliminary results obtained with pure hydrolyzates from *Populus tremuloides* show the same favourable tendencies. More work is however needed to optimize the coupling between pretreatment and cyclization.

This work opens the way for an improved technology in the utilization of hemicelluloses (pentoses-rich) to produce furfural as a by-product within an overall refining strategy for biomass and bioenergy.

Acknowledgments

The authors are indebted to the CQVB, FCAR, NSERC and EMR for financial support in the different phases of the biomass/bioenergy program.

The technical contribution of J.P. Lemonnier is acknowledged.

References

1. Heitz, M., Capek-Ménard, E., Koeberle, P.G., Gagné, J., Chornet, E., Overend, R.P., Taylor, J.D., Yu, E., Fractionation of *Populus tremuloides* at the Pilot Plant Scale: Optimization of Steam Pretreatment Conditions Using the STAKE II Technology. Bioresource Technology, (1991), 35, 23-32.

2. Kwarteng, I.K. Kinetics of Acid Hydrolysis of Hardwood in a Plug Flow Reactor, Thayer School of Engineering. Dartmouth College, N.H. 03755, U.S.A.; Ph.D. Thesis, (1983).

3. Sproull, R.D., The production of furfural in an extraction coupled reaction system. Purdue University, Lafayette, Ind., U.S.A., Ph.D. Thesis, (1986).

4. Abatzoglou, N., Koeberle, P., Chornet, E., Overend, R.P. and Koukios, E.G., Dilute Acid Hydrolysis of Lignocellulosics: an Application to Medium Consistency Suspensions of Hardwood using a Plug Flow Reactor, The Can. J. Chem. Eng., (1990), 68, 627-638.

5. Dubois, C., Selectivité des réactions chimiques en série: étude de la cyclisation catalytique des pentoses en milieu bi-phasique. Université de Sherbrooke, Sherbrooke, Qué., Canada; M.Sc.A. Thesis, (1991).

Costs of Biomass Pyrolysis Derived Liquid Fuels

A.V. Bridgwater
Energy Research Group, Chemical Engineering and Applied Chemistry Department,
Aston University, Aston Triangle, Birmingham B4 7ET UK

An analysis of capital and operating costs is presented for fuels that may be produced by pyrolysis processes, from which production costs have been estimated for several liquid fuels. The results suggest that the current lack of interest is not just due to adverse economic viability, but also the status of the technology, concerns over environmental problems and utilisation of the product. As with gasification, lack of long term operation, operating experience and perceived reliability problems also contribute to the relatively poor image of the technology.

Une analyse des coûts en capital et des coûts de fonctionnement est présentée concernant les carburants qui peuvent être produits par les procédés de pyrolyse et à partir desquels les coûts de plusieurs carburants liquides ont été estimés. Les résultats indiquent que le manque actuel d'intérêt ne tient pas seulement à la viabilité économique douteuse, mais aussi au statut de la technologie, aux préoccupations environnementales et à l'utilisation du produit. Comme dans le cas de la gazéification, l'absence de fonctionnement longue durée, les expériences de fonctionnement et les problèmes perçus au chapitre de la fiabilité contribuent aussi à l'image plutôt négative de cette technologie.

Introduction

Pyrolysis produces gas, liquid and char: the relative proportions of which depend very much on the pyrolysis method and reaction parameters. Much of the present interest in pyrolysis currently centres on the liquid products due to their high energy density and potential for premium liquid fuel substitution. Several liquid fuels can be produced directly or indirectly - oil or bio-oil and slurries of charcoal with water or oil. In order to carry out the cost analysis it is helpful to briefly review the technology and products **(1)**.

The main products of pyrolysis are:

- Liquid. "Oil" or bio-oil has been produced in yields of up to 80% wt on feed (dry basis) from flash pyrolysis. While early small scale commercial operations have given similar high yields, a more conservative figure of 50% wt is used in the calculations in Table 1. The liquid has an elemental analysis similar to source biomass and an HHV of 20-25 MJ/kg. The product can be used directly as liquid fuel oil or upgraded to a hydrocarbon fuel.
- Solid char can be produced at yields of up to 30% wt on a volatile-free basis (up to 50% with high volatiles) depending on the process. Low char yields accompany high liquid yields. The char has an HHV of 30 MJ/kg and the product can be used as solid fuel directly, or briquetted, or slurried with water or oil to give a liquid fuel. Cost analyses for both slurry fuels are included.
- A fuel gas is also produced:
  EITHER an MHV fuel gas from indirectly heated processes at yields of up to 80% wt on feed from flash pyrolysis at high temperatures; or up to 20% wt from low temperature flash pyrolysis. It has an HHV of 15-20 MJ/Nm^3 and the product can be exported or used in-plant for process heat, feed drying or power generation. This is not considered further as the process is assumed to be energetically self sufficient by using this byproduct gas.
  OR an LHV fuel gas is produced from partial gasification processes at yields of 80-140% wt on feed. It has an HHV of 4-8 MJ/Nm^3 and again the product can be exported or used in-plant for process heat, feed drying or power generation. This is not considered further as the process is assumed to be energetically self sufficient by using this byproduct gas.

Liquid Products

Yields of liquids from pyrolysis can be influenced by the rate of reaction, with fast or flash pyrolysis at lower maximum temperatures giving the highest liquid yields, with up to 75% wt being reported. This liquid product may be readily burned and has been employed for this purpose. Problems have, however, been reported in its use, and special precautions may have to be taken in handling, storage and combustion. For these reasons, pyrolysis liquids cannot be readily assimilated into a conventional fuel marketing infrastructure, although there is adequate evidence that they can substitute for fuel oils in several applications.

The discount necessary for these lower quality fuels to be adopted in place of conventional fuels could be up to 20%, but this is insignificant compared to the current range of oil prices in the market place as shown later. Upgrading is necessary to give a product that is compatible with conventional fuels, but this is expensive. The products that are considered in the economic analysis are listed in Table 1 with explanations for selection, and significant parameters and properties.

Costs of Pyrolysis Derived Liquid Fuels

Of major importance in implementation of pyrolysis technologies, as with gasification technologies, is the economics of production. The relevant parameters of the costs of production are summarised in Table 2, with capital costs shown in Table 3, and process yield in Table 1. Production costs for each product are summarised in Table 4. These are based on directly attributable costs with no allowances for environmental benefits such as low sulphur, or fiscal benefits such as replacement or reduction of imports. Costs include drying and using surplus process gas, and are compared to equivalent fuel oil prices on a heating value basis. The 1989 average (scheduled and typical) medium fuel oil price is £2.26/GJ, and cover a range from £1.53 to £3.02/GJ. Costs in **bold** are those below this market value. This is probably the nearest product to bio-oil in terms of handling. The 1989 average (scheduled and typical) light fuel oil price is £2.84/GJ, and cover a range from £2.08 to £3.63/GJ; and the corresponding figure for heavy fuel oil price is £1.95/GJ, and cover a range from £1.41 to £2.52/GJ. The 1990 premium for low sulphur (<0.3%) fuel oil is around £50/t, and up to £100/t for large quantities in some European locations. The equivalent fuel oil price is then about £3.5/GJ. Costs underlined in the Table are below this figure, although refuse derived products may not meet this specification and are therefore excluded. Other fiscal credits may be attributable to biomass derived fuels which would further enhance their attractiveness, but more R&D is necessary to quantify these effects and relate them to locations and circumstances. Two Figures are included to show the effect of feed cost - Figures 1 and 2.

Discussion

Liquid fuel production from waste with its high disposal credit clearly offers the more attractive opportunities for commercial implementation, with oil and char-water slurries giving similar production costs. Char-oil slurries offer a significant advantage due to the higher yields of products, but there are more uncertainties in their production and use which require resolution. Both feed cost and product yield are major economic factors in minimising product cost. Both wood and agricultural wastes offer opportunities for oil and char-oil products at higher capacity plants and if the feed cost is sufficiently low. Char-water slurries are less interesting in these situations. A major uncertainty is the value to be attached to the product. It might be compared to a light fuel oil with a scheduled price in the UK as high as £3.63/GJ which opens up wider opportunities for wood, agricultural wastes such as straw, and energy crops. Alternatively if the product is compared to heavy fuel on the open market, and also requires a discount for successful sales, then only larger capacity high disposal credit waste fed processes have a chance of success. On the market side, therefore, conventional liquid fuel product prices and their movements are a dominant influence on the rate of implementation and degree of success of these technologies.

Upgrading Oil to Hydrocarbons

Upgrading technology is not well developed with most attention being paid to either hydrogenation or zeolite processing to give synthetic gasoline and other hydrocarbons. The third alternative is slurrying with water or oil to give a liquid fuel substitute. This is only at an early stage of development, with few details available on technology complexity, additives or costs. Some estimates of the production costs of upgraded liquids as hydrocarbons (gasoline) are indicated in Table 5, to give some idea of the performance of the three main technologies for producing gasoline from biomass by flash pyrolysis and liquefaction. These technologies are at a much less advanced stage of development.

Conclusions

The production cost analyses presented show the short term potential for energy production from wastes, particularly where there is a substantial credit associated with disposal of the waste, as feedstock contribution is particularly significant. Why, then, has so little been done to gasify these wastes, or any attractively priced biomass resource?

The answers are firstly that considerable attention has been paid to waste conversion throughout the world, and secondly the lack of success is due to many factors, including technology, economics, environment, politics, and social attitudes. A comprehensive assessment of these factors and their interactions cannot be considered here, but some of the salient points are identified.

- There <u>are</u> still technical problems to be resolved in reliability and performance, but these are not insurmountable and there is evidence in several operating examples in Europe and North America where problems have been satisfactorily resolved. A strong RD&D programme is considered essential to support these technologies and remove uncertainties.

- The economic attractiveness has been reduced by low conventional energy prices which has removed the pressure to resolve technical problems and develop better systems. This is compounded by the perceived uncertainty of these newer methods of energy production, with a preference by industry and government to invest in low risk ventures in the absence of any significant economic advantage.

- A further factor in the perception of pyrolysis products arises from comparisons with conventional fuels, i.e., fuel oils. The lack of understanding of the nature and utilisation constraints of these new fuels will inhibit their use in the short term unless significant changes occur in the world.

In a more unstable environment, the incentives to find alternative energy sources will cause more serious attention to be placed on these technologies. The pressures will come from the environmental lobby who expect cleaner and safer waste disposal methods, and the energy lobby who will react to an unstable energy supply situation when it arises. There seems little doubt that there will be considerable opportunities for thermochemical conversion of biomass and solid wastes in the future particularly if there is pressure from increasing conventional energy prices and also from environmental concerns over waste management practices. There is need for technology assessment and, if appropriate, support for long term demonstration. This will help ensure that the technologies are sufficiently developed and mature to provide reliable service when the occasion demands.

Acknowledgements

Some of the data reported in this paper were generated for the Energy Technology Support Unit at the UKAEA, Harwell, UK, as an update of earlier work **(5)**. Data on upgraded products was generated for the New Energy Vectors Programme of the EEC DGXII R&D Programme in Non-Nuclear Energy. All opinions and statements are the views of the author and in no way reflect any views of the UKAEA, the UK Department of Energy or the EEC.

References

1. Bridgwater, A.V., Beenackers, A.A.C.M., and van Swaaij, W.P.M., "Biomass Pyrolysis and Liquefaction: Status and Opportunities in the European Community", Proceedings of Euroforum New Energies, EEC Conference, Saarbrucken, October 1988
2. Bridgwater, A.V., "Economic And Market Opportunities For Biomass Derived Fuels" in *Gasification and Pyrolysis*, Ed G-F. Ferrero et al., (Elsevier Applied Science, 1990).
3. Bridgwater, A.V., and Double, J.M., "Production Costs Of Liquid Fuels From Biomass", in *Advanced Transport Fuels*, Ed G. Imarisio and J-M. Bemgten, EEC 1990 in press
4. Bridgwater, A.V., "The Economics Of Transport Fuels From Biomass", these proceedings
5. Bridgwater, A.V. "A Review of Thermochemical Conversion Technologies", Report to ETSU, Department of Energy, 1989.

Table 1
Liquid Products from Flash Pyrolysis

	Composition	HHV MJ/kg	Yield, wt % on daf feed
Crude pyrolysis liquid (oil or bio-oil)	-	22.5	75
Char-water slurry	60% wt char (a)	18.0 (b)	50 (c)
Char-oil slurry	20% wt char (d)	24.0	80 (e)

Notes
- a this is a probable upper limit for a pumpable slurry
- b based on a char heating value of 30 MJ/kg
- c based on 30% wt char yield (volatile free) on daf feed and 60% wt solids in the slurry
- d this is a probable upper limit for a pumpable slurry
- e based on 50% wt oil yield and10% wt char yield on daf feed and 20% wt solids in the slurry

Table 2
Typical Economic and Financial Parameters for Pyrolysis Base Cases

Pyrolysis technology	Entrained flow or fluid bed
Pyrolyser efficiency (see text)	58 to 93%
Feedstocks	Refuse, straw and wood
Feedstock heating value	20.00 GJ/t daf
Number of shifts (for 8000 h/y operation)	4
Project life	10 y
Scheduled operating hours per year	8000
Availablility (Actual operating hours / Scheduled operating hours)	80%
Actual operating hours	6400
Throughput	5 te/h, daf basis
Capital cost (including working capital)	See Table 3
Feedstock cost after processing	as stated
Utilities cost	£0.24/GJ produced
Yearly maintenance cost, fraction of capital cost	0.025
Yearly overheads, fraction of capital cost	0.080
Total cost of labour per shift	£25 000/y
Nominal cost of capital	10 %
Inflation rate	5 %

Table 3
Capital Costs of Pyrolysis Processes for Bio-oil, Char-water slurry and Char-oil slurry Production (2)

Capital cost of pyrolysis processes	1 t/h	2.5 t/h	5 t/h	10 t/h
Basic pyrolysis process				
Lower	£213 000	£412 000	£678 600	£1 117 800
Average	£370 000	£715 700	£1 178 900	£1 941 800
Higher	£735 000	£1 421 700	£2 341 800	£3 857 300
Target capital cost	£291 500	£563 800	£928 700	£1 529 800
(target capital cost = mean of average and lower cost)				
Pyrolysis for Bio-oil	£437 250	£845 700	£1 393 000	£2 294 700
Target plus 50% for feed preparation, feed drying and oil recovery				
Pyrolysis for Char-water slurry	£320 650	£620 200	£1 021 600	£1 682 800
Target plus 10% for slurry preparation				
Pyrolysis for Char-oil slurry	£466 400	£902 100	£1 485 900	£2 447 700
Target plus 60% for feed preparation, feed drying, oil recovery and slurry preparation				

Table 4
Production Costs by Feed Type and Scale

Scope: As in Table 2. Costs include drying and using surplus process gas
The 1989 average (scheduled and typical) **medium** fuel oil price is £2.26/GJ, and cover a range from £1.53 to 3.02/GJ. Costs in **bold** are those below this market value. This is probably the nearest product to bio-oil in terms of handling.
The 1990 premium for low sulphur (<0.3%) fuel oil is around £50/t, and up to £100/t for large quantities in some European locations. The equivalent fuel oil price is then about £3.5/GJ. Costs underlined are below this figure, although refuse derived products may not meet this specification and are therefore excluded

	Production cost, £/GJ			
	1 t/h	2.5 t/h	5 t/h	10 t/h
Pyrolysis Liquid (Bio-oil)				
STRAW				
£22/t delivered (£26/t daf)	5.34	4.19	3.72	3.43
£17/t on-farm (£20/t daf)	4.81	3.66	3.19	2.89
REFUSE				
Wet fluff £-0.60/t product (£-1/t daf) Includes disposal credit of £10/t raw refuse	2.94	**1.79**	**1.32**	**1.03**
WOOD				
£30/t daf	5.70	4.55	4.08	3.78
£20/t daf	4.81	3.66	3.19	2.89
Char-Water Slurry				
STRAW				
£22/t delivered (£26/t daf)	6.15	4.82	4.29	3.98
£17/t on-farm (£20/t daf)	5.48	4.15	3.63	3.31
REFUSE				
Wet fluff £-0.60/t product (£-1/t daf) Includes disposal credit of £10/t raw refuse	3.15	**1.82**	**1.29**	**0.98**
WOOD				
£30/t daf	6.59	5.26	4.74	4.42
£20/t daf	5.48	4.15	3.63	3.31
Char-Oil Slurry				
STRAW				
£22/t delivered (£26/t daf)	3.99	3.14	2.79	2.57
£17/t on-farm (£20/t daf)	3.60	2.76	2.41	**2.19**
REFUSE				
Wet fluff £-0.60/t product (£-1/t daf) Includes disposal credit of £10/t raw refuse	**2.26**	**1.41**	**1.06**	**0.42**
WOOD				
£30/t daf	4.24	3.40	3.05	2.83
£20/t daf	3.60	2.76	2.41	**2.19**

Table 5
Liquid Hydrocarbons Production Cost by Flash Pyrolysis and Direct Liquefaction
(3, 4)

Scale 1000 daf t/d wood @ £30/t

	Flash Pyrolysis				Liquefaction	
Product costs	£/t		£/GJ		£/t	£/GJ
Primary liquid						
Product cost	70.7		3.1		102	3.4
Primary product	------Pyrolysis liquid------				---Liquefaction liquid---	
Product value *	48.0		2.1		66.0	2.2
Upgrading technology	Zeolite		Hydrotreating		Hydrotreating	
	£/t	£/GJ	£/t	£/GJ	£/t	£/GJ
Final product						
Product cost	370	8.6	378	8.8	404	9.4
Final product	Gasoline, premium		Gasoline, blending		Gasoline, blending	
Product value *	250	5.8	225	5.2	225	5.2
Ratio Cost : Price	1.30		1.68		1.96	

* September 1990, ex-refinery, tax and duty free

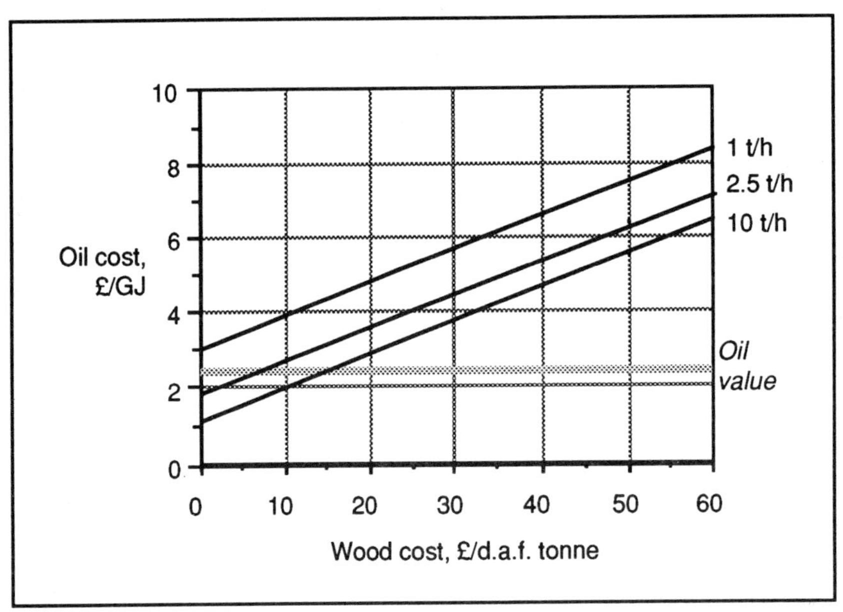

Figure 1
Oil Cost v Wood Cost

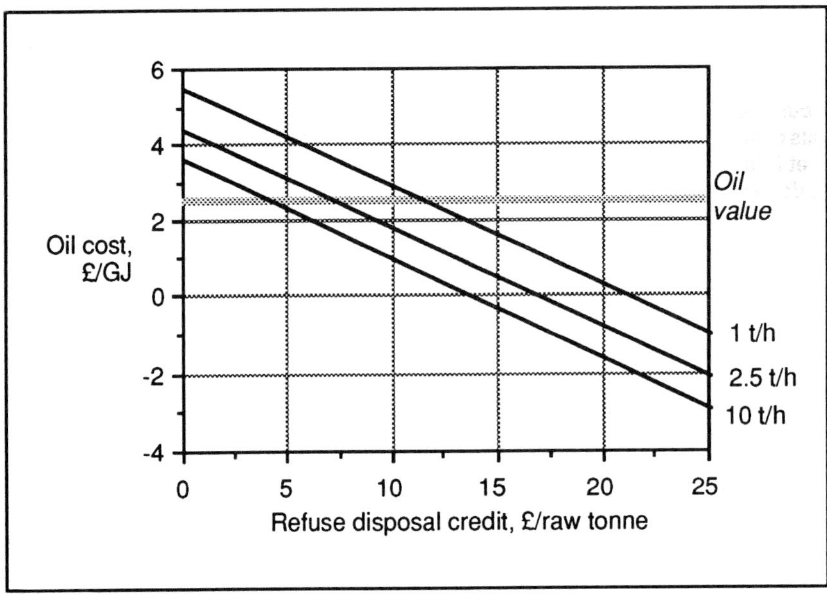

Figure 2
Oil Cost v Refuse Disposal Credit

The Economics of Transport Fuels from Biomass

A.V. Bridgwater
Energy Research Group, Chemical Engineering Department,
Aston University, Aston Triangle, Birmingham B4 7ET UK

There has been considerable interest in producing fuels from biomass and wastes since the oil crises of the last two decades which has been reinforced by subsequent environmental concerns and recent political events in the Middle East. This project was undertaken to provide a consistent and thorough review of complete processes for producing conventional liquid fuels from biomass from biomass feed at the factory gate to final product storage. It was carried out to compare both alternative technologies and processes within those technologies in order to identify the most promising opportunities that deserve closer attention.
The processes covered are indirect liquefaction by thermal gasification and liquid fuel synthesis; direct thermal liquefaction and catalytic upgrading; and biochemical conversion through hydrolysis and fermentation. Feedstocks include wood, straw and refuse. The liquid products considered include gasoline and diesel hydrocarbons that in some cases would require minor refining to convert them into marketable products; conventional alcohol fuels of methanol which has established opportunities and fuel alcohol which is as yet unproven in the market place; and bioethanol.
Results are given both as absolute fuel costs and as a comparison of estimated cost to market price. Generally the alcohol fuels are more attractive in comparing costs and prices, but the advantage is lost in absolute terms.
In the future work programme about to start, the direct liquefaction processes of flash pyrolysis and pressure liquefaction will be confirmed and extended to cover a wider range of individual processes rather than the generic processes modelled to date. In addition, there is still considerable analysis to be carried out on results from the model, including production cost analyses and sensitivity studies.

Les crises du pétrole des deux dernières décennies, auxquelles sont venus s'ajouter les préoccupations environnementales et les récents événements politiques du Moyen Orient, ont suscité un intérêt considérable pour la production de carburants à partir de la biomasse et de déchets. Ce projet avait pour but de fournir une revue complète et sérieuse de tous les procédés de fabrication de carburants liquides conventionnels à partir de la biomasse, depuis la charge d'alimentation à l'usine jusqu'à l'entreposage du produit fini. Il a été mené afin de comparer les techniques et procédés possibles dans ce domaine en vue de déterminer les possibilités qui sont les plus prometteuses et qui méritent une plus grande attention.
Les procédés étudiés sont la liquéfaction indirecte par gazéification thermique et synthèse de carburant liquide; la liquéfaction thermique directe et l'amélioration catalytique ainsi que la conversion biochimique par hydrolyse et fermentation. Les charges d'alimentation comprennent le bois, la paille et les déchets. Les produits liquides étudiés sont les hydrocarbures d'essence et de diesel qui, dans certains cas, devraient être légèrement raffinés pour les convertir en produits commercialisables; les alcools carburants conventionnels à partir de méthanol dont les possibilités sont déjà établies et l'alcool carburant, qui n'a pas encore fait ses preuves sur le marché, ainsi que le bioéthanol.
Les résultats sont donnés en coûts absolus du carburant et en comparaison du coût estimé sur le marché. En règle générale, les alcools carburants sont plus intéressants selon la méthode comparative, mais ils perdent cet avantage si on applique la méthode des coûts absolus.
Dans le cadre du programme de recherche sur le point de commencer, les procédés de liquéfaction directe de la pyrolyse-éclair et la liquéfaction sous pression seront revus et élargis pour comprendre une plus vaste gamme de procédés spécifiques plutôt que les processus génériques conçus jusqu'ici. En outre, des analyses considérables restent à faire sur les résultats du modèle, y compris des analyses de coûts de production et des études de sensibilité.

Introduction

The objective of this project is to examine the full range of technologies for conventional transport fuel production from biomass and construct a robust techno-economic computer simulation so that an objective and consistent evaluation of the alternative technologies may be performed. The feeds, technologies and products are summarised in Figure 1. This paper summarises the methodology and scope of the project and provides more extensive results. Details of the methodology have already been published **(1, 2)**. The programme is referred to as AMBLE - Aston Model of Biomass to Liquid Energy.

Modelling and Simulation

The general process routes outlined in Figure 1 have been analysed and subdivided into discrete process steps for modelling. Each process step is a self contained operation which includes all operations and processes necessary for it to function. These have been built as large as possible to minimise complexity and maximise ease of use and flexibility.

Process steps

The overall process is considered to be made up of a number of process steps, through which the material being processed passes sequentially, and by combining the results of the models of the process steps making up the complete plant, then the performance of the whole process can be determined. The process steps modelled are listed in Table 1, with comments on the source of data and limitations. Each process step is modelled as an integrated unit that considers all the inputs and outputs of feed, reagents, utilities and products and byproducts as depicted in Figure 2. The mass and energy balances that are performed in each step are as complex as is necessary for a robust model to be derived of the required accuracy that is valid over the full range of variables that might be encountered by the user. A process is synthesised as a sequence of such steps as depicted in Figure 3.

Reception, Storage, Handling and Pretreatment

The front end processing is made up of physical handling steps, size control through milling, grinding pelletising and screening, and drying with different scales of operation having different equipment specifications. This part of the process is more subject to local circumstances than conversion or downstream processing and is therefore rather less reliable for specific plants, although good "typical" performances and costs will be produced by the model.

Conversion

Robust empirically modified equilibrium models of gasification have been adapted for inclusion in the programme. These have been validated for each of the gasifiers included in the model and the example of the Lurgi CFB gasifier is included as Figure 5 to demonstrate the good agreement obtained **(2)**. Similar agreement has been achieved for the IGT and ASCAB/Framatome/Stein gasifier. The direct liquefaction processes are much less well developed and are not conducive to the same approach to modelling as gasification. The current models are, therefore, empirical and based on published data, supplemented where relevant with direct inputs from the researchers **(3)**. The results are, therefore, less robust.

Product Upgrading and Refining

Data on the design, specification and costing of process steps for the production of liquid fuels from synthesis gas has been provided by John Brown Engineers and Constructors as basic data and algorithms. This information has been used to model the steps **(4)**. In all cases the data is from commercially developed or demonstrated processes and is therefore considered to be reliable. The modelling of direct thermal liquefaction processes has been based on data derived in-house. In all these cases the quality of the data is relatively poor and the uncertainty of the resultant model is, therefore, higher.

Process Specifications

Complete process plants are modelled from feed delivered at the factory gate to a marketable product in storage. The size range considered is 100 to 1000 daf t/d feed input: larger sizes are considered unrealistic, and smaller sizes will suffer excessively from diseconomies of scale.

Process Synthesis and Programme Output

The processes are constructed by linking process steps in logical sequences based on specification of feed, product and, where relevant, conversion technology. Although there are 47 individual process steps, these can only be combined in a limited number of ways. These are depicted in Figure 6 for direct liquefaction through pyrolysis and liquefaction and, for comparison, indirect liquefaction in Figure 7. The results from running the programme are output in three formats:

1. Summary output of process performance and product cost,
2. Main results which gives more extensive information on process performance, utilities requirements and cost analysis,
3. Complete mass and energy balance for a complete process. Examples are given in **(4)**.

Results of the Program

Results of the AMBLE programme are shown in Figures 8 to 17. Most of these graphs show the effect of feedstock cost on the cost of the liquid fuel product, but Figure 17 shows the effect of plant throughput on product cost. In each of the graphs of feedstock cost against product cost, the grey lines show the approximate current values of the feed cost and product cost. Table 2 shows the base case data used in producing these graphs. In every case, the product cost calculated at the current value of the feedstock cost is in excess of current liquid fuel prices. Only in the case of ethanol produced from wheat does there seem to be a possibility of a currently viable process, but this is discussed in more detail below.

Gasoline from Wood by MTG (Figure 8)

In the case of gasoline being produced via gasification, methanol synthesis and MTG synthesis, the Lurgi gasifier gives the most economic results. The Lurgi gasifier does not give quite such a good yield of methanol as the ASCAB gasifier, due to the higher content of methane in its product gas, which acts as an inert in the methanol synthesis and which must then be purged, representing a loss of efficiency. However, the Lurgi gasifier uses considerably less oxygen for gasification, reducing both the utilities costs and the total production cost. The IGT gasifier produces a gas with a very high hydrocarbon content. This results in a very poor yield of methanol and hence gasoline. This might be rectified by reforming and recycling of a fraction of the purge gas in the methanol synthesis step, but on the straightforward comparison with the Lurgi and ASCAB cases then the IGT gasifier can be seen to be inefficient for the production of synthesis gas.

Gasoline From Wood by MOGD (Figure 9)

This route, similar to the route previously discussed except for the substitution of the MTO/MOGD process for the MTG process, gives very similar results, as might be expected. The main difference is that the gasoline costs slightly more to manufacture than that from the MTG route, due to the increased complexity of MTO/MOGD compared with the MTG plant.

Diesel From Wood by MOGD (Figure 10)

Costs are similar to gasoline from wood, but slightly lower.

Diesel from Wood by SMDS (Figure 11)

The Shell SMDS process gives a cheaper product than that from MOGD. This is because the Fischer-Tropsch technology converting synthesis gas directly to fuels is less complex than the three stage process of methanol synthesis followed by MTO and MOGD synthesis. Again, the Lurgi gasifier gives a cheaper product due to the reduced oxygen consumption, despite the lower yield of diesel.

Methanol from Wood (Figure 12)

Methanol production gives the same basic pattern of results as the other indirect liquefaction processes already described, with Lurgi gasification giving a lower product cost than ASCAB gasification, and IGT gasification giving very high prices due to the low product yields. Methanol production cost is less than for hydrocarbons, because of the reduced process complexity. The disadvantage of methanol is that it is not currently used as a motor fuel, and would require a new market infrastructure. Two methanol values are quoted - the current estimated cost of production, and the current market price.

Fuel Alcohol from Wood (Figure 13)

Fuel alcohol - a blend of methanol with higher alcohols produced by a modified methanol synthesis process - gives a pattern of results similar to methanol production, but at a higher product cost, due to the increased capital cost of the Fuel Alcohol Synthesis plant. There seems to be no fundamental reason why the Fuel Alcohol Synthesis unit should be more expensive than a methanol synthesis unit, but the process is still in development, which might explain the higher quoted capital costs.

Direct Thermochemical Conversion of Wood to Gasoline Blending Stock (Figure 14)

The product costs predicted by the model for the direct thermochemical routes are considerably lower than for other routes converting wood to fuels. There is little to choose between them, the pyrolysis and zeolite upgrading option being the cheapest at lower feedstock costs due to a lower capital cost, but losing out to the pyrolysis and hydrogenation route at high feedstock costs due to the poorer yield of product. The liquefaction route gives slightly higher costs than the pyrolysis routes due to the higher fixed costs. While the product costs from these routes are the cheapest predicted by the model, it should be remembered that the product is not suitable for direct use as gasoline, and some further upgrading or blending would be required to make use of it as a motor fuel. Also, these routes are the least developed technically, so there are greater uncertainties associated with the predicted costs.

Methanol from RDF (Figure 15)

Similar results to methanol from wood, but at a higher level of cost due to the lower value as a feedstock of RDF. RDF cannot be gasified at the operating conditions specified in the IGT gasifier model - this is because the high steam content of the gasifying agent and the low calorific value of the RDF do not allow the energy balance to be satisfied in an autothermal process - additional heat would be required to allow gasification to take place. The low heating value of the refuse means that a greater proportion of the chemical energy of the feedstock has to be sacrificed to achieve the high operating temperatures of the ASCAB gasifier than is the case for wood. The Lurgi gasifier, operating at a lower temperature, is not so susceptible to this effect, so the difference between the performances of the two gasifiers is greater than in the case of wood feedstock.

Methanol from Straw (Figure 16)

The low initial moisture content of straw means that the economic performance of a methanol plant based on straw as feedstock is better than one using wood. However, these figures are based on a 1000 te/day plant, and it is unlikely that more than 500 te/day could be supplied to a plant using straw feedstock.

Effect of Throughput (See Figure 17)

The effect of throughput of the plant on economics has been tested for the production of methanol from wood. The other processes will follow the same general trends. The effect of scale is less important at high throughput, as would be expected, and the effect of scale is greater on the IGT gasifier based process because of the low yield of product.

Fuel Prices

Typical liquid fuel prices in September 1990, after some stability had set in after the August Middle East crisis, are given in Table 3, apart from methanol for which a current state-of-the-art production cost is given to represent a new venture rather than the depressed market price (6). The figure for fuel alcohol is estimated from methanol costs as there is no commercial production and no market.

Comparison of Prices

Absolute cost of fuels at 1000 t/d daf feed rate with feed costs at typical values and taking the best figures from each case described above are summarised and ranked in Table 4. A comparison of production costs with current prices is summarised and ranked in Table 5. In the case of methanol, the current production cost is taken as would apply to a new state of the art plant as the prices are depressed due to a current surplus of capacity. Hydrocarbon products are credited with refined fuel values although they would require some tertiary processing to meet fuel specific

Discussion and Conclusions

Specific Results

The most attractive processes from comparing production costs to product values are generally the alcohol fuels which enjoy a higher market value from their non-fuel uses and sometimes artificial prices. As fuels they are less attractive as replacements due to the difficulty of assimilation into a developed market infrastructure, although as fuel extenders they are valuable and probably deserve a credit higher than their fuel value because of environmental benefits.

The only exception to the status of alcohol fuels in comparison of costs and prices is direct production of highly aromatic gasoline by direct liquefaction of wood through pyrolysis and zeolites but this is the least developed and most speculative of all the processes. For hydrocarbons the direct liquefaction processes appear to offer considerable advantages, although the state of development is less advanced. Of the indirect thermochemical routes, the MTG appears to be the most attractive for gasoline and the Shell SMDS for diesel. Both these processes also have the advantage of producing high-quality products: MTG gasoline is highly aromatic and therefore has a good octane number, while SMDS diesel has a cetane number far in excess of current specifications, which would allow new engines tuned for this fuel to be much more efficient than current diesels. The SMDS process also has the advantage of producing jet fuel of high quality.

This final analysis should be viewed cautiously since product values are very uncertain and can change significantly over short periods of time, and feedstock prices are also subject to substantial variation with time and location, and this is often the major cost item in the production costs. There is still considerable analysis to be carried out on results from AMBLE, including production cost analyses and sensitivity studies, as well as continued development to improve accuracy and extend the range of technologies considered.

General Results

A unique process simulation and evaluation package has been completed that provides consistent comparisons of the full range of liquid fuels production processes from biomass. The package has been validated and also gives good absolute values. There is considerable work to be completed in analysing data and carrying out sensitivity studies as well as extending the model to the the less certain technologies of direct liquefaction.

Acknowledgements

The support of the New Energy Vectors Programme of the European Community Research and Development Directorate DG XII is gratefully acknowledged. The following organisations have made valuable contributions:

- John Brown Engineers and Constructors Ltd of London, UK for product synthesis and upgrading process performance and costs.
- Parsons Whittemore of Croydon, UK for technical advice on the feed handling and pretreatment steps,
- Lurgi GmbH of Frankfurt, Germany for discussions and model validation,
- Shell of The Hague, Netherlands for discussions and model validation,
- Stein Industrie and ASCAB of Paris, France for discussions and model validation.

References

1. Bridgwater, A.V., and Double, J.M., "Liquid Fuels from Biomass: Process Assessment and Modelling", presented at the First EEC contractors meeting of the programme *Production and Utilization of New Energy Vectors*, 6-8 June 1988, Saarbrücken, West Germany.

2. Mehrling, P., Reimert, R., "Synthetic Fuel from Wood via Gasification in the Circulating Fluid Bed", in Beenackers AACM & van Swaaij WPM, "Ad-

vanced Gasification", Reidel Publishing, Dordrecht, Netherlands, 1986, p73-113

3. Beckman, D., Elliott, D.C., Gevert, B., Hornell, C., Kjellstrom, Ostman, A., Solantausta, Y. and Tulenheimo, V., "Technoeconomic Assessment of Selected Biomass Liquefaction Processes", (IEA), VTT Report 697 (VTT, 1990)

4. Bridgwater, A.V. and Anders, M., "Production Costs of Liquid Fuels by Indirect Coal Liquefaction", in *Advanced Transport Fuels*, Ed G Imarisio and J-M Bemgtem, EEC 1990 in press

5. Bridgwater, A.V. and Double, J.M., "Production Costs Of Liquid Fuels From Biomass", in *Advanced Transport Fuels*, Ed G Imarisio and J-M Bemgtem, EEC 1990 in press

6. Sheridan, B., ChemSystems Ltd, Private Communication 214 December 18, 1990

Table 1
List of Process Steps and Sources of Data

Reception, storage and handling

101	Reception of wood
102	Reception of RDF
103	Reception & storage of straw
111	Bulk storage of wood
112	Bulk storage of RDF

Pretreatment

201	Size control of wood chips
213	Straw size reduction
221	Drying of wood.
222	Drying of RDF
261	Grinding of liquefaction feedstock
291	Buffer storage of wood
292	Buffer storage of RDF
293	Buffer storage of straw

Conversion

301	Framatome gasification
302	Lurgi CFB gasification
303	IGT Renugas gasification
313	Flash pyrolysis
321	Liquefaction
398	Dummy gasification
399	Generic oxygen gasification

Product synthesis

401	Gas compression
421	Fischer-Tropsch synthesis - SMDS
431	Methanol synthesis
432	Fuel Alcohol synthesis
441	MTG synthesis and distillation
442	MOGD synthesis and distillation

Product upgrading

501	Hydrogenation of pyrolysis oils
502	Hydrogenation of liquefaction oils
511	Zeolite upgrading with flash pyrolysis

Product refining

601	Distillation of crude methanol
611	Refining of upgraded oils

Table 2
Base Case Data

Technical Data

Feedrate	1000 t/d daf
Wood feedstock moisture content	50% wet basis
Straw feedstock	17% wet basis
RDF moisture	33% wet basis

Economic and Financial Data

1	Project Life, years		10
2	Days per year of operation		340
3	Real Interest Rate, %		10.00
4	Inflation Rate, %		4.00
5	Feedstock Cost, GBP/daf t		30.00
6	Labour Cost, GBP/man year		15000
7	Maintenance, % CapCost/year		2.50
8	Overheads, % CapCost/year		10.00
9	Required real ROI, %		0.00
10	Utilities costs:	LP steam, GBP/t	5.00
		HP steam , GBP/t	10.00
		Process water, GBP/t	1.00
		Cooling water, GBP/t	0.04
		Fuel gas, GBP/GJ	3.00
		Hydrogen, GBP/t	1500
		Oxygen, GBP/t	40.00
		Power, GBP/MWh	30.00
		Feed fines, GBP/t	10.00
		(NB GBP = £ sterling)	

Table 3
Wholesale Prices and Production Costs of Fuels, September 1990

Product	Gasoline	Diesel	Medium fuel oil	Heavy fuel oil	Methanol	Fuel alcohol
Wholesale price, tax free, £/te	260	183	99	79	80	N/A
Production cost, tax free, £/te	-	-	-	-	140	180 *

* Estimated equivalent current cost of production
Current crude oil cost at this time was $35/bbl

Table 4
Estimated Fuel Cost
Basis: 1000 t/d daf feed, feedstock costed at typical cost

Rank	Product	Feed	Route	Uncertainty	Cost, £/GJ
	Natural gas				*3.0*
	Diesel				*4.1*
	Methanol				*5.6*
	Gasoline				*5.8*
1	Methanol	straw	gasification	low-moderate	7.0
2	Gasoline	wood	pyrolysis + zeolites	high	8.6
3	Diesel	wood	gasification + SMDS	low-moderate	8.8
4	Gasoline	wood	pyrolysis + hydrotreating	high	8.8
5	Methanol	wood	gasification	low-moderate	9.1
6	Gasoline	wood	liquefaction + hydrotreating	high	9.4
7	Fuel alcohol	wood	gasification	moderate	9.8
8	Methanol	RDF	gasification	moderate	9.9
9	Gasoline	wood	gasification + MTG	low	12.4
10	Gasoline	wood	gasification + MOGD	moderate	14.6
11	Diesel	wood	gasification + MOGD	moderate	14.6

Table 5
Comparison of Fuel Costs to Prices
Basis: 1000 t/d daf feed, feedstock costed at typical cost

Rank	Product	Feed	Route	Uncertainty	Cost/Price
1	Methanol	straw	gasification	low-moderate	1.17
2	Methanol	wood	gasification	low-moderate	1.46
3	Gasoline	wood	pyrolysis + zeolites	high	1.48
4	Fuel alcohol	wood	gasification	moderate	1.56
5	Methanol	RDF	gasification	moderate	1.58
6	Gasoline	wood	pyrolysis + hydrotreating *	high	1.69
7	Gasoline	wood	liquefaction + hydrotreating *	high	1.81
8	Gasoline	wood	gasification + MTG	low	2.14
9	Diesel	wood	gasification + SMDS	low-moderate	2.15
10	Gasoline	wood	gasification + MOGD	moderate	2.52
11	Diesel	wood	gasification + MOGD	moderate	3.56

* includes consideration of lower quality product

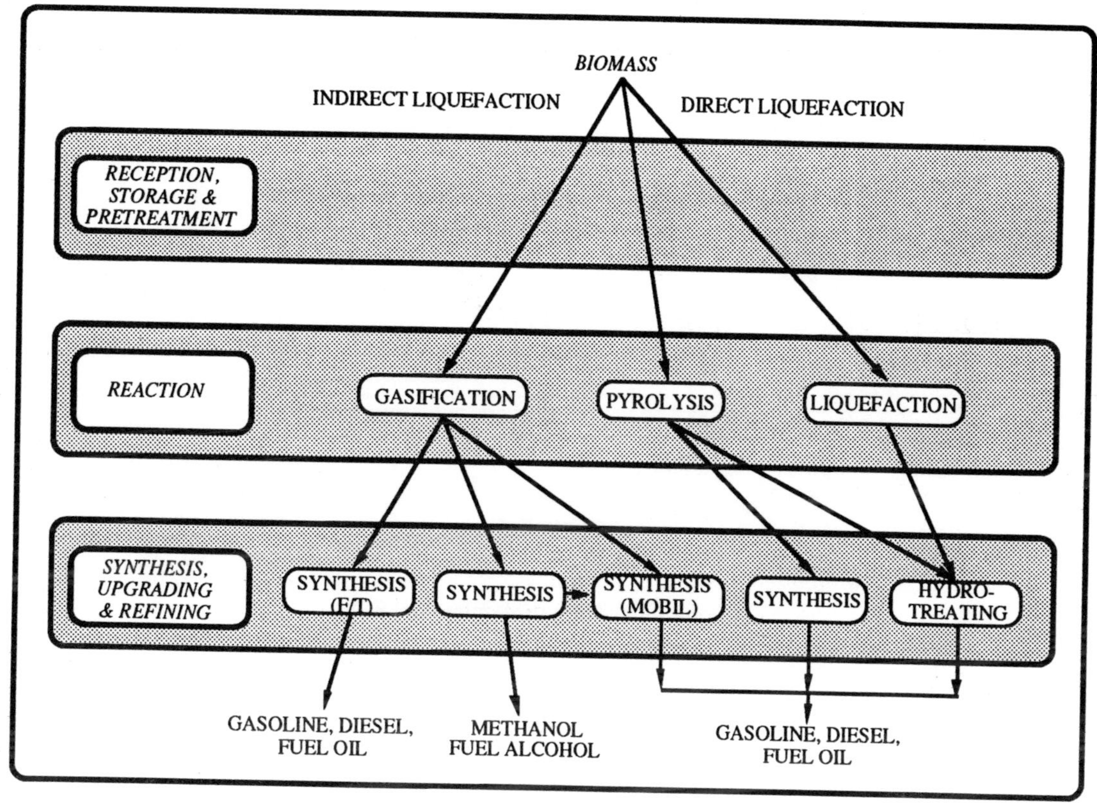

Figure 1
Thermochemical Process Routes to Liquid Fuels

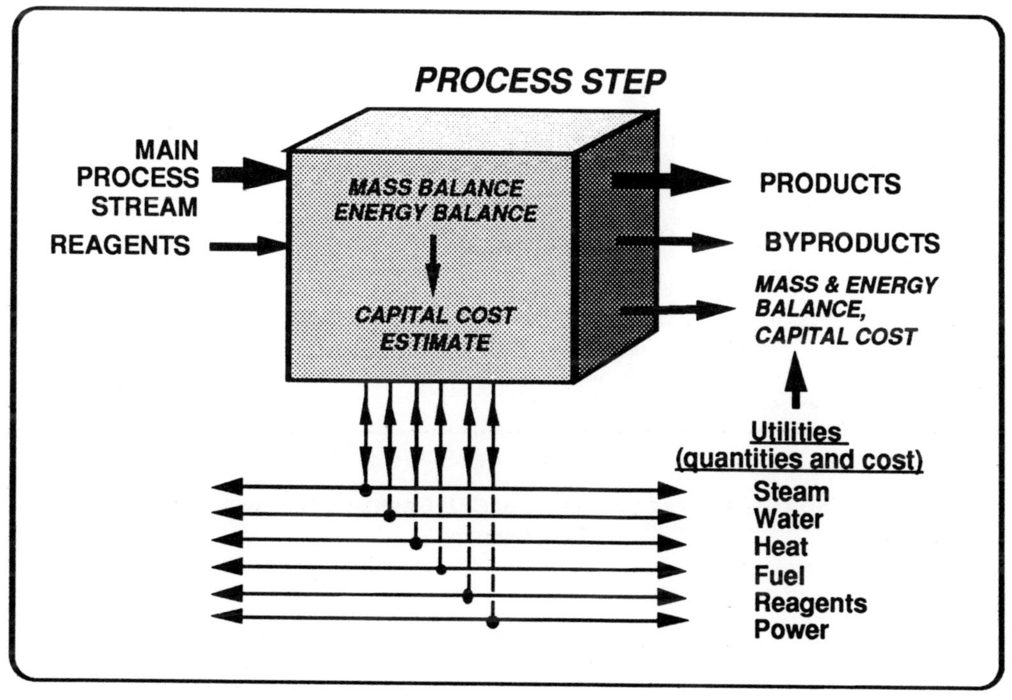

Figure 2
Conceptual Diagram of a Typical Process Step

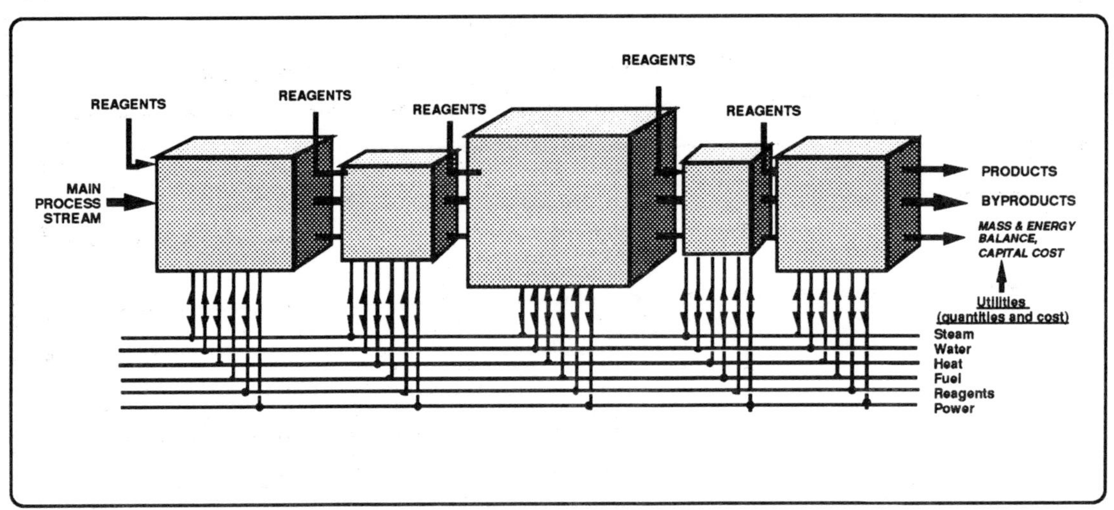

Figure 3
Process Synthesis from Process Step

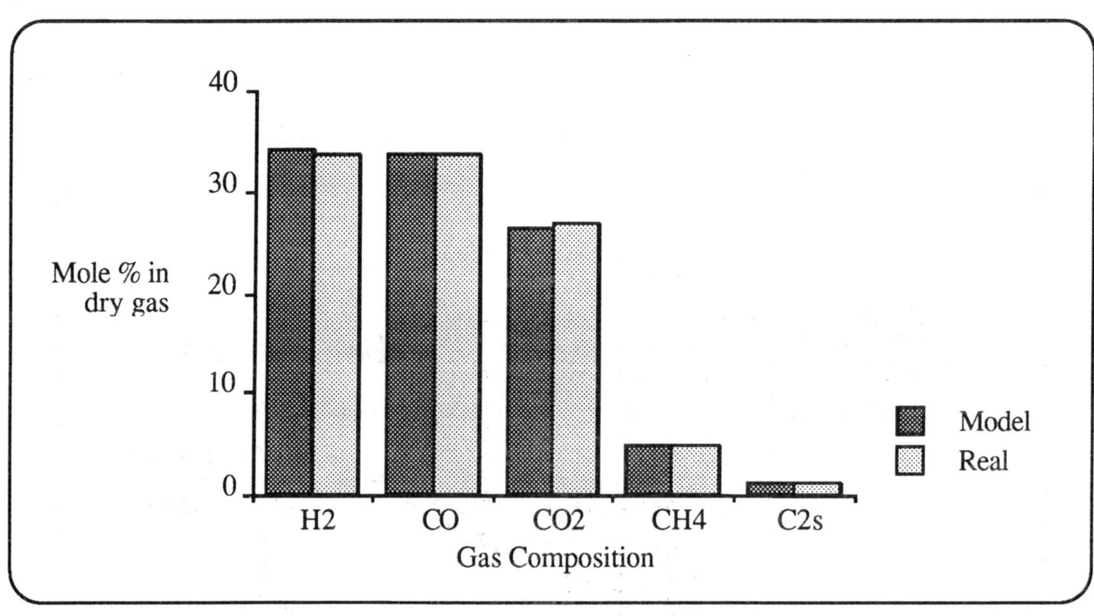

Figure 5
Comparison of Results from the Lurgi CFB Gasifier Model with Achieved Performance

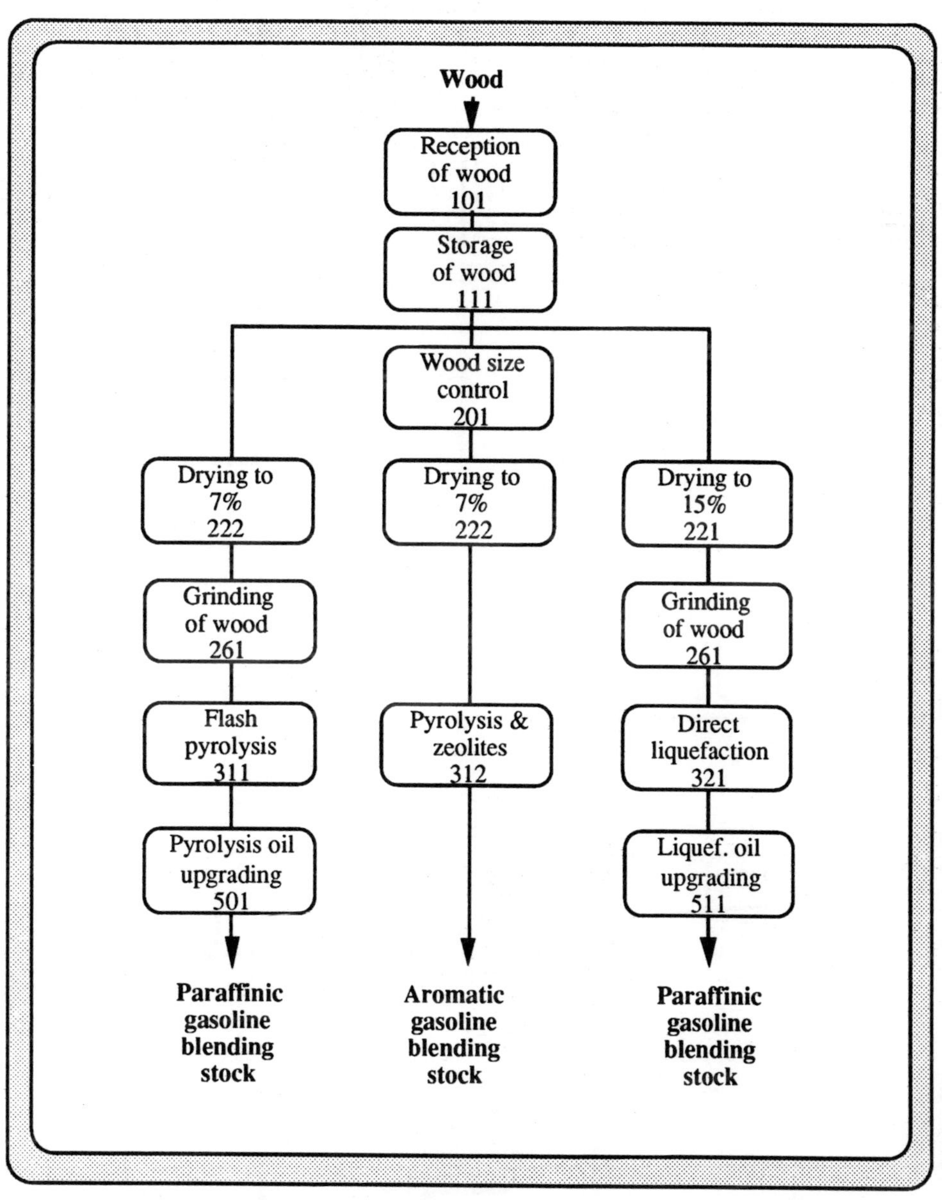

Figure 6
Direct Thermochemical Liquefaction Routes by Pyrolysis and Liquefaction

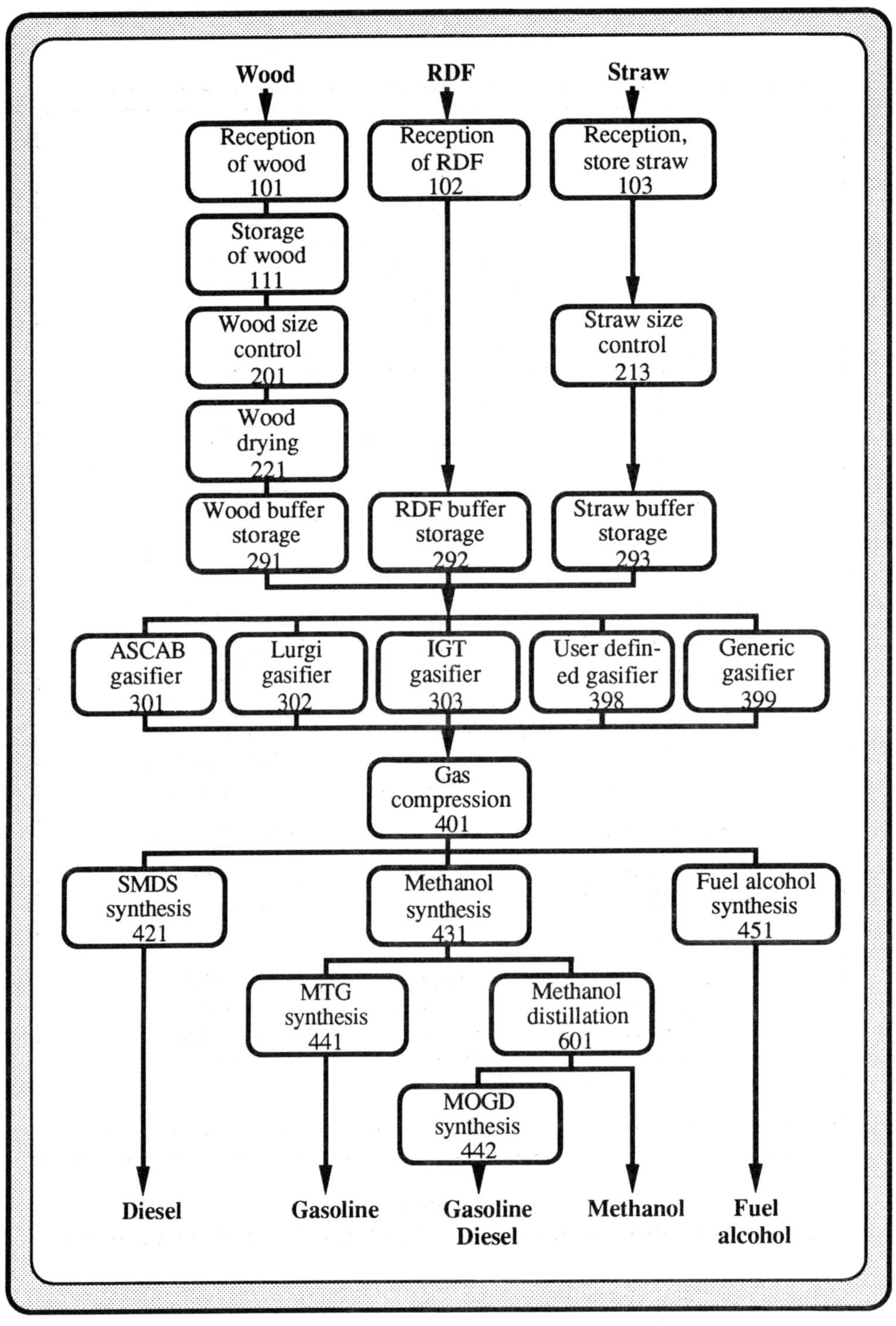

Figure 7
Indirect Thermochemical Liquefaction Routes

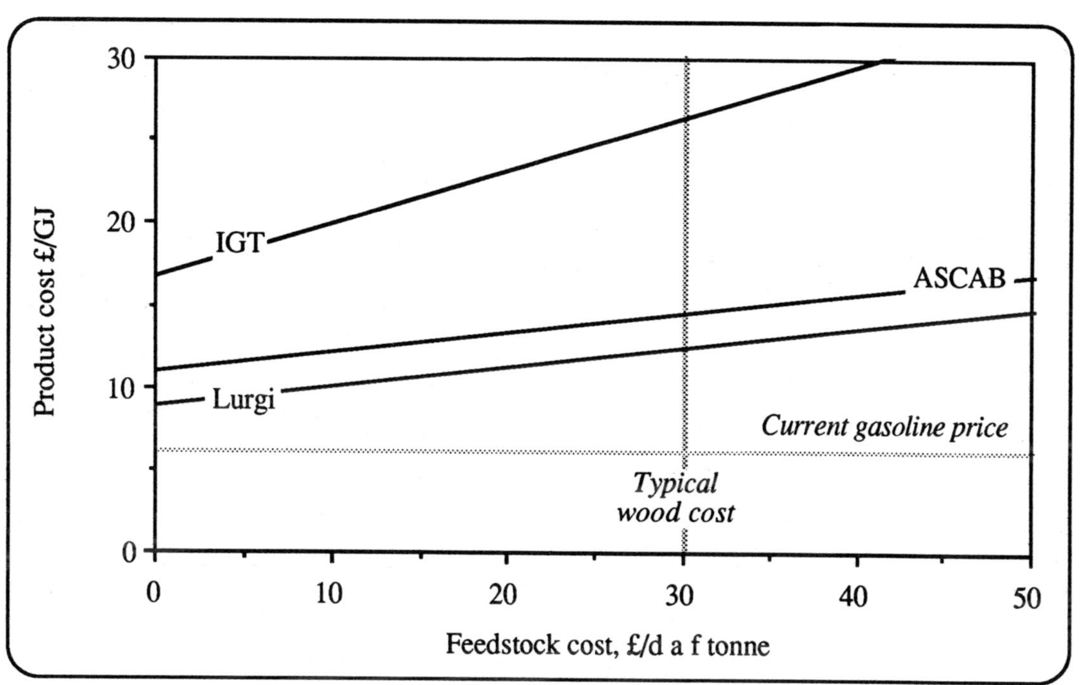

Figure 8
Gasoline from Wood by the MTG Process

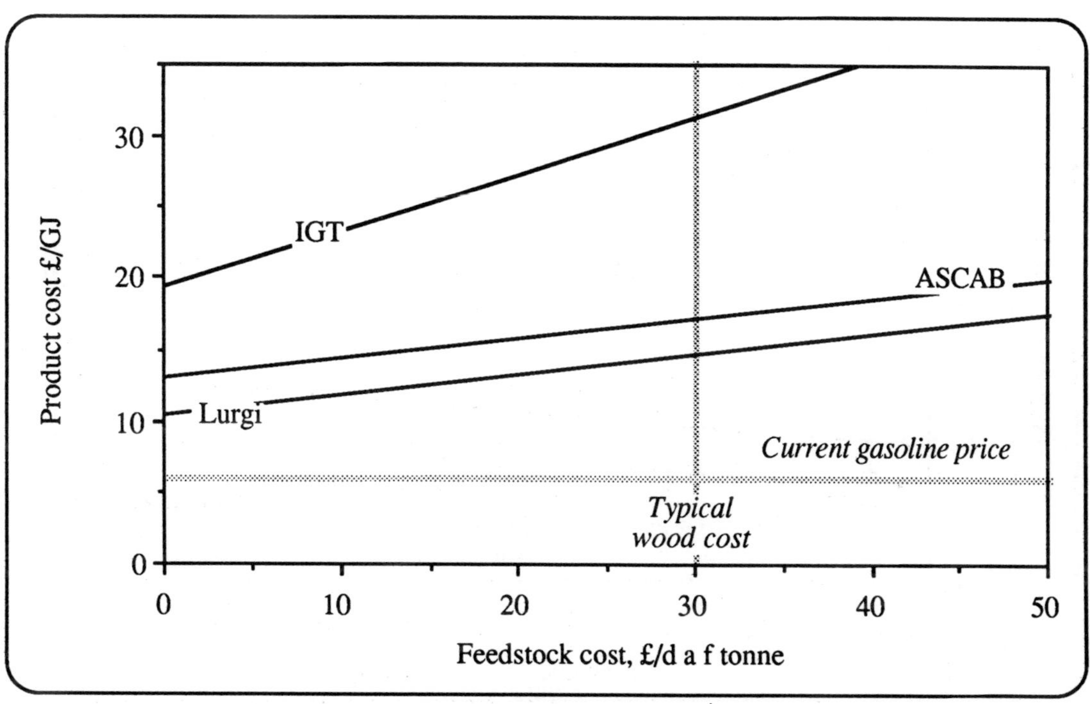

Figure 9
Gasoline From Wood by the MOGD process

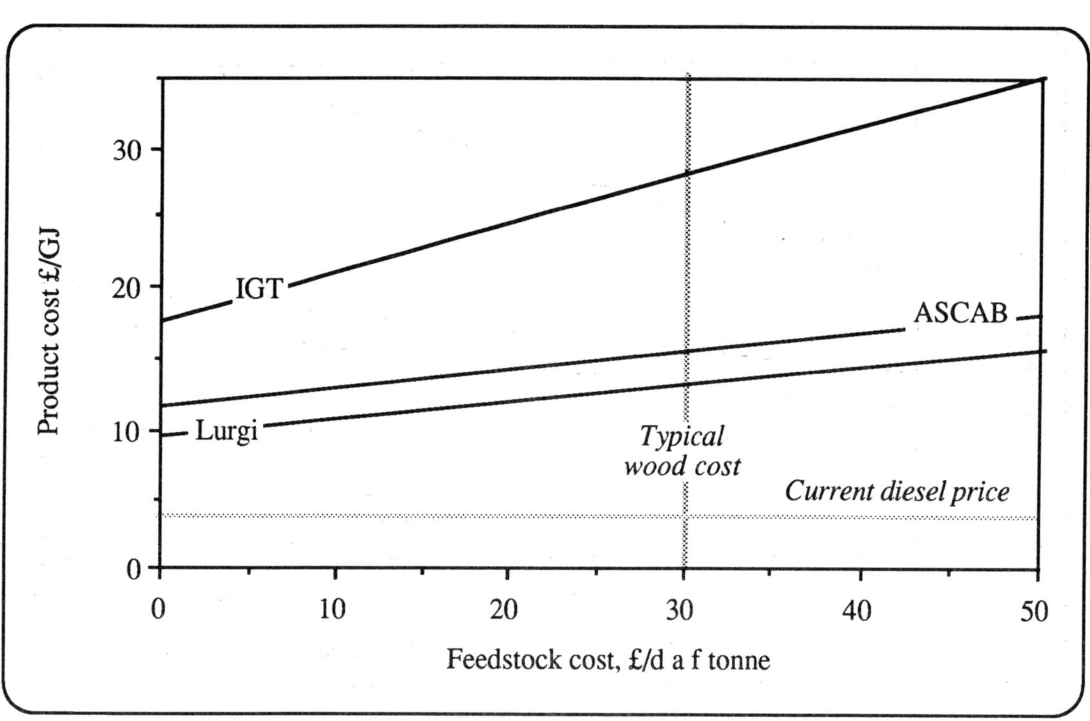

Figure 10
Diesel From Wood by the MOGD process

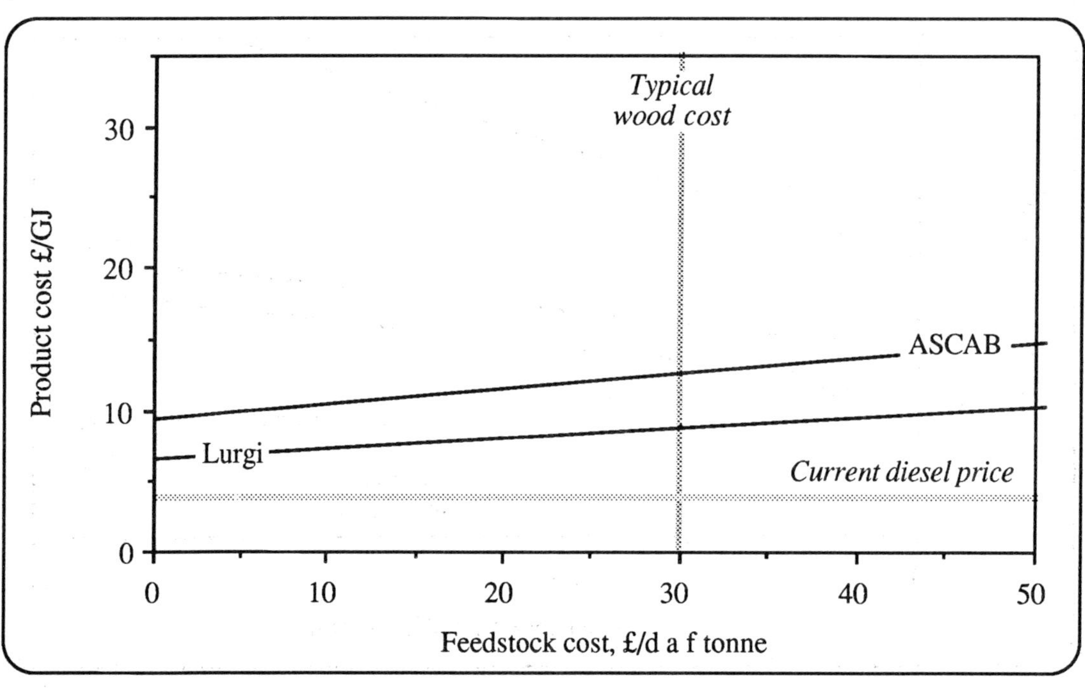

Figure 11
Diesel From Wood by the SMDS process

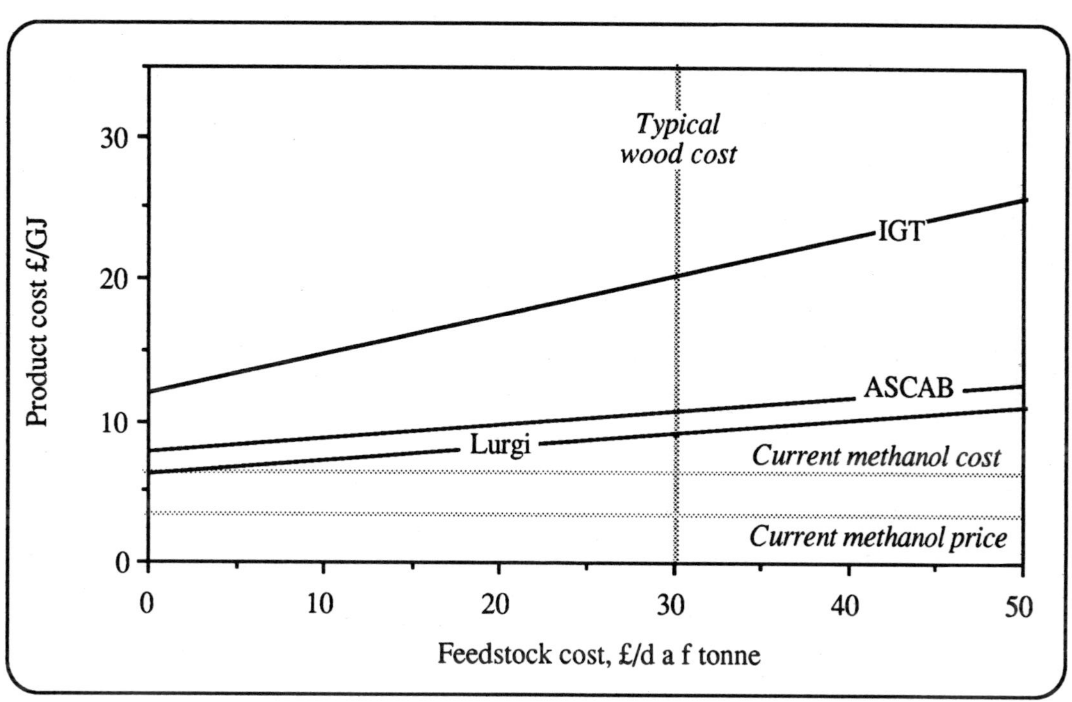

Figure 12
Methanol from Wood by Gasification

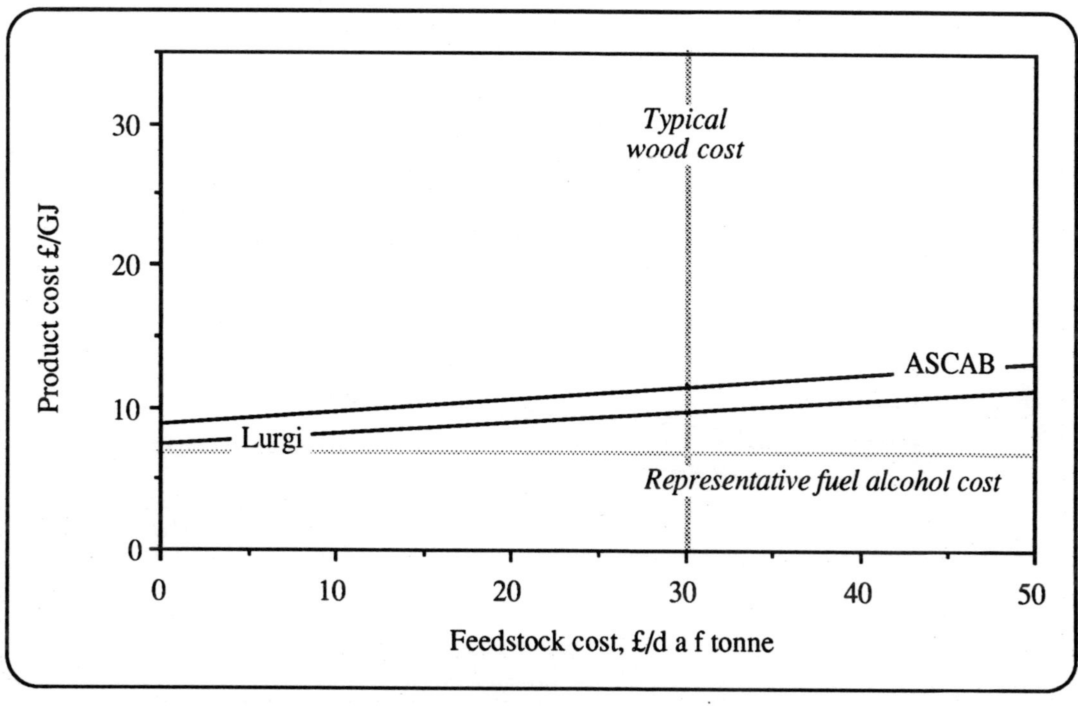

Figure 13
Fuel Alcohol from Wood by Gasification

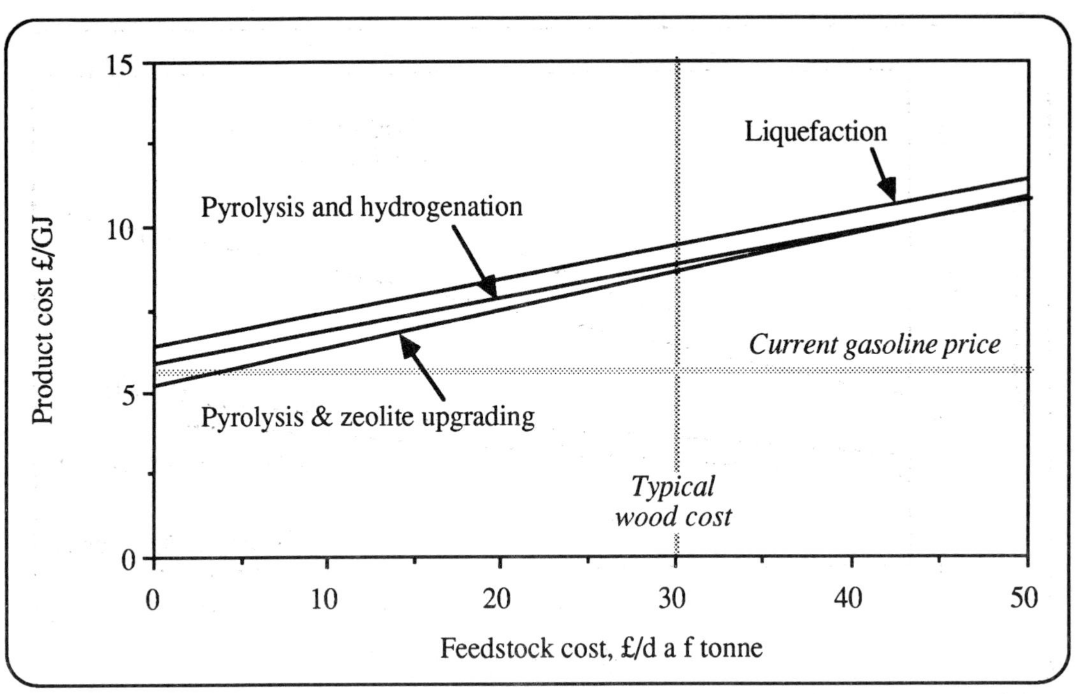

Figure 14
Direct Thermochemical Conversion of Wood to Gasoline Blending Stock

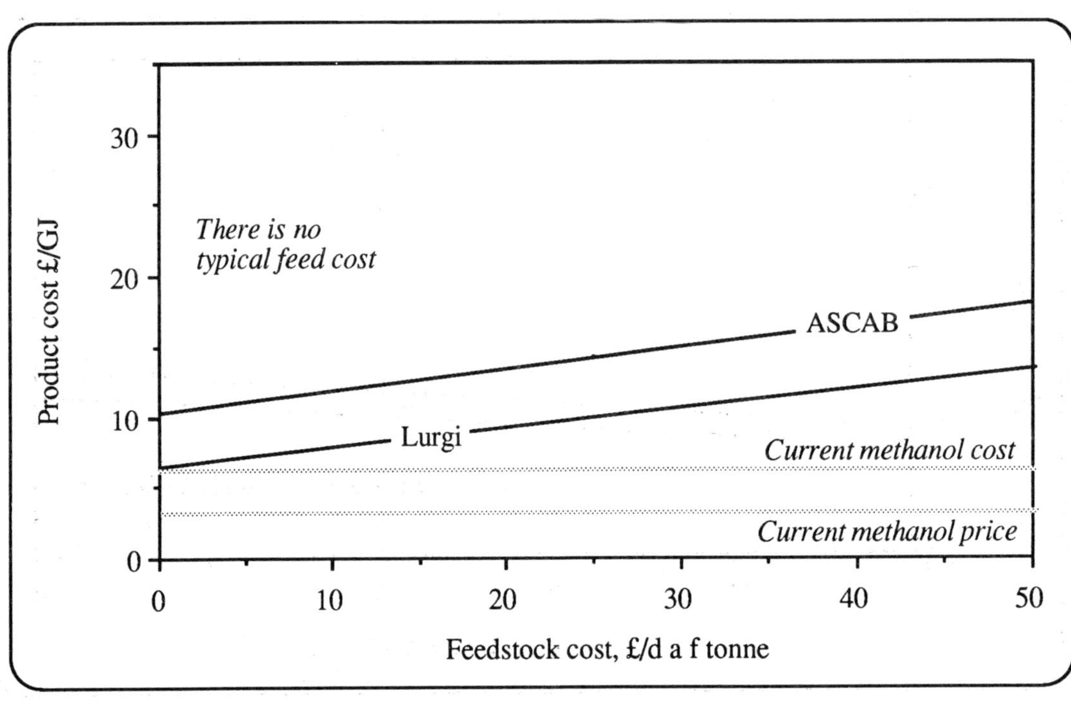

Figure 15
Methanol From RDF

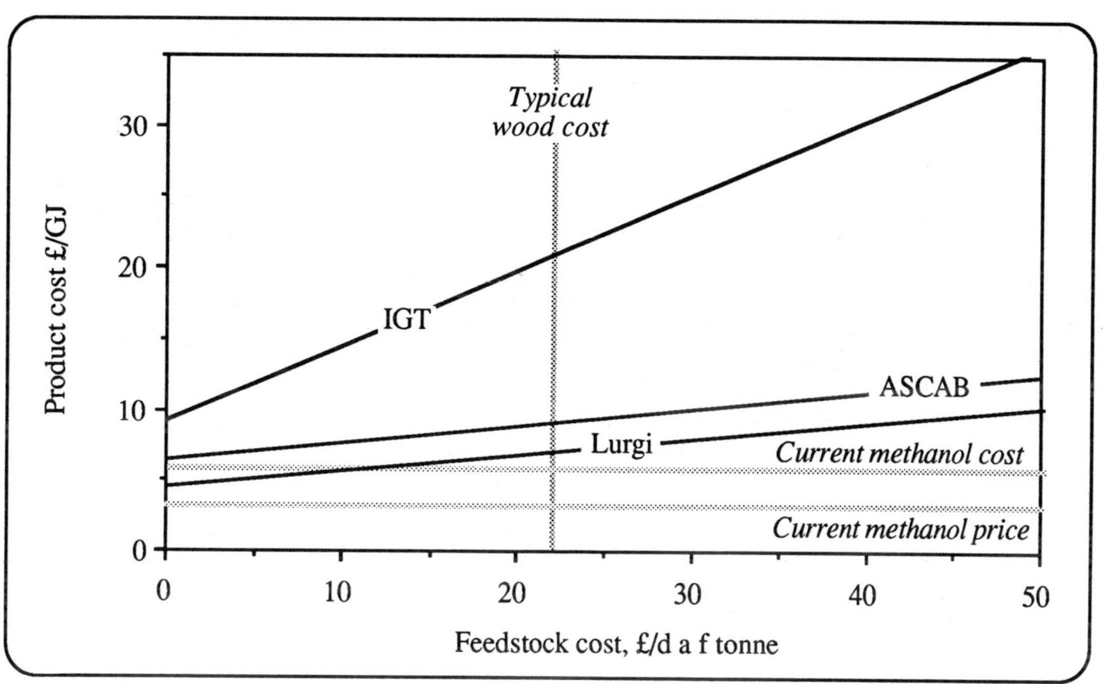

Figure 16
Methanol From Straw

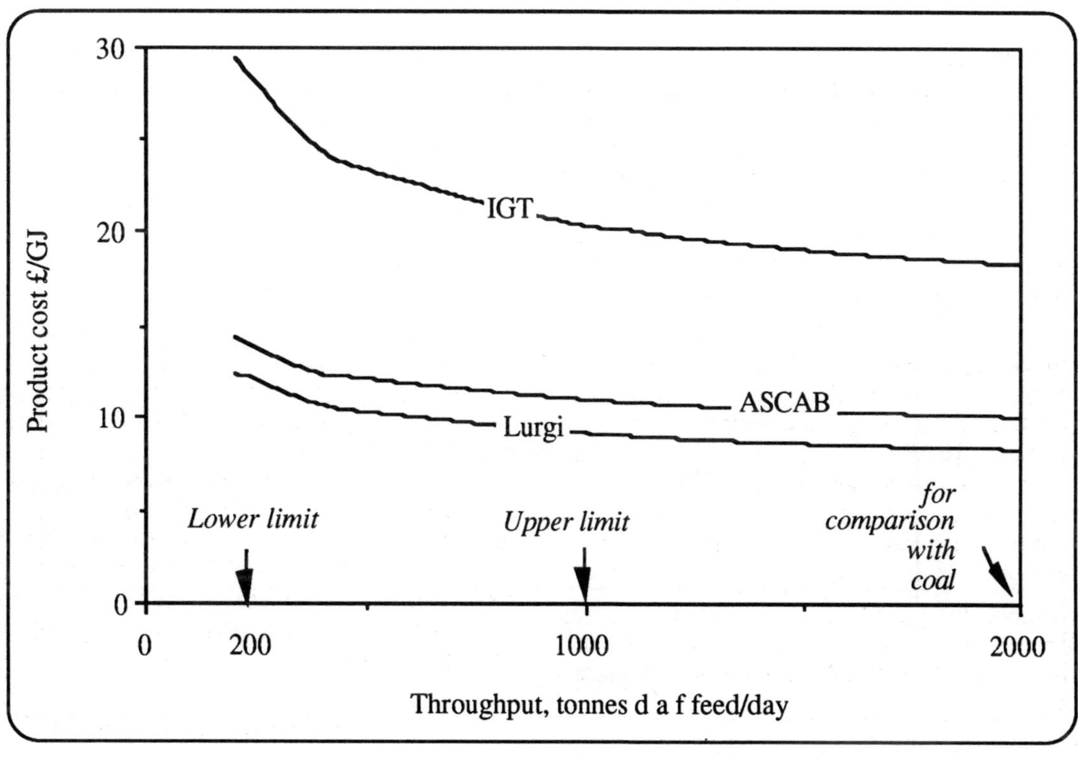

Figure 17
Economy of Scale for Methanol from Wood

Commercialization of Fast Pyrolysis Products

G. Underwood, Vice President
Red Arrow Products Inc., Manitowoc, WI, U.S.A.

Acknowledgements

I would like to express thanks to Energy Mines and Resources for the invitation to speak and for their continued faith in the future of fast pyrolysis. We have benefitted immensely from the funding of Canadian research into biomass fast pyrolysis. In particular, we have benefitted from our cooperation with Ensyn Engineering Associates Inc. with whom we continue to jointly develop RTP fast pyrolysis technology. I would also like to express appreciation to Ensyn for their perseverance in their efforts to develop RTP. As most scientists are aware, research and development can be very trying at times, especially when it is done in the context of a small business. Finally, we thank the State of Wisconsin for supporting our development efforts with funding from the Oil Overcharge Fund, Waste to Energy Program.

Background of Wood Pyrolysis Derived Products

The first commercial wood-derived pyrolysis product was charcoal. Charcoal has been used for gunpowder manufacture, cooking, heating for hundreds of years. Some kilns today differ little in appearance and functionality from those of a hundred years ago. In the 1800's, condensation of the gaseous effluent from the kilns to generate useful products was implemented commercially. Production of methanol and acetic acid were commercially feasible for several decades. However, after World War II, the economics of commercial pyrolysis of wood became less favorable because of the availability of low cost petroleum. Outdoor home cooking largely sustained the charcoal manufacturing industry, and persists as the major market for wood pyrolysis products today.

During the early part of this century, Cliffs Dow in Midland Michigan was a principal charcoal and condensate producer. They developed separation techniques for a couple of wood pyrolysis products for use in the food industry. For undisclosed reasons, Cliffs Dow never did commercialize production of these flavoring substances in North America, but the technology was transferred to a European company and is still used in Europe today.

Another byproduct of the charcoal manufacturing process is pyroligneous acid, which is a crude condensate that sometimes includes both an aqueous and an oily, water-insoluble phase.

Pyroligneous acid was the forerunner of today's commercial smoke flavorings and is still used today in some instances. For example, in Japan pyroligneous acid is classified as wood vinegar and is used in treating casings. In one region in Spain it is used as a table vinegar. In the U.S., it is classified as an artificial smoke flavoring.

Over 30 years ago, Red Arrow pioneered the technology to produce natural liquid smoke using a slow pyrolysis process. Pyroligneous acid differs from natural smoke flavorings primarily in the lack of ability to brown meat. This distinction is the primary reason why smoke flavorings and not pyroligneous acid have become such popular replacements for the use of vaporous smoke in meat processing today. In the U.S. this replacement with liquid smoke is about 75% of all smoked foods. Another reason why smoke flavorings have succeeded in supplanting smoking is the increased safety of the food supply in the virtual elimination of polycyclic aromatic hydrocarbons (PAH) originating from wood pyrolysis. Many of these are carcinogens and are non-polar, hence water insoluble. The manufacturing of smoke flavorings which Red Arrow pioneered involves as a crucial processing step, the phase separation of the water insoluble tarry phase. This tarry phase contains the PAH while the water-soluble phase contains the acids, carbonyls, and phenols which comprise the essentials of a commercially useful smoke flavoring.

Some of the other advantages of smoke flavorings over natural vaporous smoking are reduced emissions, elimination of charcoal disposal, improved product consistency, microbiological control, antioxidative effects, reduced smokehouse maintenance and cleaning, and they make possible the employment of advanced food processing technology.

As stated previously, the browning ability of aqueous smoke flavoring is a primary reason why smoke flavorings have succeeded where pyroligneous acid has not. The reason for this is essentially the presence of reactive aldehydes, the principal browning agent in smoke flavorings. Conventional charcoal production processes, from which pyroligneous acids are typically produced, involves very slow pyrolysis resulting in low yields of the desirable aldehydes. In these processes, acids are still produced, hence the term pyroligneous acid, but yield of the reactive aldehydes is low. Without the ability to brown foods, the advantage of pyroligneous acid over vaporous smoke is severely reduced.

Liquid Smoke Developments

Smoke flavorings are complex mixtures containing over a thousand identified compounds. By way of proximate analysis, a typical aqueous smoke flavoring will contain about 25% organics and 75% water. The organics are 11% acids, 12% carbonyls and 2% phenolics. The acids are useful in control of Listeria, Staph., molds, fungi and Salmonella. The phenols are the principal flavoring substances and are highly effective natural antioxidants. The carbonyls are effective antimicrobial agents as well as the source of browning. There are other classes of substances such as lactones and nitrogenous bases which contribute substantially to the overall smoke flavor. Some of these are organoleptically detectable at low ppb concentrations.

A consistent need in the meat industry is a consumer-driven preference for less flavor and more brown color. Processors have continued to seek more color, yet due to the flavor intensity of commercial smoke flavorings have been unable to achieve the desired effect. Recently, Red Arrow launched a line of smoke flavorings with reduced overall flavor to attempt to capitalize on this idea. Another specialized application which requires high browning is the application to casings which are used to manufacture sausages and hams. All major types of casings are potential markets for high browning, smoke flavorings: cellulose, fibrous reinforce cellulose, collagen, and natural casings. Some of these casings are on the market while others are still in the conceptual stage due to cost, technological and marketing considerations.

Fast pyrolysis is a technology which can be applied to the development of low flavor, high browning, liquid smoke products. While there have been a number of motivating forces driving fast pyrolysis technology over the years, the direct linkage of the technology to a marketable product has been lacking. Extensive collaborative effort between Ensyn and Red Arrow, facilitated by the commitment of Canadian and Wisconsin government program funding has brought this technology to the continuously operational pilot plant stage at a scale of about 2.5 tons/day. This RTP plant has been operating since November, 1989, producing liquid smoke, food flavorings and boiler fuel from biomass.

Future of Fast Pyrolysis Products In the Food Industry

There is a future in the food industry for fast pyrolysis-derived smoke flavorings. The improved yield with fast pyrolysis over slow and the resulting economics could translate to lower manufacturing costs. The application to casings as mentioned earlier is one emerging food processing technique in which browning is more important than smoke flavor. Another area is improved browning in a variety of food processing operations. Certain smoke flavorings may be used to accomplish this where others cannot, but it is the ability to brown under these milder conditions that result in a successful application. Dextrose and other reducing sugars are used in food processing to enhance browning via what are known as Maillard reactions. These are very complex and not completely understood reactions characterized by initial reaction of sugars with amino groups followed by subsequent reactions resulting in brown melanoidin pigment formation. In some instances smoke flavorings may be substituted for the reducing sugars allowing the reactions to proceed more rapidly and at lower temperatures. It is in these situations where fast pyrolysis products may be able to succeed where conventional products have not.

The carbonyls resulting from wood pyrolysis in particular are highly effective browning agents. Their reactivity with amines is well documented in the literature. These carbonyls are highly effective in obtaining brown reaction color even at room temperature. Their usefulness in smoke flavorings in food processing operations such as bacon manufacture where temperatures do not exceed 128 F is well known. It is in these applications of smoke flavorings where fast pyrolysis products will likely be used in the future.

Conversion of a fast pyrolysis liquid to a product containing functionally useful amounts of active browning carbonyls is the major hurdle which must be overcome to be successful. Of course there are also governmental regulations and procedures which must be complied with to produce a marketable product. So the road to commercialization of a product of this sort in the food industry is difficult indeed, notwithstanding the promising state of the technology as it exists today.

One of the aldehydes produced in potentially useful quantities from the pyrolysis of wood is hydroxyacetaldehyde. Quantification of it in fast and slow pyrolysis liquids has been reported by a number of researchers. It is not an approved food additive, but it may have some other applications as a specialty chemical.

In the chemical industry, hydroxyacetaldehyde may have some uses in polymer production where specialty polymers with alcohol functionality is desired. It is readily convertible to ethylene glycol should the high-yield, large-scale production of hydroxyacetaldehyde from waste cellulose become a reality. Hydrogenation of hydroxyacetaldehyde from wood pyrolysate liquids to produce ethylene glycol has been achieved as shown in the literature. An additional product of this hydrogenation could be propylene glycol from acetol the yield of which from wood is about 1%.

There are a number of approved flavoring chemicals present in wood pyrolysis products. But while fast pyrolysis may be used in the manufacture of smoke flavorings in the future, chemicals production from wood pyrolysis is less certain. Production of commodity chemicals will not likely be feasible for a few years. Specialty chemicals has been one of the strategies driving the research and development efforts in fast pyrolysis in recent years. At present it is not economically feasible to produce chemicals from biomass without an integrated "refinery" approach. Certainly, there is no one chemical for which a process can be operated commercially. Nevertheless, it is possible to produce a few chemicals whose combined revenues may support a pyrolysis plant. Among the 50 or so FEMA GRAS (Flavor and Extract Manufacturers Association, generally recognized as safe) chemicals which may be of commercial interest in the flavoring industry are acetic acid, formic acid, acetol, pyruvic aldehyde, diacetyl, furfural, 5-methyl furfural, 2,6 dimethoxyphenol, 2-methoxyphenol, eugenol, isoeugenol, maltol, and cyclotene. The markets for each of these vary widely. The state of separation technology and it's cost are the major considerations for the development of viable production operations for flavor chemicals.

In conclusion, the future of fast pyrolysis in the specialized area in the food industry niche that Red Arrow is involved in is promising. We have come a long way from the early days of "gram" scale fast pyrolysis research to the 2.5 ton/day RTP pilot scale reactor in continuous operation at our plant today. Scaleup to a larger RTP system is a next logical step should widespread commercial acceptance of products from it become a reality. Reactor system improvements to reduce maintenance and downtime are ongoing. One major expense which will have to be incurred in the future is the complete engineering design of systems in the range of 25 tons/day which are needed for industrial production. We are convinced that within the next few years these larger scale systems will be available so that the full potential of fast pyrolysis technology can be realized.

Discussion:

D. Scott (University of Waterloo): I wonder if you could give us some idea of what the total market is in the food industry for this fast pyrolysis or for the smoke flavourings in terms of organic content rather than what you sell.

G. Underwood (Speaker): Well, broadly, the various data that has been published, was available in the review back in 1981-82 and since that time the market has grown. So, I am estimating somewhere (between three manufacturers in North America), it is about a \$20 million U.S. market world-wide. That includes all of the products that would be derived from the water soluble fraction and not just the water soluble fraction itself.

There is in the neighbourhood of about four million gallons of water soluble slow pyrolysis products now manufactured by the three that are manufactured in North America. There is about four million gallons of product, about forty million pounds per year and it is about 25% total organics.

N. Bakhshi (University of Saskatchewan): From what kind of a wood or a feed material are all these materials which you have tried or which you are using?

Speaker: What kind of wood species?

N. Bakhshi: Right, yes.

Speaker: The hardwood species. We use typically oak, hickory, beech, birch and maple - a lot of our product is oak and maple. Also there are some other speciality products. We have one customer in British Columbia that absolutely insisted that they had to have a squamish alder smoke. Supposedly they had logs floating down the river that they were going to use and then ship the sawdust to us, but we have not been able to produce that one yet. However, we produce a mesquite smoke flavouring too, and some of the other manufacturers produce an apple or a cherry smoke, but very minor products.

H. Pakdel (University of Laval): What is the market and value for maltols which you showed in the last slide?

Speaker: Maltol? We have one but I guess it depends on whether it is synthetic or natural. We have arguments that you can use to say if you extract it out of here it is natural, but it would be anywhere from \$50 to \$200 a pound generally. I do not know what the total market is, but I have received indications from one of our customers, that if we could produce the material, they could probably sell 10,000 pound per year and that is probably only a small part of the market.

I just cannot recall, offhand, what the estimate is for that. However, I am sure it is in the order of 1 to \$10 million a year - in that magnitude. It is a flavour enhancer. It is generally used at 1 to 10 parts per million.

H. Pakdel: Do you have access to the information on the maltol market and the present value?

Speaker: I have some estimates which were compiled a couple of years ago, but they are not really complete. I could do a little more work and I am doing that now through another one of our customers. I am trying to find out what the market for a couple of those mentioned are.

H. Pakdel: That is usually from hardwood?

Speaker: Yes, that is from a hardwood product. In Europe, I think they use beech.

Economics of Flash Pyrolysis of Peat

P.B. Fransham
Encon Enterprises Inc., Calgary, Alberta

Introduction

Encon Enterprises Inc. was incorporated in 1989 to commercialize the fast pyrolysis of peat moss. To this end, a demonstration plant has been constructed and tested. The primary focus of the testing program was to evaluate the technical and economic feasibility of a peat pyrolysis plant in central Newfoundland. The plant has been described elsewhere **(1)** and while certain design limitations were identified, the conclusions obtained from the research indicate that the technology is viable. Data obtained from the demonstration plant and also from laboratory investigations **(2)** have been used as input into an economic model. The overall economics are excellent and the project should be of interest to venture capitalists. Project scheduling has been completed and if funding can be secured a commercial project should be under way by the middle of 1991.

Economic Model

A manufacturing model consisting of eight schedules was written using Lotus 123. The model is comprehensive and has been extensively tested in a variety of business applications. The eight schedules are as follows:

1) Income Statement
2) Cost of Goods Manufactured
3) Changes in Financial Position
4) Production and Sales
5) Raw Material and Labour Costs
6) Variable Operating Costs
7) General Administration Costs
8) Salary and Wages

The model can be set up on a monthly, quarterly, or annual basis. Usually a month by month analysis is carried out until the plant has reached steady state operation. Annual analyses are then performed until the required time period is reached. In the case of a peat pyrolysis plant, a monthly analysis was performed until the end of the fifth year, followed by annual statements until year ten.

A second model (also written in Lotus 123) was subsequently developed to summarize the data from the first model into a more concise format. The second model synthesized the data into annual statements that included depreciation, taxes, inflation factors, discount factors and internal rate of return over a ten year period. By varying parameters such as capital cost, revenue, operating costs, and raw material costs, it was possible to evaluate the sensitivity of Internal Rate of Return (IRR) to changes in each of the parameters. The summarized data contained in Model 2 is presented in Table 1. The following discussion highlights some of the inputs to the model and important conclusions.

Production Profile

A 50 tonne per day plant using dry peat moss as a feedstock would produce the following basic pyrolysis products:

char	20 tonnes/day
liquids	20 tonnes/day
gas	10 tonnes/day

The production profile shown in Table 1 lists activated charcoal as one of the products. Research carried out during the past two years has shown that the char can be upgraded to a middle grade activated charcoal after a 40% burn-off. Activated charcoal production uses raw char as a feedstock and therefore char production decreases as activated charcoal production increases. Activated charcoal production is projected to reach maximum levels by 1995.

The main pyrolysis plant is slated to operate on a three shift, five day week. Normally Saturday would be reserved for plant maintenance and in the case of significant repairs, Sunday could also be assigned to maintenance. Experience has shown that heat exchangers and pumps require regular maintenance and an adequate supply of spare parts will need to be on hand. The activated charcoal plant will operate on a 40 hour work week. Should demand for Encon's activated charcoal increase after 1995, then additional production capacity can be added via a second and if necessary a third shift.

Capital Costs

Table 1 lists the major capital components and the year in which the costs will be incurred. Inspection of the capital cost components shows that the bog development is estimated at $750,000, process equipment at $2.3 million, with the remainder of the $3.5 million costs slated for salary, building, and start up. Assuming a project start date of June 1991, the majority of the capital costs are incurred in 1992.

Bog development costs and harvesting costs were estimated from existing data **(3)** and adjusted to 1990 cost figures. Major suppliers to the demonstration plant were requested to provide quotations for a 50 tonne/day plant and hence the capital costs for the full scale plant are considered to be accurate to within 10%.

Operating Costs

Operating costs have been partitioned between: plant labour; peat costs; factor overhead; and selling and administration. Once the production profile has reached steady state in 1995, factory overhead accounts for approximately 50% of the operating costs. Included in this category are all energy costs, maintenance, and activated charcoal production costs. Since activated charcoal forms an integral part of the production profile, the numbers presented in Table 1 cannot be used to determine the economics of the pyrolysis conversion component. Peat cost include the cost of harvesting, screening and drying to under 10% moisture. All related harvesting and capital costs are blended into the overall peat cost.

Raw peat is a viable energy source and could be used to fuel the dryer, pyrolysis plant, and activated charcoal plant. The energy cost used in the analysis assumes a $4.00/MBTU energy cost (the local cost of #6 fuel oil) and is very conservative. Analysis has shown that peat can be harvested for on site combustion for about $1.00/MBTU. or about 75% less than the prevailing fossil fuel cost. Peat is low in sulphur and ash and when dry is an excellent fuel source. Peat would be a viable alternative energy in Canada if it were not for the high cost of transportation to energy consumers. On site combustion and/or pyrolysis to liquid fuel, overcomes this limiting factor, lessening the dependence on fossil fuels, and improves project economics.

Price Structure and Revenue

Conservative pricing has been used for establishing the value of the product streams. Energy prices have been set according to the local cost of energy and activated charcoal prices from current import data. Laboratory testing established the higher heat value (HHV) for each product streams and the energy value was calculated on the basis of energy per unit weight. The char has an estimated value of $100/tonne, oil at $12.00 per barrel, gas at $1.00/MCF, and activated charcoal at $1.50/kg. At full capacity the plant should have an annual revenue of $2.6 million.

Cashflow Calculations

Cashflow has been calculated by carrying forward losses incurred during the early years to offset taxes in the later years. A positive cash flow is obtained in 1993, but no taxes are payable until 1998. The estimated net present value of the investment over ten years is $68366. before taxes and $569440. after taxes. The internal rates of return for the tax out and tax in cases are 16.4% and 15.63% respectively.

Sensitivity

A sensitivity analysis was performed on the major cost and revenue sectors. Parameters used in generating the base case listed in Table 1 were varied from -20% to +20%. The outcome of the sensitivity analysis is shown in Figure 1. The Internal Rate of Return (IRR) is most sensitive to variations in operating costs and revenue and least sensitive to variations in peat costs and capital costs.

Since these numbers were generated using conservative product values and operating costs, the rate of return is considered to be the minimum expected from the project and actual rates of return should be in the 20% range. Significant operating cost reductions can be afforded by using peat for drying, heat input to the pyrolysis reactor, and for steam activation. It is estimated that the operating costs could drop by 15% by using peat as the main energy source and the rate of return will jump to 22%. The present value of the energy is about $20.00 (CDN) up from $12.00 used in the analysis and the IRR is presently close to 22% after tax. It is evident from the analysis that peat pyrolysis in Newfoundland is a viable economic entity and Encon is optimistic about the possibility of constructing a plant in Bishop's Falls, Newfoundland.

All rates of return have been calculated assuming no Government involvement in the project. The philosophy used in the analysis was to determine the real economics of the project rather than those including subsidies, loans or outright grants. If some Government support is forthcoming, then the rates of return to the private investor become increasingly attractive. It is generally conceded that an IRR in the 20% range is sufficiently high to attract investment capital and given the economics presented above, investment capital should be forthcoming.

Project Sponsors

The following is a list of the project costs as of October 1990 and the contribution by Government agencies and private enterprise:

Project Costs	
Capital Costs	$400,000
Materials and Supplies	$300,000
Engineering and Labour	$300,000
TOTAL	$1,000,000

Sponsors

Energy Mines and Resources	$400,000
National Research Council	$230,000
Department of Energy	$90,000
Culmen Enterprises Inc.	$280,000

Bridge and receivable financing was provided by the Royal Bank of Canada.

References

1. Fransham, P., Fast Pyrolysis Demonstration Plant. R&D Contractors Meeting on Biomass Liquefaction, Published by CANMET, 1990.

2. Scott, D.S., Internal reports and laboratory data

3. Noval Technologies Ltd., Bishop's Falls Milled Peat Mining Project. Canada/Newfoundland Conservation and Renewable Energy Demonstration Agreement. Report Number CR 85-70, March 1985.

Discussion:

C. Roy (Université Laval): Your presentation reminds me of some aspects of my PhD work which was on pyrolysis of peat, actually, more than a decade ago. You will be interested to hear that Esteban Chornet, with whom I have been doing my PhD, and I, have published a very extensive paper on pyrolysis of peat. I would be glad to give a copy to you if you give me your card. I have, however, a couple of comments here on your hypothesis regarding your program. Quite a bit, relies on the revenues generated by the activated charcoal and to another extent, on carbon black.

Now, I am sure you are aware of the Norit which is currently producing activated charcoal from peat and with actually quite a great profit. I think their price is even in the range of $3,000 per ton. There has also been many more unsuccessful attempts in making activated charcoal from peat. The one very important parameter being the level of peat. The ash content of the Norit Plant, and I think get their peat in Holland, is less than 3%.

Now, finally, I don't know what the ash content is of the peat you are considering, but this is something you must take into account, because it is going to keep you away from a certain type of activated charcoal, like the granular charcoal which requires a lot of cohesion. As I have said, there have been unsuccessful attempts in Québec, even trying to identify and find a proper bog for activated charcoal making. So, this is one thing, regarding carbon black, you are bringing a number of 25 cents per pound I think. I suspect that the carbon black you get from peat will be more of an Austin grade type of carbon black, the one which is produced from coal in Virginia. I think that your value is probably twice what is probably the real market for this carbon black. So, I am just warning you to keep this in mind for your own information.

P. Fransham (Speaker): I only put that on as an example, kind of taking from you what the value adding was. We actually don't anticipate getting into the carbon black market, because I don't think it is a good enough quality. It would be a tremendously hard sell to convince people to take it.

C. Roy: It would be, at the best case, as a filler and I think you could get 12 cents a pound, not 25 cents a pound.

Speaker: The ash runs usually under 5% in this particular bog where we are. So, I don't know how that is going to impact. The ash content in the char though is up around 10%, by the time you concentrate that 5% up. So, is that going to be problematic? We would agglomerate the char, maybe even with the pyrolysis oils and tars, before we would activate and it is a question of whether we could keep that particular particle and give it enough cohesive strength to actually be a granular and tap that higher value added market. We still have to do some research in that area.

Norit, with their high value activated charcoal only represents about 4% of the Canadian imports. The U.S., largely through Calgon, represents about 75% of the import. So, we are not pricing with Norit, we are pricing against Calgon and the big U.S. suppliers. Ironically, about half the imports go in through Western Canada and half go in through Ontario. Where they are distributed after that we haven't yet ascertained. It would appear that there would be a market in Western Canada and also a market in Eastern Canada, but with Free Trade that could well kill us. This is where we have to make an assessment, really, on the impact of Free Trade.

C. Roy: Just to complete, the last difficulty with activated charcoal, it is an industry which is a captive one, in the hands of a few - as you have said, there is Calgon, there is Norit, to illustrate - but there are just five or six of them world-wide. It is extremely difficult to tap into this market, because you get into the regeneration type of service. So, they have linked and bound customers to them with an established relationship over the years. I would, therefore, suspect a better strategy would be perhaps to sell the material to those companies so that they activate it.

Speaker: Yes, we haven't really developed a full strategy on how we are going to break into that market, but I can appreciate that it is not going to be an easy sell and that is why we started off in one with very small sales. We will not build up to our peak peat volume until years five or six for exactly that reason.

D. Scott (University of Waterloo): Since the activated carbon is such a big part of the economics of Peter's presentation, I might make a comment on the quality.

There are multitudes of uses, of course, for activated carbon and one has to target some market they are after.

So, we targeted the water treatment market. In fact for the kind of material that you would want to remove from contaminated water, i.e., the large molecules that give bad taste, bad odour, bad colour, so as to make it potable. In fact, this activated charcoal that Matt Siska produced from peat or from hog fuel, is the equal of any commercial product, in decolourizing and sweetening ability.

Table 1. Cashflow summary

ENCON ENTERPRISES INC - PYROLYSIS PLANT

Year	1991	1992	1993	1994	1995	1996	1997	1998	1999	2000	TOTAL
	1	2	3	4	5	6	7	8	9	10	
Production profile											
Char tonnes/year	0	0	1984	3066	1318	1318	1318	1318	1318	1318	12958
Oil BBL/year	0	0	19277	33057	33057	33056	33056	33056	33056	33056	250671
Gas MCF/year	0	0	36372	62372	62372	62370	62370	62370	62370	62370	472966
Activated charcoal kg/yr	0	0	190000	684000	1368000	1368000	1368000	1368000	1368000	1368000	9082000
Capital costs: ($CDN)											
Plant design	10000	0	0	0	0	0	0	0	0	0	10000
Bog selection	10000	0	0	0	0	0	0	0	0	0	10000
Development plant	10000	0	0	0	0	0	0	0	0	0	10000
Equipment purchase	80000	0	0	0	0	0	0	0	0	0	80000
Plant equipment	0	1500000	0	0	0	0	0	0	0	0	1500000
Building construction	0	100000	0	0	0	0	0	0	0	0	100000
Harvesting equipment	0	500000	0	0	0	0	0	0	0	0	500000
Field ditching	0	41000	0	0	0	0	0	0	0	0	41000
Perimeter ditching	0	60000	0	0	0	0	0	0	0	0	60000
Site location/prep/erect	0	50000	0	0	0	0	0	0	0	0	50000
Progress assembly	0	800000	0	0	0	0	0	0	0	0	800000
Final surface preparation	0	91400	0	0	0	0	0	0	0	0	91400
Start up	0	50000	50000	0	0	0	0	0	0	0	100000
Total capital ($CDN)	110000	3192400	50000	0	0	0	0	0	0	0	3352400
Operating costs: ($CDN)											
Labour	0	0	113481	191637	234468	234468	234468	234468	234468	234468	1711926
Peat costs	0	0	145488	249488	249480	249480	249480	249480	249480	249480	1891856
Factory overhead	0	0	238153	644793	1096233	1096233	1096233	1096233	1096233	1096233	7460344
Selling and admin	193256	278849	328176	328176	367176	367176	367176	367176	367176	367176	3331513
Total op costs ($CDN)	193256	278849	825298	1414094	1947357	1947357	1947357	1947357	1947357	1947357	14395639
Prices:											
Char $CDN/tonne	100	100	100	100	100	100	100	100	100	100	
Oil $CDN/BBL	12	12	12	12	12	12	12	12	12	12	
Gas $CDN/MCF	1	1	1	1	1	1	1	1	1	1	
Activated charcoal $CDN/k	2	2	2	2	2	2	2	2	2	2	
Total plant revenue											
Revenue:	0	0	751096	1791656	2642856	2642842	2642842	2642842	2642842	2642842	18399818
Financial assumptions											
Inflation factor (%)	0	0	0	0	0	0	0	0	0	0	
Discount rate nominal (%)	11	11	11	11	11	11	11	11	11	11	

ENCON ENTERPRISES INC - PYROLYSIS PLANT

Year	1991	1992	1993	1994	1995	1996	1997	1998	1999	2000	TOTAL
	1	2	3	4	5	6	7	8	9	10	
Tax calculation:											
Revenue ($CDN)	0	0	751096	1791656	2642856	2642842	2642842	2642842	2642842	2642842	18399818
Op costs ($CDN)	193256	278849	825298	1414094	1947357	1947357	1947357	1947357	1947357	1947357	14395639
Depreciation	11000	330240	335240	335240	335240	335240	335240	335240	335240	335240	3023160
Profit ($CDN)	-204256	-609089	-409442	42322	360259	360245	360245	360245	360245	360245	981019
Loss carried forward	-204256	-813345	-1222787	-1180465	-820206	-459961	-99716	0	0	0	-4800736
Taxable income	0	0	0	0	0	0	0	260529	360245	360245	981019
Total taxes	0	0	0	0	0	0	0	78159	108074	108074	294307
Unburden cashflow	-292256	-3141009	211038	712802	1030739	1030725	1030725	1030725	1030725	1030725	3674939
Total taxes	0	0	0	0	0	0	0	78159	108074	108074	294307
Net cashflow	-292256	-3141009	211038	712802	1030739	1030725	1030725	952566	922652	922652	3380634
Cum cashflow	-292256	-3433265	-3222227	-2509425	-1478686	-447961	582764	1535330	2457982	3380633	
NPV (unburdened)	683666										
IRR (unburdened) (%)	16.4										
NPV (after tax)	569440										
IRR (after tax) (%)	15.63										

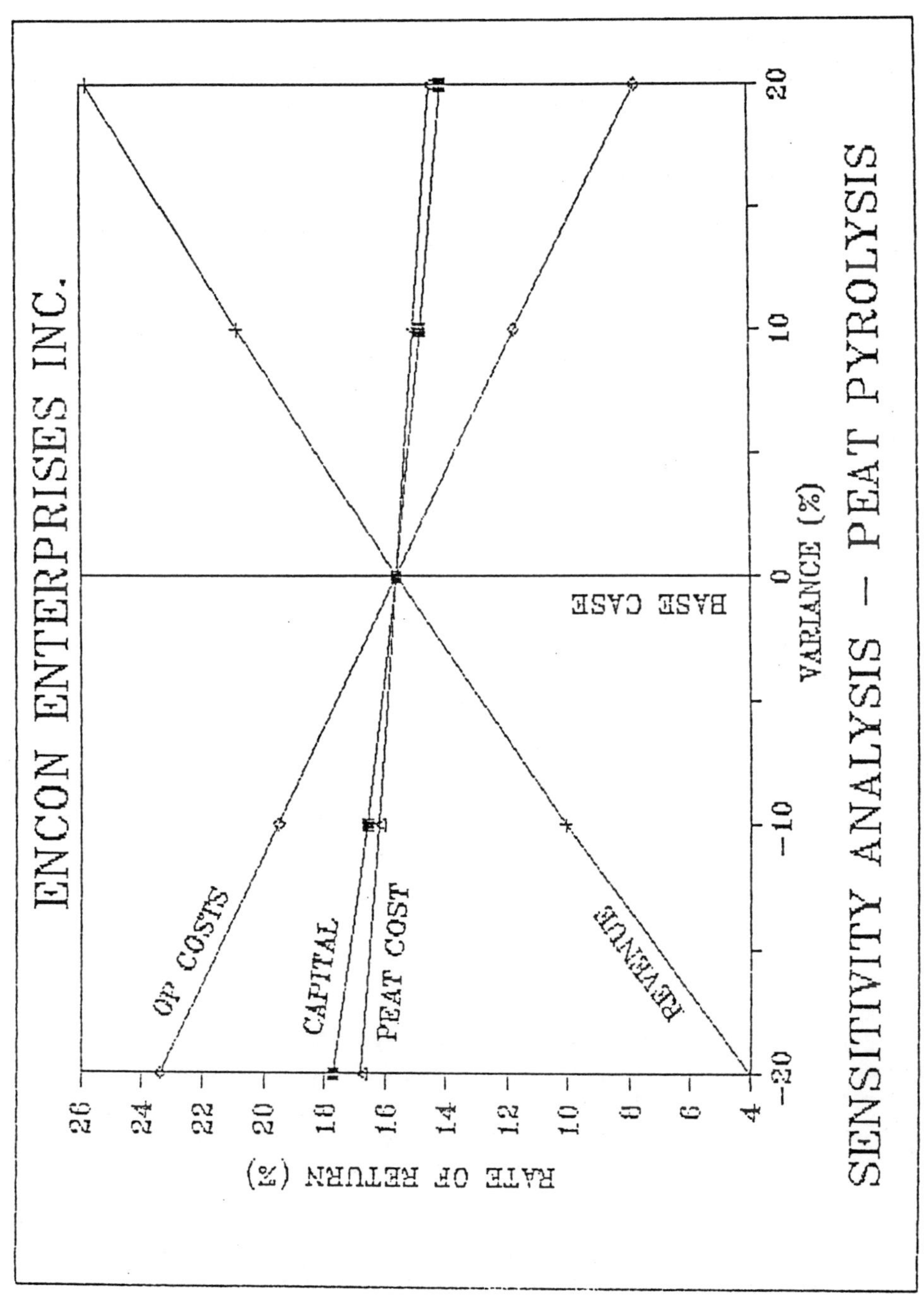

Figure 1 Sensitivity Analysis

Scale-up of the Ablative Fast Pyrolysis Process

D.A. Johnson, W.A. Ayres and G. Tomberlin
Interchem (N.A.) Industries, Inc., 8016 State Line, Leawood, Kansas 66208

Interchem (N.A.) Industries, Inc. is a member of the Pyrolysis Materials Research Consortium (PMRC). This Consortium was formed in 1989 to permit and encourage private commercialization of technology financed by government funds from the U.S. Department of Energy/Solar Energy Research Institute (DOE/SERI). Five companies were selected to join the consortium; Allied-Signal, Corp., Aristech Chemical Corporation, Georgia-Pacific, Corp, Plastics Engineering Company and Pyrotech/Interchem, Inc. The DOE/SERI has conducted research and development in the areas of (1) breaking down biomass material via fast pyrolysis into constituent chemicals; (2) processing or separating the chemicals into products or feedstocks; and (3) using the feedstocks to make products such as phenolic resins and aromatic hydrocarbons. Titles to the technologies developed by DOE/SERI in this area have been acquired by Midwest Research Institute Ventures in order to license these technologies to the Consortium members.
Pyrotech/Interchem's contribution to the Consortium is to scale-up the ablative fast pyrolysis process. During the Spring of 1990, Interchem started construction of a scale-up to a 36 dry ton per day facility for the production of fuel oil and charcoal. The equipment, referred to as a Petroleum Synthesis Unit (PSU) was designed to be mounted on two skids so it could be entirely built in the shop requiring minimal field installation. Construction and installation of the PSU was completed in September, 1990.

La société *Interchem (N.A.) Industries, Inc.* est membre du *Pyrolysis Materials Research Consortium* (PMRC). Ce consortium a été créé en 1989 en vue de rendre possible et d'encourager la commercialisation privée de technologies financées à même les fonds gouvernementaux provenant du *Solar Energy Research Institute* du *Department of Energy* (DOE/SERI) des États-Unis. Cinq sociétés ont été choisies pour se joindre au consortium, soit l'*Allied-Signal, Corp.*, *l'Aristech Chemical Corporation*, la *Georgia-Pacific, Corp.*, la *Plastics Engineering Co.* et la *Pyrotech/Interchem, Inc.*, Le DOE/SERI a effectué des travaux de recherche et de développement dans les domaines suivants: 1) dégradation de la biomasse par pyrolyse rapide, en vue d'obtenir des substances chimiques; 2) traitement ou séparation des substances chimiques en produits ou en charges d'alimentation; et 3) utilisation de ces charges d'alimentation pour fabriquer des produits comme des résines phénoliques et des hydrocarbures aromatiques. La *Midwest Research Institute Ventures* s'est portée acquéreur des titres de propriété des techniques mises au point par le DOE/SERI dans ce domaine, en vue d'en conférer les droits d'exploitation sous licence aux membres du consortium.
La *Pyrotech/Interchem* s'occupera, dans le cadre des travaux réalisés par le consortium, d'accroître l'échelle du procédé de pyrolyse rapide par ablation. L'*Interchem* a entrepris, au printemps 1990, la construction d'une installation d'une capacité de 36 tonnes de matière sèche par jour, destinée à produire de l'huile combustible et du charbon.

Site Selection

A site was chosen in Southern Missouri where the disposal of waste wood is an ongoing environmental problem. Currently the waste wood is being burned in Tee-Pee Burners or dumped in large piles. With the increasing environmental pressures and regulations, local mill owners are seeking viable alternatives for wood waste disposal. Fast pyrolysis offers a unique solution to this disposal problem. The process reduces air emissions over 70% from the current technology of burning in Tee-Pee Burners. In addition, there are no groundwater discharge problems which occur when the waste wood is simply piled in large areas and allowed to decay.

Interchem plans to showcase this facility as a new method whereby mill owners can dispose of their waste wood in an environmentally safe and economic manner.

The Process

The following steps outline the overall process for scale-up facility. A flow diagram of the process is located in Figure 1.

Wood Handling System

The sawdust is transported to the system in two different ways; (1) by live bottom truck from a storage silo and (2) by pneumatic conveyor from the flooring plant to the live bottom truck (see Figure 2).

Pyrolysis System

The sawdust enters the hopper which has a 1-1/2 ton capacity. The bottom of the bin has a vibrator which delivers the wood to the PSU feeder system. The wood will be discharged through a rotary valve which is used as a pressure seal. The wood is then dropped into an eductor which receives inert gases from the PSU system. These gases pneumatically transport wood fuel to the PSU for reaction.

The PSU system involves the use of a very high throughput reactor in which entrained biomass particles enter tangentially into a vortex tube at one end and exit tangentially from the opposite end. The vortex tube wall is radiantly heated by energy released from the surrounding refractory wall which is heated as some of the char and gases are oxidized. Gas and oil vapors are produced as the biomass slides along the inside surface. These gases then go to the Condenser.

The gases and char from the recirculation system enter the combustion chamber of the PSU. In this chamber, substoichiometric combustion of the residual wood particles and gases takes place. These exit the combustion chamber and go to a cyclone for char removal and then to the Burner.

Condenser

The gases from the pyrolysis unit enter the Condenser at process temperatures and are cooled to approximately 85°F via a heat exchanger. The vapors are condensed to liquid oil and returned to the sump. Approximately three gallons of oil per minute are discharged to the storage tanks and the remaining vapor stays in the recirculation system and becomes part of the heat transfer media which eventually enters the combustion chamber of the PSU.

Burner

All exit gases are passed through the burner which is lined with refractory material. Complete combustion takes place in the burner and very low levels of carbon monoxide, nitrous oxide and particulates are all that remain.

Oil Handling System

The oil is pumped to one of two 5,000 gallon storage tanks. These tanks are diked on a sloping floor and are covered by a roof. During the loading of the tankers, the truck will be backed into the diked area so that any potential spill is contained.

Char Handling System

The char is pneumatically conveyed from the cyclones into a covered truck-trailer. This is done with a nitrogen overlay and the air is recycled.

The PSU system (see Figures 3 and 4) will process 13,000 dry tons of waste wood annually to create approximately 1.4 million gallons of fuel oil and 1,900 tons per year of charcoal.

The oils produced by the process are oxygenated fuels with a BTU value between 90,000 - 95,000 BTU per gallon. They can be used as a replacement for No. 6 Fuel Oil or upgraded to a No. 2 Fuel Oil substitute. One of the primary advantages of wood pyrolysis fuel oil over conventional No. 6 Fuel Oil is that it contains no sulfur. With the passage of the new Clean Air Act, air emissions by companies are going to be tightened considerably especially in relation to SO_2 emissions.

The char produced by this facility will be sold to local charcoal briquetting manufacturers in the vicinity of the facility. Many of the major charcoal briquette producers are located in this area and are always looking for suppliers which can deliver char with standard characteristic.

Economics

Each pyrolysis unit will cost $750,000 (U.S) to develop, build and install. This cost includes the materials handling system, pyrolysis system, oil handling system, char handling system and all trucks to deliver the wood, oil and charcoal. The only costs which are not included in this price include the land on which the unit is built and a building to house the unit.

Future Products and Activities

Further development is underway to increase the product lines derived from the pyrolysis of waste wood. As part of the PMRC activities, research and market development is ongoing in the area of phenolics replacement in adhesives and engineering plastics. As this development continues, there should be a market created for a pyrolysis P/N substitution of phenol in these products. In addition, the oil produced by the pyrolysis system can be converted into carbon black and the charcoal can be upgraded into activated carbon which is a growing market. All of these commodity chemicals are products which our company is actively exploring.

Interchem has plans underway to build up to twenty similar facilities. This scale facility is an ideal size to be located at mill sites since they are skid mounted and mobile.

Plans are also underway to scale-up the unit by 8 times in size. The New York State Energy Research and Development Authority has just awarded a $500,000 research contract to Integrated Woodchem, Inc., a wholly owned subsidiary of Interchem, to develop a 288 dry ton per day waste wood to chemical facility in New York State. This one year contract will fund site selection, engineering design and permitting for the plant development. Upon completion, Integrated Woodchem will build a $25 million plant to convert 100,000 tons per year of waste sawdust and wood chips into valuable commodity chemicals.

Discussion:

K. Oehr (B.C. Research): Could you say something about waste treatment of an aqueous phase, if there is any, from the process and how you deal with that and how it effects economics for the whole plant?

D. Johnson (Speaker): We are not expecting a whole lot in the aqueous phase. What we do have, we just make activated carbon from it and you can start treating it through.

P. Fransham (Encon Enterprises): So, you use the activated carbon to treat the ---

Speaker: Sure, you just take one of your products.

P. Fransham: What is your record - something like 3/4 of a million dollars? It was just near the end of your talk; was that your capital cost?

Speaker: It is the capital cost of the unit.

P. Fransham: Of the 36 tons per day?

Speaker: The 36 tons per day plant includes the dryers, the pyrolysis unit, the condenser, the tanks, the trucks, the whole kit and kaboodle.

D. Beckman (Zeton Incorporated): Our group has been looking at the ablative reactor process, and one of the concerns we have was scaling it up. Could you explain to me how in your 36 ton per day plant you control the temperature, the wall temperature in the reactor, which, I believe, is critical to that process?

Speaker: You control the combustion process in the unit, in the combustion unit around it, and you can control that by the amount of gas and char you pump in.

D. Beckman: But are you getting a feedback for your wall temperature?

Speaker: Once we get it up and running we expect to be able to keep it at a constant temperature. But it is not a unit where you are going to be closing it down on weekends. We will run it 24 hours a day, seven days a week so that we can maintain temperature.

P. Laborde (CQVB): Could you tell us how the consortium is working and if it is involved in the research and development, and what is the involvement of each industry in terms of effort, money or *in kind* consideration?

Speaker: The general way the consortium works is, and yes, I do believe it is really working, as an industry - I think Jim has spoken from the research and the government side of it and as an industry member, we are all really excited about it. Basically how it works is that we will meet twice a year for an update on research and there is also a business committee that will look at other things. The research that SERI does will be looked at. There are also presentations by most of the member companies. Most member companies are all doing some type of *in kind* research, depending upon where our interests lie. In our case it is on a scale-up. Part of what we will do is provide oils now in larger quantities to all the members to work on.

As an industry member that is to my benefit, because now some of these other companies decide, oh yes, we have made this nice adhesive and we are going to go to market or resins or whatever products they are making. Since they have already used our oil to test, it now provides us a market. I think it works the same way, so there is a lot of interaction between the companies, we are not really competing with each other. The other nice thing about it is, as industry members, and I sit on the business committee, we can sit down and go back and talk to SERI and DOE and say, well, we now, as industry, have picked up this research and we are really running with it. It is becoming more and more proprietary inhouse as every company is doing their own thing.

There are also these really exciting areas over here that we haven't looked at yet, and maybe you could try to go out and start doing the R&D back on these sides again. Then we will start picking that up and because I think with the pyrolysis, it is what you have been hearing the last couple of days, there is so much available as far as the chemicals and everything else goes, you know, different markets and different things that can be made from the oils that we haven't looked at yet.

N. Bakhshi (University of Saskatchewan): You showed us something about the combustion of this oil; is that oil by itself?

Speaker: No, it is the oil that comes out of the pyrolysis unit. In fact, that was oil out of SERI's unit.

N. Bakhshi: There are no problems as far as combustion is concerned?

Speaker: No. Well, I mean there are some mechanical problems that you have to handle, but there are no problems with the burner itself. So basically, we are looking at minor retrofits to existing boilers that are using number six.

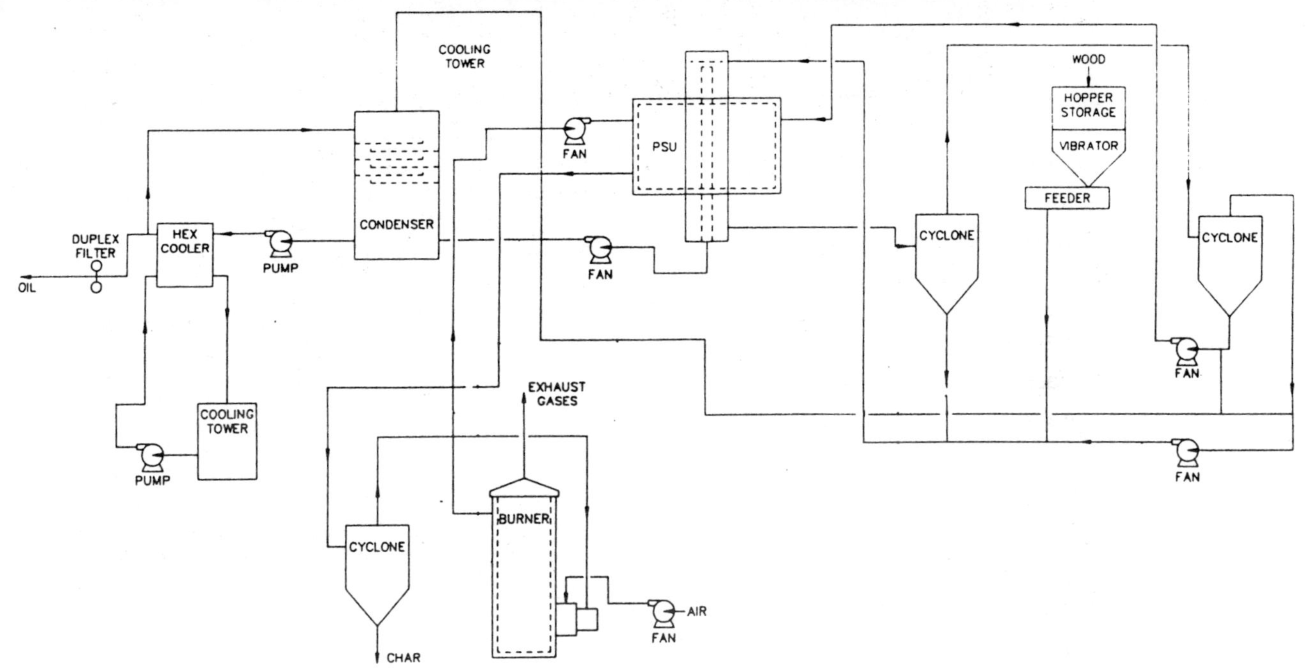

Figure 1. Process Flow Diagram.

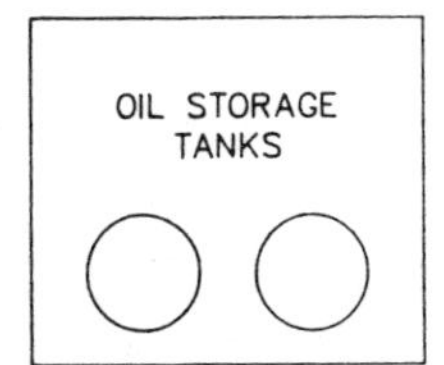

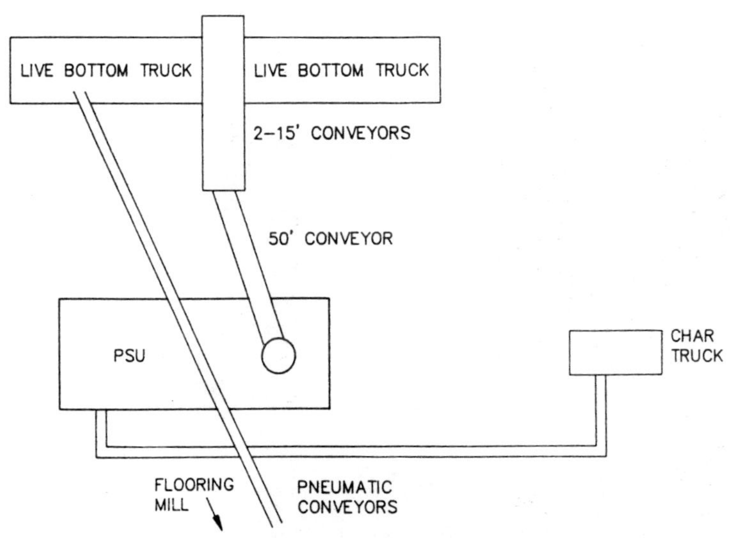

Figure 2. Layout of the Pyrolysis System (not drawn to scale).

Figures 3 and 4. Photographs of the Petroleum Synthesis Unit.

Some Problems of Technology Development for the Small Innovators

A. Wong
Arbokem Inc., Vancouver, Canada

Many of the fundamental problems encountered by the small innovators (including independent inventors and small technology development companies) are related to the lack of funds for continued research and development. The situation is particularly acute at the technology demonstration stage. There are significant deficiencies in the present means of funding research and development, and eventual commercial implementation of new technology, for the small innovators.

Government R&D Funding

Under the present economic situation, most Federal Government programs are based on cost sharing, typically in the range of 10 to 50%, in the form of repayable and non-repayable contributions. But the small innovator is left with a basic problem with such a means of public funding assistance. The small innovator has to find the remaining funds to support his research and development activities. The company has to take on other (frequently non-related) business activities just to sustain itself, and can not devote its maximum energy to research and development.

In reality, the principal beneficiaries of the present government funding assistance programs are the larger companies, universities and industry research institutions. With the exception of universities, all others are sufficiently wealthy to provide their own funding for research and development. If they do not spend money on technology research and development for their own business prosperity, then they deserve to be put out of business by the lag in technological competitiveness.

Tax and/or cash rebate is a sensible approach to encourage company research and development. But it might be better structured to induce a company to meet a certain minimum qualification level such as 1% of the gross sales or income for technology research and development. In Canada, many resource industries are spending much less than this 1% level. Should they be given any tax rebate?

Venture Capital

Another source of funding for technology research and development is venture capital fund. The Vancouver Stock Exchange has a reputation for raising money in such a fashion. Venture capitalists are difficult people to deal with. Venture capitalists force the small innovator to sell his body and soul. The venture capitalists do not care about technology or technology development. They are only interested in making a fast dollar - the faster the better. In many cases, the small innovator is eventually enslaved by the venture capitalists.

Large Established Companies

Can the small innovator go to an established larger company for financial support for his technology idea? It is not easy. The larger company with its own research department is probably the most difficult to convince. The most common reasons of disinterest include:

- It is not invented here
- We have our own research jobs to protect
- Acceptance of an outside idea from a small innovator would make us look bad within the Company

The small innovator also has to be wary of any apparent interest shown by the Company research department. It can be a ploy to extract new ideas from the small innovator. These corporate researchers can not be trusted. The small innovator is left in limbo and wastes time in this pursuit and perhaps even loses control of his novel idea to others. It is a jungle in the realm of technology research and development.

In another scenario, the small innovator's idea is so attractive that the large company could decide to buy it to keep it out of the market. The innovation might damage the market of the company's present line of products.

Banking Institutions

Banks never lend money for technology research and development, or even implementation of advanced technology. In general, banks only lend money to those persons or companies with money. The small innovator is usually not in this class of people or companies.

In a recent study commissioned by the Canadian Bankers' Association, small businesses are blamed for their lack of understanding of the role of the bank in company financing. Perhaps the banks should examine their own, sometimes despicable, behavior towards small businesses. The study pointed out that the banks are not in the "risk" business. Small-scale technology innovators need not apply. The record shows that banks are often driven by greed to finance high-risk junk bonds, leveraged buy-outs, and Third World loans. When the banks stumble with such loans, the banks shift a large share of the write-down burden to the small businesses.

The banks are generally more comfortable with "proven state-of-the-art technology". But "proven state-of-the-art technology" is in fact yesterday's technology. Such an orthodox approach guarantees that at the slightest downturn in the economy, the enterprise with the classical technology will be adversely affected. This outcome should not be surprising to the banks. The enterprise with the "proven state-of-the-art technology" has the same cost structure as its competitor with the same technology. Yet, when new advanced technology is proposed, the banks could not be interested in project financing. Such is the fact of modern banking practices.

Commercial banks like independent studies made by reputable large engineering or market study consulting companies. The problem here is that the small innovator has to disclose his technology and other trade secrets to these consulting companies. These consultants had generally spent nothing on technology research and development; and yet they have virtually unrestricted access to novel technologies. Moreover, the study exercise is frequently worthless because these consultants are generally not knowledgeable with the particular leading-edge technology. The small innovator is forced to buy "the consultant's name". It is an unavoidable financial burden on the small innovator with limited financial resources.

Government Intervention

It is futile to try to reform the banks. It is not possible to banish the venture capitalists. The Government could undertake a more effective role to foster Canadian technology development and comercialization.

In order to partially alleviate the cash-related problems of the small innovators in bringing his technology to the demonstration stage, the government might organize the formation of a "Technology Development Bank (TDB)". The TDB would be different from the existing Federal Business Development Bank (FBDB). The FBDB lends money, popularly advertised as the lending bank of the last resort, to small businesses such as shopping mall cookie shops, FBDB lacks the emphasis for the realization of innovative technology in Canada.

Unlike the FBDB, the TDB would have a very different mandate. It should be a financing institution with a specific goal to assist the small innovator in the demonstration of advanced Canadian-made technology. Some of the criteria for making high-risk (with high-benefit to the Canadian economy) loans might include:

- Lending could be made at a high debt-to-equity ratio, at the prevailing Bank of Canada interest rate.
- The loan could be in the form of debentures or redeemable preferred shares. Giving loan guarantees and allowing private banking institutions to reap the profit with no risk is not a good way to do things. The guiding principle for all should be "no risk, no profit".
- In order to avoid direct competition with commercial banks, private banking institutions would be given the first option to provide similar loans to the small innovator at the prevailing fair-market interest rate.
- Lending should be restricted for the purpose of technology demonstration, but not commerical operation.
- If the technology demonstration is ultimately successful, the small innovator would be on his way to sell the technology commercially, with much less difficulty. The loan can then be repaid to replenish the TDB reserve.
- The small innovator must prove that he is already taking extraordinary financial risk to develop his own technology. In other words, he has to have sufficient faith in his technological innovation.

Concluding Remarks

The small innovators (including independent inventors and small technology development companies) in Canada are faced with cash-related problems in technology research and development. The problem is particularly acute in the technology demonstration phase.

Present government funding assistance programs are oriented mainly to assist the larger organizations such as established companies with other business activities, universities and industry research institutions. Unfortunately, the answers for the small innovator are often not found in venture capitalists, larger companies with their own research departments, or traditional banking institutions.

At least for the technology demonstration stage, the Government could provide vital and timely assistance in the formation of a "Technology Development Bank (TDB)". The TDB would be empowered to make loans and to purchase debentures or preferred shares for technology demonstration projects when the small innovator has no other reasonable financing means available.

Because each innovator has his own particular funding requirement, it is not possible to suggest any one particular route to secure technology development financing.

WORKSHOP

Chairman: J. Robert, Chief
Bioenergy Supply Technology
Canada Centre for Mineral and Energy Technology
Energy, Mines and Resources Canada

INTRODUCTION

This workshop was held following the R&D Contractors Meeting on Biomass Liquefaction at the Delta Hotel, Ottawa, October 23-24, 1990. The key theme of the workshop was to decide where we are going, what direction we need to take and what future collaboration would be possible, with the European Community.

DISCUSSION:

J. Robert (Chairman): I guess, there is basically four groups of people that have been represented here: industry, such as the Dows and the Petro chemicals and Petro Canada, Esso, to the university groups, from Saskatchewan to Waterloo and Laval and then the government people. Each one of you have your own need. I would like to find out what, specifically, is our ministry's need, right away. Can they see any applications or technology about where we are going? Petro Canada - I do not want to single them out, but I would like to stimulate some comments.

A. Silva: I talked to Ed Hogan and I would like his reaction to the meeting.

E. Hogan: Maybe I will start off with one of Aldolfo's major points. In the thermo-chemical area we are kind of proud of ourselves these days in the fact that we can take a lot of these waste materials and treat and convert them into various end products and we are kind of congratulating ourselves at how great we are in the treatment of waste. One of the points Aldolfo made that we have to look at and I do not think we have yet to any great extent, is are we going to be polluters ourselves and what kind of commissions are we going to have regarding air quality and water emissions, etc.

I know, we have looked at this to some extent. I know Christian is now doing some work on some gain tests and that sort of thing. But I think we have got to go a few stages further to find out what exactly are going to be some of the environmental problems that, we, ourselves face.

We have been looking at paralysing rubber tires and petroleum sludge for example and also sawdust waste. We are getting concerned within E.M.R. that we may find ourselves in the same predicament as some industries that were trying to treat waste, and now we have become the polluters. With some of the regulations within Environment Canada these days and the Green Movement, we could find ourselves that way

C. Roy: Well, of course the quality of emissions is important and we are not going to transfer one problem into another problem, so that is certainly a major aspect. So, it seems that some processes can already cope with the aqueous phase. For instance, in case of pyrolysis the condensate is certainly a problem. What I am hearing is that this is being taken care of by some processes here like the pyrotech. It seems with the activated-charcoal, I am glad to hear that this is being handled.

In terms of air emission, I do not see a big problem there for most processes because there is no sulphur, which is a good beginning and there is no chlorine. That is basically in the water phase, which is more my concern. Now, when pyrolysis is being applied to other substrates, as was mentioned here of petroleum sludge and municipal sludge, that is another business.

To carry on, I think that the one point which I noticed, was the enlargement of pyrolysis applications in other fields. We started strictly in wood and now we are slowly getting into other areas. I think this is the way to go. Not to say that wood should not be carried on, I think this area is very interesting. But there has to be some of the effort and some of the money spent also in the environment area, because this is where the future is. Also wood waste - that is a good beginning - but let us remember there are many specific wastes. It will be a big, big problem in coming years.

Municipal sludge which we have been discussing, contaminated soil, and I think incineration is not going to survive through all the regulations which are coming and which are tougher and tougher. Landfilling is not the solution any more and that is what is happening basically in B.C. There is stuff between the sea and the mountains; the price of land is drastically increasing.

So, it is a wonderful opportunity for the pyrolysis community to tap into this market and continue building on the wood area, which has to evolve now in other sectors.

I am also very glad to see here a good balance between fundamental research and applied research. With some researchers perhaps like myself, having one foot in industry and one foot in university - which is in the middle.

But I take it, it is absolutely necessary that government continues to support fundamental R&D. That is because applied research evolved from basic research and this would be from me, as a university professor, the message and I know it has been tough over the last years for fundamental researchers to get money from Canada.

I know most bureaucrats are sympathetic with fundamental research, but I think we have to say here strongly and in the face of politicians, that it is important to maintain a fundamental program.

C. Johnson: Could I just comment for a second on Ed's question, I am involved in permeating several plants right now with the ablative pyrolysis - Jim does not get an aqueous phase, so we do not have to worry about it. So, I am not expecting one in the scale-up, but let's put it like anything else, I do not rule anything out one hundred percent until I see it work. As far as air, there are air emissions, and you are going to have particulates, you are going to have carbon monoxide and you are going to have knox. It is also very easy to get an air permit though, because if you just burn the wood, in the system I showed you where we do not have any air control on it at all, you are still reducing your emissions by 60%.

So, you can go in and you can tell them - well, I have got a TP burner out there now, and if we go in, and if I don't do anything else but flare the gas off, I am still going to reduce air emissions over current, which makes you look great.

We had our permits in California in thirty days, which for California is unheard of, because of those kinds of arguments.

Most of our other systems, other than this first one, all have air filters on them. So, now you are cutting the air emissions somewhere between 85 and 90 percent plus over what a current burn system would be. So, I think those kinds of things really help you look at it. The only other discharge you are going to have out of an ablative process, is you start needing larger quantities of cooling water as you start getting into things. So, you will have a thermo-discharge that you need a permit for and those are pretty well the only things that we really need to worry about.

R. Larue (PR Euromart): We specialize in the transfer and the sale of technology, non-government. We are doing this as a private business. I seem to attend a lot of shows and a lot of conferences like these. There is one thing that I find appalling and it is that I come over here as a guy that has not got a clue what biomass is. You are all the experts. You come out with unbelievable technologies, things that the whole world needs. Well, the government, you know, is providing money to develop this technology. It is exciting.

There is one thing that I have been wanting to say for a long time, every time I attend a meeting. I think what should be done is between governments and I am talking at the international level.

We have people from EEC here and we have people from North America. One thing that should be done, is there should be some type of way to set up business. So, you have invited business people over here to see if they can use the product themselves and you would have to get hundreds of them.

I think that the first thing that should be done, is maybe set up an independent group whereby you would have businessmen, you would have government and you people, with all of this technology that all of your governments own around the world. Put them together and somehow put out a publication that lets the other guy know, I have got this over here, or I have got that for sale. Then maybe the government won't have to put out so much money or they might even just match the money that you can raise on you own.

J. Roberts: Well, that is the intent - I was trying to get some collaboration going between Europe and Canada through such agencies as the International Energy Agency, the CEC, etc. We have a group here like the Encons and the Ensyns that represent the middle man, between the universities and the research institutions and the commercial people. Do they have anything that they see in what direction it's going, or any specific needs other than finance. Or have you changed your tune here?

P. Fransham: As sort of the small intermediate businessmen, as Christian Roy put out the other day, there is this gap between big business and the small guys. I guess we serve a purpose in that you have to have somebody who is going to go out and prove to industry that this is going to work. That is actually the toughest individual fact facing somebody in sort of my position, who is now trying to transfer the enthusiasm, the results, and make really the sale of this technology to somebody.

That is where, I think in some ways, we need help. It is very hard as a researcher and an engineer and what not, to be the promoter also. I could not have done this without the government's support because I just did not have the financial resources behind me. So, this has really been me taking on something on behalf really, on a lot, of EMR. I think we probably have to keep a certain level of government in there.

The one thing that you have to have though, is you cannot give total freebies out any more. The industry has to put something up and has to share some of the risk. Otherwise, you do not have the drive to make it go. It is surprising when you have your money involved in something how dedicated you become to see its completion because there is an economic advantage.

On the other hand then, I guess when the day comes to share the profits, government should somehow share in those down stream profits too, which they actually do and this is where it is kind of hidden.

When Encon finally gets this plant up and running, when you look at the cut that the government takes in the way of income tax from all of its levels, the money is paid back several times over. The government money, in the way of taxes, always comes right off the top. So, I would not hammer too much on this aspect that we have to get the government off our backs.

I think government has to continue to fund this. Recognizing that a large return through economic growth, they are going to see that return and maybe that is how government should analyze things. Okay, I am going to invest $1 million dollars now, but over the next ten years, I am going to collect this much in taxes. When you start looking at the rate of return there, you realize that it is not that bad a deal for most government agencies.

Chairman: Bob, have you got anything to add from your experience?

R. Graham: There are just three things that I wanted to say. I wanted to echo what Christian Roy said, and I think it is very important for companies like ours to have the backing of the university groups that you have been funding. Not just university groups, but the B.C. Researchers and the University of Sherbrooke, in our own case, have been tremendously valuable. We cannot afford to develop the analytical expertise and to purchase equipment.

These groups, I think someone was asking, in a very positive way, the value of the work that is going on, develop an understanding of bioenergy and some of the fundamentals. They can return it back to engineering companies that are trying to develop and optimize processes.

So, I think, as far as direction, I hope that the governments, the economic community and the Canadian government, continue to fund fundamental research in those areas that can be of assistance to us.

The second thing, I think that Canada has clearly developed good pyrolysis technologies and where we can cooperate with Europe is in the implementation and the application and developing uses for some of these fast pyrolytic oils or vacuum pyrolytic oils. There are certainly a lot of scenarios in Europe that are more favourable for the implementation than in North America on the bioenergy side and on the environmental side, but I think that is coming in North America.

A third comment. I agree with Peter as well, that someone somewhere along the line, has to assist these technologies in being demonstrated. As far as Ensyn is concerned, all that interest has come probably from outside of Canada and it is coming from Europe now in demonstration and certainly with Red Arrow, a private company.

Why have we been able to do this - I am not talking about just Ensyn, I am talking about the whole community in Canada which has done such good research on process development, we believe, over the last ten years, and it seems so hard then in Canada, in our own backyard, why are we not implementing? Why are we not demonstrating here?

Chairman: Well, that is what we are trying to find out. We are at a very opportune time. We have the technologies developed that would treat waste. All waste has to be treated and that is taken for granted now, and we happen to get energy out of it to boot.

So, we have got two good things going for the technologies but I think the industry itself, can probably, through seminars such as these, see further applications of it. They have not made contacts, and this is probably the collaboration that will happen between industry and individual researchers or universities, that will further stimulate the take-up and the commercialization of their technology.

What we are seeing is extremely long times for this to happen. How does it operate In Europe? Tony or Dr. Grassi.

Dr. Grassi: I shall try to explain a little bit about our strategy to implement the biomass activity, in such a way, we hope to grow fast enough so we can satisfy everybody.

We have been active as I mentioned yesterday, for about 15 years. It was rather at a modest level, but in any case, we accumulated over a long period, enough results, and several are at the experimental pilot stage. With such a large base of information, we asked ourselves how could we now proceed to speed up all the processes?

First of all, biomass could make a contribution to five large-size markets. One is the heat market, one is the transportation market, one is the chemical market, one is the electricity market and also, in our consideration, the pulp for paper production interests us.

Considering these five largest markets, we could say four since transportation is difficult; everyone in the last year has seen the difficulty for bio-methanol and it will take some time and it is difficult to introduce it as a product.

The second is chemical. The chemical business is a very rich one and so they can afford to pay for oil at $100 a barrel and so at the present time it is difficult. Of course, it is a modest dimension in comparison to the other, but it is a very hard balance. However, there is some potential contribution and we have now the first example and I think many others will follow in the future.

There are two markets, one the heat market and the other the electricity market where I think a product like pyrolytic oil could really make an important contribution. Pyrolytic oil can be utilized directly for the production of steam. You do not need to go to a refinery and this is a tremendous advantage.

Of course, we are starting to do some experiments perhaps or some mixing at the refinery, a small percentage seems feasible - a small percent. Perhaps improving the quality, we could go up to 10 percent. We do not know yet, it is too early - perhaps when we find the electricity market. So, we say that we have to discuss with the market people and see if they are ready to consider this possibility. We discussed it with several authorities in Europe and in particular with two represented at this meeting.

If it was imposed at this time, it would be much more strict in the future. It is a product that could be really good fuel for electricity production. We considered that after methane, perhaps it would be the best fuel from an environmental point of view. Then if we start to discuss this with the authorities like market people and convince them we are able to produce, in an economic way, a competitive way, taking into account the premium for the sulphur absence and other characteristics and, if it is possible, we can utilize it directly and it can be an economical product that we are able to absolve in huge amounts then immediately you would find everything is possible.

Industry is starting to develop technology and the technological transfer has already begun. Technology must be provided from industry. It is not something you achieve in a day and I do not think there is any need for a conference on technological transfer. If the market is opened now everything will be financed. It has to be economical of course.

Although the first exercise in Europe was to have a serious discussion with the authorities where possible. In some countries, like France, who are big investors for nuclear, it is not possible at this time. Maybe in years to come they will be able to consider alternatives. In particularl in southern Europe because of our situation where the electricity market could be extremely optimized for biomass in general and we are ready for this kind of product.

Although, we started from a discussion of the market to see where we are able to open doors, we immediately have to consider industries - and large industries are coming. We have the transfer of technology and we have cooperation. We need R&D, increased research development and system development. I think if the market is ready to accept it, there would be a tremendous growth of all activity including collaboration. We have already started and our results are very important because small and medium size industry and also big industrial organizations are coming.

On the other hand, because there is a general movement in our direction, we are getting much more consideration from the political level and this is due to the fact that we are no longer a small group of people at the scientific level. When we have more industries, I also think researchers receive more consideration and perhaps we will have more money to increase the spectrum of the technological development that we need to take us from our starting phase. So, this is how we think and what we are doing. We hope to get our first results within the next few months and perhaps at this level there could be a real change of attitude and also as to the financial means to speed up all the processes to get this kind of bio-fuel.

First, there will be help for electricity and after that if we are patient, I think there will be much more consideration given for a real fuel substitution.

A.V. Bridgwater: One of the differences in the EEC funding and Canadian funding, is that the research is done under the heading of a framework program which lasts for four years. There is either one or two causes for proposals within that four years which are then evaluated. Among the number of criteria for evaluating the proposals is international collaboration and that is normally dependent on at least two European countries who must collaborate and that is an important criteria in the assessment.

Another criteria is that any proposals that have an industry, university or industrial research institute collaboration are also more favourably dealt with. That is one way of ensuring that industry is actively involved in the development and the R&D.

Inevitably, of course, and correctly there is more fundamental work, but I think that one of the innate differences between European research, is that European research seems to be a rather shorter term and rather more applied research or you can take it towards solving shorter term problems within Europe.

Another way of getting industry involved, like here, is to try and invite industries to contractors' meetings and to seminars and workshops. Dr. Grassi has set up a number of networks of experts in which, again, both the experts to the Commission and industry who are known to be interested or sympathetic, meet and consider and formulate plans and strategies and produce working documents, recommendations and publications. There are many mechanisms for getting industry involved in the R&D programs in order that they become acquainted with the developments in hand and help to implement the R&D being carried out.

Chairman: Maybe we can have some ideas from the academia on this. Where do you see the bio-oils are going? Should we concentrate more on our R&D, more infractions, separation, upgrading? We are looking for directions too.

As you know, we have this request for proposals, but we evaluate them on pretty well the same basis, but I think we want to see Europe; your need's answered, because you are looking at it from a broader perspective than we are at times. Is there anyone from industry?

D. Scott: I have a few comments I would like to make. I think it is quite obvious to those of us who have been developing processes from the laboratory and trying to get them out into a demonstration scale, Canadian industry has no interest, really, in biomass as a source of energy. This is only to be expected, because after all, lots of fuels are plentiful. They are cheap in Canada and they will limit their use of them only when they are forced to do so.

Therefore, the only way in which Canadian industry is going to support this technology transfer or basic research or demonstration and finally, full scale plants, is either through the waste disposal activity or because some government regulation is forcing them to do so.

On the other hand, Europe has a very keen interest in biomass as a source of energy. In some areas of Europe, like the Mediterranean countries, any and all forms of alternative energy are fair candidates for examination and research. This is, by no means, true in Canada. In fact, you can see already that the developments are following, in a sense, this natural split that we in Canada are turning our attention to the speciality uses, because neither Ensyn nor Encon are really in the energy business.

We are concentrating on chemical recovery, because in the short term, this, I think, is where the future lies for the Canadian biomass developments. Europe, on the other hand, and I have some small involvement there, is obviously developing the energy projects much more actively than we are in Canada.

We do have a common and equal interest in waste disposal. So, I would suggest that if research approaches are to be followed, that we might take this as the direction we are naturally heading in anyway. The Canadian researchers and developers would concentrate on the speciality uses, the chemical uses and the Europeans concentrate on the energy uses. Then we will mutually share our results and have a wider scope in this way.

Chairman: Are you suggesting some sort of delineation, formally, or just on an ad hoc basis?

D. Scott: Well, this would not be a bad thing, but since this becomes highly political, I am not really able to comment.

Chairman: I understand. We have industry, at last.

A. Silva: I think I would like to mention something because I agree with Prof. Scott.

I mentioned before to Ed that what I learned here was that people are sometimes doing work for the fuel energy consumption and I do not see any future for pyrolysis or biomass to produce fuel. We have a lot of fuel in Canada and cheaper. We saw economics just now, on how cheap it is to produce gasoline and diesel out of the oil that we have. Now we have Hybernia coming, we have Nova Scotia, we also have another project maybe later on and there are more than what we need really here in Canada.

I see biomass, and I agree with Prof. Scott in two main areas. Really, for chemicals, and I do not know much about it, and the other one is, something that we intend to be the leader of in the world, and it is related to environmental problems.

I think with the technology that you are developing, you can help the industry, in general, and there you have a lot of industry interested in cleaning up the mess that we have. A mess that we have already and which is going to continue into the future. There is so much to do. We have contaminated soil; we have contaminated water; and we have contaminated air. When we talk about producing something from biomass, you would have to look at the environmental impact and there are two areas of environmental impact.

The quality of the material that you are producing - if you are producing the materials, that product, whatever it is going to be, has to satisfy some environmental guidelines.

Gasoline, for instance, is an example from the oil industry that we have become accustomed too. I was born and everyone was born with gasoline beside them in some way. We saw that it was very friendly; it is not any more; it is bad.

Water - we are consuming water from lakes and rivers and we are putting back those things again, but it is not the same quality.

We also have problems in our refinery and every refinery has problems and in every other industry there are problems - contaminated soil.

Look at the gas station beside your house - and the woman the other day at the seminar - the woman who had a close count of something like 19,000 underground tanks and half of them are leaking. It is not the oil industry alone; it is everyone who participates in that. We have had a lot of problems there and pyrolysis can help us with this. So, the environmental impact is quite important and you have the technology to help here.

Let me give you another example of how you can help. They use lubricants. There is a lot of lubricants that everyone uses in the car. What do we do with that? Where do we send it? There are some outfits who can recycle it but it is not enough. Something else can be done.

I think that EMR helped technology - they can do some hydrocracking or hydrotreating to recycle that material. Christian Roy mentioned the cleaning of oil sludge, for instance. It was very interesting to learn what was done with Ultramar. If that would work, I think it is a good idea to pursue that. It is an area that every refinery produces - you mentioned one percent, but perhaps in some refineries it is more than one percent of the total capacity and work needs to be done there too.

So, when we are talking about biomass to produce fuel, I worry about it. I do not think there is a future in Canada for this. Perhaps in Europe because they do not have as much energy as we have in Canada.

Chairman: Ian Maars, do you have any perspective from the chemical aspect at all - from the government monitoring? I do not want to single you out, but you know, I think you have made comments before and you know the technologies.

I. Maars: Yes, I am not too close to this issue. I just meet once a year with Ed and his people to review some of the projects that are proposed. I would agree with Dr. Grassi, on chemicals. I do not see any way that these processes are going to be economic. Waste disposal to me is the big one. As Adolfo says, energy we don't need, but recycling of plastics and the whole recycling industry - here I think that technology has a lot to offer. If we are going to go into pyrolysis, then go all the way and burn the hell out of it and recover the energy.

Chairman: Dick Wilson, he is an industrialist, so he is more credible.

D. Wilson (Dow Chemical): Well, I don't know if I am more credible. I have a number of comments. I think there are two shades to almost everything that has been said.

First of all, I think pyrolysis should have a role in the reduction of municipal solid waste. The combustion is a bad word these days, incineration, the Green Peace have got the acronym "NOPE" - "No Place on Planet Earth" and rapidly promoting this hysterical, single molecule of fear and forgetting the fact that these heavy metals are going to be buried in the ground, whether in the form of the gross garbage or in the form of ash.

Nevertheless, the big argument could be made for pyrolysis that you have an opportunity of recovering some chemical values. Ian Maars has put a finger on the problem that I see in part. We have talked about the feasibility, for instance, of collecting ethylene glycol, which chemically and technically is certainly feasible and I will give you an example.

For instance, when you make ethylene glycol and with conventional technology, you do a little test by putting some sulphuric acid in it and then cooking it up and the reason for this is to detect some parts per billion level of impurities that are going to effect the polyester process. If you are making ethylene glycol, you do a UV analysis, which again is looking at parts per billion. I was in that business for about 15 or 20 years and never did figure out how to take those materials out of the product after they were made. We tried all kinds of strategies to stop making them in the first place.

So we have some horrendous problems of recovering purity chemicals from these types of highly oxygenated products. I think that we have to look harder at it. As I mentioned to Don Scott, when I talked to my cohorts in the chemical speciality business, the general reply is "Dick Wilson, get lost, get out of our hair. Look at the problems compared to what we have got now."

That is a fact. There are problems, but we have to start looking.

Getting back to the pyrolysis and municipal solvent waste, I have another comment. I think that if you go to this RDF process, the "Refuse Derived Fuel", and get the metals recovery out of it, then you do the pyrolysis. You now have the ability to have this energy source partly because you have got this good volume of oil and it can be moved to a use point. Now, cost of fuel, that fuel, compared to the others, is going to be high. There is no question about it. However, if you roll the total costs into $100 a ton disposal cost of the MSW and then back that into the fuel cost, that helps some. In the year 2000 those costs are going to be there. So, I do not think it is completely out of the question in specific areas.

The question of water from the pyrolysis process has come up and in almost all cases, you are going to be feeding some water with the process and you are going to be making some.

In Ontario, we have the municipal industrial strategy for abatement. In the case of my own organization in Sarnia, we generate 750,000 data points a year to satisfy that analysis.

Now, this is a complex operation, with 13 different chemicals being made in the one plant and about four or five sewers. Nevertheless, the cost that everybody is running a pyrolysis unit and trying to extract some of the chemicals with aqueous systems, and/or generating some cooling water, you had better not ignore that one, because that is soon going to be duplicated in the air field as well.

In this one plant, we spent over $2 million in capital and have an environmental lab the size of this room, and the cheapest instrument in it now is approximately $90,000. Dow Chemical is, I think, one of the leading chemical companies in safety. When I look at these chemical plants, the pyrolysis unit, the pilot unit, and the costs are estimated at $150,000 or $200,000.

Also, let me tell you, I could not make those things for $500,000 and pass the Safety, Health & Loss Prevention Review that we would have to have. We have these plants of potential gas leaks in enclosed buildings and that is not going to fly forever folks, because the Departments of Labour will not buy it and the costs are going to go up a long way.

E. Chornet: I think we do not have to be that negative about things. I remember, and I am old enough to remember, how things happened in 1970 and my impression is that biomass has been used in Canada in increasing amounts as a source of fuel in the pulp and paper industry by direct combustion of waste. The amount of biomass that has been recycled in the fuel industry has doubled in ten year's time. We generate seven percent of our primary energy needs in Canada from biomass today and this mostly comes from the pulp and paper, burning combustion of residual biomass which is generated during the process.

So, my lesson or what I have learned from this, is that if industry can see an advantage in using the biomass and this happened during the 70's when the oil prices went up, then industry will use it for a source of energy, either through direct combustion, which is an admittible situation, or it will go into pyrolysis if we need to move it around.

I do not see, however - and I agree with Don Scott, the generalized situation in Canada in which biomass will become a great supply of energy needs, either for transportation or for fuel. I do not see, for instance, how in Quebec we will produce electricity from biomass.

We have a sad story concerning this, in view of the biosyn project. This happened when we could not produce methanol because the price went down and we then suggested to Hydro Quebec to shift the gasification unit, which was ten tons per hour at that time, to an electricity generation process. Hydro Quebec was not interested because it was much more costly than establishing a dam in the north and just carrying the electricity to the south. So, that is the situation for electricity, and I do not see any change for awhile.

Now, as to where I am going with this, my first point is that if the residual biomass has a negative cost; if there are environmental constraints that say you cannot leave it there because it becomes waste; and this becomes a social pressure created by this problem, then industry will use it.

The pulp and paper industry has done it magnificently. Most pulp and paper mills today are self-sufficient in energy. They do not import oil from Petro Canada or other things - they use their own waste.

Now there are laws in Canada and they are common, at least in Quebec, that say when you cut a tree that you cannot leave 20 percent of that tree on the ground because that impedes the reforestation and there are very good arguments to prove this.

When this happens people are forced to do something and the only large market is energy.

Now plastics, et cetera, encountered the limits of having very pure feedstock and so on and so on. This may develop as a social pressure by putting a tax on waste. I think we are joining the municipal waste or biomass waste from the forest. Ultimately we will have to pay for it but I think slowly we will be forced to develop along the lines I mentioned.

The second point has to do with chemicals. I have participated, as I have indicated, in some industrial ventures that are profitable - for instance a sugar company that uses biomass to produce a chemical. We were interested in bio-gas and we processed it in two years. We optimized the bio-gas fractionation scheme for them which was not put in operation in Brazil. It was not put in operation for political reasons, however, the economics were very clear.

With Domtar we are now getting into some structural products. So, I see special niches that are going to be created, that require a lot of imagination from scientists, managers and decision-makers and perhaps that is the way things go. You cannot have a program with massive influence on things, you have to go slowly with showcases. Perhaps, in the chemicals and the pulp and paper area we are going in this direction.

An area that would be very good in pulp and paper, particularly for Europeans, less for us, would be to use residual biomass to produce better fibre. There is a crying need for this.

Rather than using 70 percent of the tree, could we use 80, could we use 85, could we use 90 percent? This will have an immense benefit for industry and in fact for everybody. I am convinced there are niches or specialities that will develop and I am very optimistic about this.

P. Fransham: Now, I approached NOVA, and I started off commercializing the fast pyrolysis process and I went back to NOVA and I said, "Look, we can handle your plastic wastes". Based on some data that Don provided me with, I said that we can get somewhere around 75% styrene back out of styrene waste. The comment back was, "Ah, we already send 80 percent styrene to the dump because it is not worth cleaning up". Therefore, until the waste has such a high value because of disposal cost, which is what Dick was saying, there is not going to be much incentive for the waste producer.

It is still easier to send it to a landfill or a selected dump or to do something else with it. I agree, that maybe it is going to be some sort of tax on the waste similar to the way we are now going with tires. I also think that tires are probably the tip of the iceberg. You now have a tipping fee on tires to ensure that they come back.

Putting taxes and more taxes on, is, at least, from an environmental standpoint, one way to meet with more acceptability than say something as straight as a goods and services tax where people are willing to pay generally for the environment but not on a personal basis.

Responding to Dick's other comment about building pyrolysis plants for $150,000.

What we built, we built to ASME standards for a relatively low sum so that we passed all of the boiler inspection, all of the electrical codes and everything and it did not cost an arm and a leg. So, I would just like to take a little bit of slight at your comment and say that we can build quality products and quality engineering without an enormous price.

C. Roy: I will try to answer the question you raised earlier regarding your money and how it might be spent. I will try to give an answer as an academia on that aspect.

I remember when starting this business about ten or twelve years ago - you will remember that we were perhaps 15 runners as process developers. Now, many have disappeared and we have not heard much of the incineration guys or gasification which is also not as significant a segment as it used to be ten years ago.

After these very difficult years, over the last five or six years, I think what we are left with now is perhaps three or four groups in Canada doing process development and these groups have been surrounded by a larger number of fundamental researchers. The know how of Canada in pyrolysis resides in this small group of people. I think, by all means, it is important to protect and keep these researchers alive both in the applied field and in the fundamental field.

How to spend it? what is the budget - $3 million a year? how to split it?

I cannot answer this question here. Perhaps you could give me more information later. I think what someone from the University would like is to have two or three years of sustained important investment in R&D, because these contracts on a project which only last one year and then your in a cash flow problem and you have to wait for the next contract and you end up with a gap of three or four months. I understand from Tony's comment that they have a different approach in Europe and it is for a longer period of time.

Matching that would still be a cost share program for applied research. I think this has to remain there but I would like to see it extended a little bit on the longer period of time and consider spending a larger amount of money for those runners who are still alive and who do worthwhile research.

So, I think that if you have a comment on this particular point, I would appreciate it. I do, however, think both groups have to be kept. Now, does this $3 million budget include the bio-technology sector as well?

Chairman: Yes, it does. It is for the whole bioenergy sector. It is not a question of how to spend it - we can spend it. It is as to what direction should we be looking at. Should we be looking at fundamental research into separation or co-products et cetera? Or are we really trying to get R&D done for industry to pick it up. To take it further, we have to pass through the activities such as the Ensyns and the Encons and other people and yourself, such as with the pilot plants, getting industry collaboration.

We do have the requirement of industry collaboration for any of our funding. It is a fundamental direction and we have been at it for 10 or 12 years now and we are at the point where we are almost pre-commercial enough in all aspects. I am looking for some future direction, because we basically want to get it into industry.

K. Oehr: I have two points that I would like to make. One, thing I would like to compliment you on and that is, it seems to me, and I talked to Ed and he has suggested there is an emphasis now on viewing proposals at any point in time, regardless of these RFP dates that used to be sent, say once a year, where you get a whole slew of proposals come in where people are trying to come up with ideas up against a deadline.

I do not see a lot of merit in that verses taking good proposals as they come. So, you should always have some cash in the till to take a good idea whenever it walks in the door. You get better proposals that way and it takes the pressure off people to feel like "Oh, my God, I have got to get a proposal out by October 31st" verses "Well, when I am ready, I have got it, I have got a really good concept." I am going to come to you and you are ready, you can review it in a rational period of time and you are going to get a better idea.

If I understand properly, the way you review proposals, that is the way you would like to go.

Chairman: Well, it is the first year we have tried it and I got a lot of flack for doing that because the proposals were not coming in, because there was no deadline. So, they started coming after me and saying you are going to elapse a lot. I think now we have straightened it out and I think we are getting cooperation now because there is a limited funding and it is a first come, first serve basis. Once you people in the research area know that it is first come, first serve, if you come in too late with a good idea and the till is empty, it is going to have to be deferred to something else. We are, though, trying it for this year that way and we hope it proves successful. Hopefully, we will get the quality of our proposals up instead of last minute rushes.

K. Oehr: The other idea is, it seems to me a lot of proposals are like bottom-up proposals, where people walk in the door, out of the cold, and say "Here is an idea, I would like to have it developed" and then you say, "Yes, that sounds okay".

You should have a few more top down proposals, where say somebody like an Ensyn might come along and say, "Boy, we have got this real problem. We want this solved, but we don't have time to deal with it". There should be a request for a proposal and they can come out month by month, periodically. They do not have to come out once a year, where you say, "Look, we would like somebody out there who thinks they can handle this to have a go at it". Just give a list of specific jobs that you would like to see handled.

For example, I know Christian is interested in recovering chemicals from an aqueous liquor; he has got his hands full with a whole pile of other things. Somebody out there should be spending their time trying to process his aqueous liquor, period. Unless they know something about it and the value to him or the value to you nothing is going to happen with that liquor unless he finds a person, inhouse, that is going to deal with it.

So, I think you should try and have a few talk-down proposals where you say, "Look, here is a problem we would like handled", and it does not have to be "Send a proposal in by December 31st". Here is the problem, if you have got a good idea, send a proposal and that is it. So, you get both types of proposals coming in.

Chairman: Well, we have done this top-down proposal's approach, as you have suggested, but it is directed. We phone up people and say, "This is the problem, we would like to see it..." but I think we will become more pro-active that way in future, once we get the main proposals out, that will identify the areas, the problem areas that we will get out.

P. Laborde: Yes, I would like to make a little comment here. I am working in an operation that is supposed to be doing a transfer between university and industry in the field of bio-mass. But it seems that in industry, not everyone is interested in biomass. Biomass is not a business.

Some people are in the waste business or in the chemical business or in the energy business and so on. So, I think the way we could develop economic projects, interesting projects for the industry, is to develop some consortiums of different kinds of industry interested, one in fibres, cellulose, another in chemical, another in energy and so on. For that, I think there is a company here that has presented a consortium in this field - pyrolysis and phenolics and different kinds of products. Perhaps it is a way to do business here in Canada, in North America.

P. Fransham: I think, you know, from what I have heard over the last couple of days, is that from a process side, I think we have pretty well nailed down that fast pyrolysis, whether it is one second, two seconds or ten seconds, but something on a very short residence time, is obviously the way to maximize the liquids.

So, from a Canadian standpoint, I do not see us investing much more, because I do not think there are many more permutations and combinations that we can hammer away on in terms of process development in the way of pyrolysis.

I echo Don's feeling here, that from a Canadian perspective, if you are looking where to go and from my own economic analysis, it is very evident that fuel is not the way to go. From EMR's standpoint, unfortunately, it is not bioenergy, it is bio-chemicals. Is EMR, from a philosophy standpoint no longer really an energy company or an energy department. Does this conflict with your mandate in a way because we do not want the energy value as it is just not worth it. What we want is the chemical value and now can this still reside in EMR?

Chairman: Well, as was mentioned, it is not really energy. Our basic mandate is producing energy for the nation, but we have the technologies and I will answer the environmental aspects too. So we can use this temporary glut of energy that still is there, but we have the technologies and we still further enhance them, as far as yields are concerned, because the energy will come more prevalent sometime in the future, plus the technology itself, will be further developed and we will see the environmental benefits. So, it is not a clear cut "energy - yes, energy - no", type of thing.

P. Fransham: It strikes me from what I have understood here, that there is a fair amount of fundamental research that could just go totally into chemical separation, but there is still a lot of work in optimizing those processes. I think those processes are kind of where the pyrolysis hardware development was maybe three or four or even five years ago.

So, now it is a question of perhaps looking at demonstration plants, larger scale pilot plants and if people like myself can produce large amounts of the liquors then somebody else can take those liquors and say, "Okay, now we have got enough feedstock to do an actual pilot test on separation as opposed to a bench and laboratory scale".

Chairman: Yes, well, maybe this will be the natural transition of what will happen in the future. We are just hoping it does.

J. Diebold: Just a few comments. With respect to a potential for the pyrolysis of municipal solid waste, we reported a few years ago that when you fast pyrolyze on RDF, made from municipal solid waste, that the condensates are very different than what we got from wood. Rather than the nice single phase liquid that we were accustomed to seeing with wood, what we got was a very different condensate which resembled asphalt. This made us change our condensation training considerably. The use of that asphaltic-type material has not been established, but is something that is probably a nice area for further work.

My point is, that you cannot extrapolate these processes without some risk. With respect to Dick Wilson's comment that you need ultra pure chemicals to satisfy the market, I am sure that is true for some applications.

However, for example, in the phenolic formaldehyde area, we are finding that you do not need pure chemicals for that application and that you can use a variety or a mixture of a large number of compounds rather than a pure phenol and do quite well with the finished product.

I think that one needs to re-examine some of the historically, chemically-pure commodity markets to establish niches for impure glycols, which worked great in my radiator to keep it from freezing, but which might not be pure enough for some chemical applications. I think the impure mixtures such as gasoline or adhesive manufacturers or things of that sort, will be the entry markets for biomass products.

I did not particularly like Ralph Overend's comment in this area, that we should be setting our inferiority complex and looking for areas in which biomass products are superior to petroleum derived products. I think that is a goal that we should attempt to maintain.

A few years ago Tom Reid sat down and listed all of the variables that he felt were important in a pyrolysis reactor and concluded that there were at least a million possible combinations.

I think it is egocentric to believe that we have come across already the very best pyrolysis systems and I think we should keep an open mind that there maybe even better ways to go about this pyrolysis business.

Lastly, I think that for meaningful collaborations between programs that are funded by different countries, that the issue of who maintains the intellectual properties, is an area that cannot be dismissed.

This is an area which has changed quite radically in the United States and it is very complex in that small business can retain the patent rights to government-funded efforts, whereas large companies must pass certain requirements, for even limited retention of their rights.

I have no idea how this works in Canada, but I do see evidence in this meeting, that the individual researchers are very much interested in protecting their patent positions.

I think this is something that is going to be a key issue to these international collaborations.

Chairman: I guess we will have to look into the question of intellectual property within ourselves.

B. Graham: There are a couple of things I would like to say too. I think that we have to recognize at a meeting like this and we have recognized this at other meetings, that the research that is reported on, is normally the tip of the iceberg. I know that a lot of groups are working on things. We obviously had that from Cindy when she was asked a specific question on what she was doing. She cannot answer a question on what she is doing and for valid reasons.

There are a number of groups that are working on things other than what they have reported here. They are working in diverse areas, and so, we can't make final conclusions from the tip of the iceberg on results that have been reported in this meeting.

The second thing I would like to say, is that I still think that there can be two or three Canadian bioenergy companies, because there is no harm in Ensyn, Encon or Laval exporting technology to Europe or to anywhere else in the world and getting a royalty. Heaven forbid if a Canadian gets a royalty from a foreign company.

The Japanese do not restrict their car making to the Japan market. We too can be bioenergy companies exporting our technology to other countries. Those are just two points I wanted to make.

D. Wilson: I just wanted to comment on a point Jim Diebold made about the quality of chemicals and say he is one hundred percent right, I did not express myself clearly.

A large part of what we have is a mind set and there are some practical applications, of course, because if you want two qualities of MEG, you have got two distillation columns, another tank and another separate system and another specification sheet and all that sort of thing.

Nevertheless, I think there are some great examples here of the aromatics being used in the epoxies and I think you could carry the same thing through to urethane. So all of us in the chemical business have to open our minds and do a much better job in this area. I am glad you made that point and I apologize for not making it clear myself.

N. Bakhshi: First of all, I think the oil situation is very complex as you have just found out. All of us know that there is no simple answer to what we are discussing. My first one or two comments will be from an academic point of view. At least what I feel and I think I hope the others will agree with me, that the research which we do is our own personal research and we are interested in that. It would just be nice if some industry would get interested in our results which are produced. So, I think you cannot restrict that. I think that is the human mind, an individual mind, and that is where you generate ideas. One may generate a hundred ideas, but perhaps maybe one or two ideas might be commercialized. Ultimately, you see, that is the human mind and you cannot restrict that.

The second point for example, refers to what Jim Diebold said a little while ago. I smiled when a comment was made here that the pyrolysis work has all been done and now, let us get on with some of the other things. I mean, we are doing some pyrolytic work and we barely started using the varied catalysts. I can tell you the preliminary work simply shows that what you can make essentially will blow your mind. Really, all that I am simply saying is that we should not limit that. There are a number of other possibilities and a number of other things which will happen.

By the same token, if you look at the biomass by itself, I do not think we should restrict it. Okay, we are all working in the biomass area, but making fuels is just only one part of it. The applications, for example, let us say activated sludge, there might be some "How much does it cost per ton to dispose of that sludge".

On the other hand, if you can make say fuels or chemicals or catalytic treatment or something else which is a valuable material, why not? So here you have an application of pyrolysis where perhaps it could be generating energy or chemicals or whatever. So, I think all these things will come true.

The other comment I would like to make and I think the usefulness of meeting like this, is in generating base ideas. What I really would like to see is, perhaps the contacting of more industries and telling them so that they can come and see what is being done and what is available. Then they can show some interest. "Well, look we are interested in this area or we are interested in that area". Somebody has to do the selling job and I think we, as academics are very poor people from that point of view and we cannot do it.

Then that job, from an industry point of view, because they are interested in whatever the idea is and when they say, "Well, look, these are some of the other problems which they are interested in" then I can think as an academic, "Oh yes, you know, I am interested. Perhaps what I am doing can be applied to this particular area and so on and so forth". So, I think that is important.

The other comment, I am not sure if it was Esteban or Bob Graham who mentioned that why should we, as Canadians, not develop the technology for world wide consumption? Why should we be afraid that in Canada we are not going to do it; the world is our market, why not? I think, as Canadians, we should exploit that situation. If you have an expertise in a certain area, we should go after it.

Chairman: Now, getting feedback from industry, we realize that it is more and more important and we are looking at ways, right now, to get their input and to find out about their products and see what we can offer. So, it is a technical transfer aspect and we are coming to the point where we have to put a lot more emphasis on it. I do agree with you and Bob, let's take our technologies and if they work, let's fill them.

It is after five now and unless there are some burning issues or anyone wants to direct some general comments on the floor from any of the two days, we can conclude this contractor's meeting.

G. Grassi: I would like, again, to say my thanks to the Canadian authorities for inviting myself with our friend from the European community to participate at the contractor's meeting. I think it has been very useful for everybody who wants to know the state of the art of the situation in Canada. I will try to invite you and your colleague here to participate at our Contractor's meeting in November in Florence. So, I thank you very much.

We know very well the situation here and because in the next few months, we are starting to consider, for us at least, important integrated projects, we could discuss which kind of collaboration and eventual participation we could have with a Canadian institution or industry for implementing these important operations. It is a complete system development for many items that are under development, that perhaps could make for interesting collaborations. Thank you again.

Chairman: Well, you are very welcome and it is a pleasure having you and learning from you. We have had a lot of interchange and we look forward to something coming out of Florence.

I wish to thank everyone for their attendance. It has been a good meeting.

Author Index

Abatzoglou, N. 198
Allen, S.G. 90
Alleyne, C.S. 126
Ayres, W.A. 236
Baeyens, J. 39
Bain, R. 101
Bakhshi, N.N. 157
Barrass, G. 181
Beckman, D. 50
Bergougnou, M.A. 52
Black, J.W. 123
Black, S. 101
Blanchette, D. 109
Boocock, D.G.B. 90
Bouchard, J. 137
Bridgwater, A.V. 6, 20, 201, 209
Brown, D.B. 123
Campbell, H.W. 78
Chornet, E. 86, 129, 137, 185, 198
Chowdhury, A.Z. 90
Chum, H. 101
Czernik, S. 64
de Caumia, B. 109
Delmon, B. 32
Diebold, J. 101
Dolenko, A. 1,2
Dubois, C. 198
Evans, R. 101
Fransham, P.B. 74, 229
Freel, B.A., ... 52
Graham, R.G. 52
Grange, P. 32
Grassi, G. 3, 6, 20
Heitz, M. 185
Hogan, E.N. 16
Howard, J. 179
Johnson, D.A. 236
Labrecque, B. 109
Lapointe, J. 185
Laurent, E. 32
Longley, C.J. 179
Majerski, P. 171
Maniatis, K. 39
McKinley, J. 126
Milne, T. 101
Morrison, A.E. 179
Oehr, K.H. 181
Overend, R.P. 86, 129, 137, 185, 198
Pakdel, H. 109, 144
Peeters, H. 39
Piskorz, J. 64, 171
Plante, P. 109
Radlein, D. 64, 171
Rajai, B. 101
Robert, J.E. 16
Roggeman, G. 39
Rossi, C. 44
Roy, C. 109, 144
Scahill, J. 101
Scott, D.S. 64, 171
Sharman, R.K. 157
Skelton, E.S. 126
Soveran, D. 184
Tomberlin, G. 236
Underwood, G. 226
Vidal, P.F. 129
Wong, A. 241